NAVFAC

Naval Facilities Engineering Command

ENGINEERING SERVICE CENTER
Port Hueneme, California 93043-4370

I0034686

SP-2209-OCN

Handbook for

Marine Geotechnical Engineering

Technical Editors:
David Thompson
Diane Jarrah Beasley

February 2012

Published by Books Express Publishing
Copyright © Books Express, 2012
ISBN 978-1-78266-051-4

Books Express publications are available from all good retail and online booksellers. For
publishing proposals and direct ordering please contact us at: info@books-express.com

ACKNOWLEDGEMENTS

Much of the background material for this second edition of the Handbook for Marine Geotechnical Engineering was developed by the Naval Facilities Engineering Service Center (NAVFAC ESC) and is based on the first edition (Rocker, 1985). These materials were extensively updated and supplemented with experience from both the public and private sector by a number of ESC and contractor subject matter experts. The primary contributors for each of the chapters is listed below.

Chapter	Primary Contributor
1	Mr. Phil K. Rockwell, MAR, Inc., Ventura, CA
2	Mr. James C. Miller, MAR, Inc., Boca Raton, FL
3	Mr. Herb Herrmann, NAVFAC ESC, Washington, DC Mr. Kyle Rollins, Brigham Young University, Provo, UT
4	Dr. Daniel G. True, MAR, Inc., Ventura, CA
5	Dr. Sheng (Tom) Lin, NAVFAC ESC, Port Hueneme, CA
6	Mr. William N. Seelig, NAVFAC ESC, Washington, DC
7	Dr. Sheng (Tom) Lin, NAVFAC ESC, Port Hueneme, CA
8	Dr. Daniel G. True, MAR, Inc., Ventura, CA
9	Mr. Blake Jung, NAVFAC ESC, Port Hueneme, CA
10	Dr. Jean-Louis Briaud, Texas A&M University, College Station, TX

[This page intentionally left blank]

TABLE OF CONTENTS

LIST OF FIGURES

xvi

LIST OF TABLES

1 INTRODUCTION

1.1 OBJECTIVE

Marine geotechnical engineering is the application of scientific knowledge and engineering techniques to the investigation of seafloor materials and the definition of the seafloor's physical properties. The responses of these seafloor materials to foundation and mooring elements, as well as other seafloor engineering related behaviors and processes, are addressed in this document. This Handbook for Marine Geotechnical Engineering brings together the more important aspects of seafloor behavior and Navy Ocean Engineering problems.

The Navy installs, or may require installation of, a variety of facilities fixed to the continental shelves and slopes, to the submarine slopes of seamounts and islands, and to the deep ocean floor. Some of these facilities rest on shallow foundations resembling a spread footing or on pile-like foundations. Other may be surface or subsurface-moored types where a buoyant element is tethered to the seafloor by uplift-resisting foundations such as piles, or propellant-embedded or drag-embedment anchors. Behavior of mooring elements lying on or embedded in the seafloor is dependant on the physical properties of the materials making up the seafloor in the immediate area. In addition, scour and slope stability problems may exist or may be created by the placement of these elements.

Navy military and civilian engineers will be required to plan for, design, supervise construction of or have technical responsibility for these facilities. Geotechnical aspects of engineering problems associated with the facilities are difficult for Navy engineers to address because of the highly specialized nature of most geotechnical topics. Also, due to a general lack of historical precedence for seafloor construction, a low level of understanding of seafloor soil behavior exists. Much of what does exist is published in documents not widely distributed. The Handbook brings this information together. It is intended for use by Navy engineers who do not have an extensive background in geotechnical engineering. The Handbook is not an all-inclusive design manual. Rather, the objective of the Handbook is to familiarize engineers with geotechnical aspects of problems, serve as a design guide for relatively uncomplicated problems, and be a technical directory to more complete discussions and to more sophisticated analysis and design procedures. Although it is intended for use with deep ocean problems (nominally beyond the continental shelf or below about 600 feet), the information contained in the Handbook is applicable to problems in shallow water as well.

1.2 HANDBOOK ORGANIZATION

This Handbook has 10 chapters; an Introduction, and nine technical chapters grouped into three major sections: PROPERTIES DETERMINATION, DESIGN OF FOUNDATIONS AND ANCHORS, and OTHER SEAFLOOR PROBLEMS.

The Introduction serves as a guide to the remaining chapters. It lists generalized features for each type of foundation and anchor, and can assist the reader in selection of an appropriate foundation or anchor type based on environmental conditions and structural requirements.

The Properties Determination section, consisting of Chapters 2 and 3, discusses on-site and laboratory determination of soil properties and presents physical property models for major seafloor soil types. Chapter 2 describes the various aspects of surveying a site, including a preliminary desk stop study and survey planning through brief descriptions of remote survey equipment, shallow and deep sampling equipment, and in-situ soil properties testing equipment. Chapter 2 also contains a section on estimating soil properties for use in a preliminary design when no field data are available. Chapter 3 describes the laboratory tests performed on recovered soil samples to generate index and engineering properties data required for analysis and design of seafloor structures. Use of index properties to classify the soil and to correlate with engineering properties is also described.

The Design of Foundations and Anchors section, consisting of Chapters 4 through 7, describes the use of topographic, stratigraphic, and soil properties information necessary to predict capacities of foundation and anchor systems. Chapter 4 covers the design of shallow foundations and deadweight anchors bearing on the seafloor surface. Design of piles for use as foundations or anchors is discussed in Chapter 5. Plate-shaped anchors embedded in the seafloor are treated in Chapter 6. Chapter 7 covers the selection and sizing of drag-embedment anchors; only the resistance developed from anchor and chain interaction with seafloor materials is discussed and not the design of a complete mooring system. References 1-1, 1-2, and 1-3 can be consulted for information regarding complete mooring systems.

In the Other Seafloor Problems section three other aspects of marine geotechnical engineering are discussed. Chapter 8 describes techniques for predicting the depth of penetration of objects into the seafloor. The techniques can be used for penetration predictions with large and small objects of various shapes (such as lost hardware, instrument packages, or foundation elements) impacting the seafloor at high or low initial velocities. The procedures described in Chapter 8 can also be used to predict the force required to embed a given object to a specified subbottom depth (shear keys below a bottom-resting foundation, for example). Chapter 9 presents techniques for predicting the force or time required for breakout of objects embedded in the seafloor and discusses conditions that can have a significant effect on breakout. Analytical techniques are given for two significantly different cases – full-suction and zero-suction along with a discussion of mechanical techniques that can reduce the breakout forces and time requirements. Chapter 10 describes scour prediction techniques. It is directed primarily toward scour problems around objects on the seafloor (local scour), but includes a discussion of nearshore seasonal seafloor profile changes. Most information on scour is drawn from historical observations and model studies of nearshore and river conditions. Insight from these studies is extrapolated to conditions more likely to exist in deeper marine environments.

Each chapter has a list of references and symbols used in that chapter. Example problems, which outline design or calculation procedures, are presented at the end of each chapter that includes design procedures.

1.3 SELECTION OF FOUNDATION/ANCHOR TYPE

Chapters 4, 5, 6, and 7 each describe a different type of foundation or anchor – deadweight, pile, direct-embedment, and drag-embedment (Figure 1.3-1). Each of these foundation and anchor types has strong points or features. This section summarizes these features (Table 1.3-1 through Table 1.3-4) to provide guidance on selecting the optimum foundation or anchor type for a given set of problem conditions.

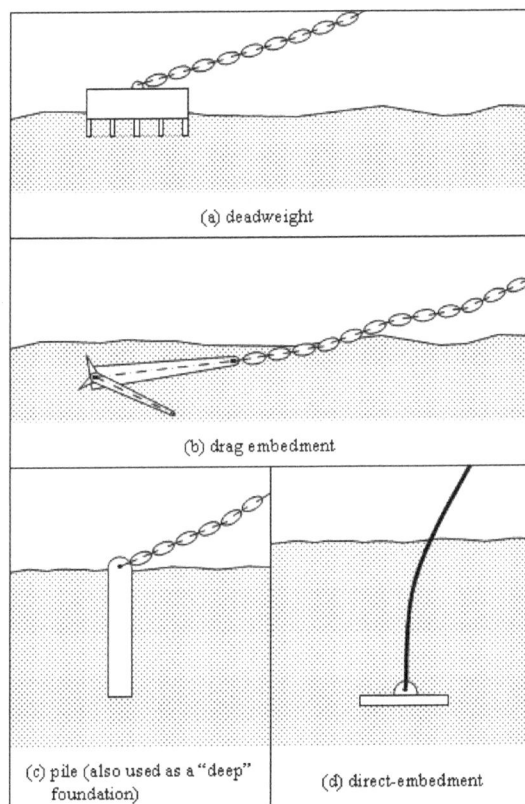

Figure 1.3-1. Simplified anchor types.

Shallow foundations and deadweight anchors are widely used in the deep ocean environment because they are simple and readily sized for most seafloor types and loading conditions. However, they do not perform well on steep sloping seafloors. In addition, deadweight anchors are not very efficient (that is, the ratio of lateral load resistance to anchor weight is low compared to other lateral-load-resisting anchor types). Table 1.3-1 lists these and other features of the shallow foundations and deadweight anchors.

Pile foundations and anchors are used where less expensive types of shallow foundations and anchors cannot mobilize sufficient resistance. A principal drawback of piles for the deep ocean is the highly specialized equipment needed for installation and the associated very high mobilization and installation costs. Table 1.3-2 lists features of piles used for foundations and anchors on the seafloor.

Direct-embedment anchors can be driven into seafloor soils by impact, vibratory, water jetting, augered-in systems. Some of the more significant advantages of the direct-embedment anchors are: (1) their very high holding-capacity-to-weight ratio and (2) their resistance to non-horizontal loading, which permits short mooring line scopes and tighter moorings. Other features of direct-embedment anchors are listed in Table 1.3-3.

Table 1.3-1. Features of Shallow Foundations and Deadweight Anchors

Features of Shallow Foundations and Deadweight Anchors
1. Simple, on-site construction feasible, can be tailored to task.
2. Size limited only by load-handling equipment.
3. Reliable on thin sediment cover over rock.
4. Lateral load resistance decreases rapidly with increase in seafloor slope.
Additional Features of Deadweight Anchors
1. Vertical mooring component can be large, permitting shorter mooring line scope.
2. No setting distance required.
3. Reliable resisting force, because most resisting force is directly due to anchor mass.
4. Material for construction readily available and economical.
5. Mooring line connection easy to inspect and service.
6. A good energy absorber when used as a sinker in conjunction with "non-yielding" anchors (pile and plate anchors).
7. Works well as a sinker in combination with drag-embedment anchors to permit shorter mooring line scopes.
8. Lateral load resistance is low compared to other anchor types.
9. In shallow water, the large mass can be an undesirable obstruction.

Table 1.3-2. Features of Pile Foundation and Anchor Systems.

Features of Pile Foundations and Pile Anchors

1. Requires highly specialized installation equipment.

2. Transmits high axial loads through soft surficial sediments down to competent bearing soils or rock.

3. Can be designed to accommodate scour and resist shallow mudflows.

4. Can be installed and performs well on substantial slopes.

5. Can be installed in hard seafloors (rock and coral) by drill-and-grout technique.

6. Drilled-and-grouted piles require more specialized skills and installation equipment and incur high installation costs.

7. Wide range of sizes and shapes are possible (pipe, structural shapes).

8. Field modifications permit piles to be tailored to suit requirements of particular applications.

9. Costs are high and increase rapidly in deeper water or exposed locations where more specialized installation vessels and driving equipment are required.

10. Accurate soil properties are required for design.

Additional Features of Pile Anchors

1. High lateral capacities (greater than 100,000 lb) achievable.

2. Resists high uplift as well as lateral load, permitting use with short mooring line scopes.

3. Anchor setting not required.

4. Anchor dragging eliminated.

5. Short mooring line scopes permit use in areas of limited sea room or where vessel excursions must be minimized.

6. Pile anchor need not protrude above seafloor.

7. Driven piles are cost competitive with other high-capacity anchor when driving equipment is available.

8. Special equipment (pile extractor) may be required to retrieve or refurbish the mooring, or new pile and pendant must be installed.

9. More extensive and better quality site data are required than the data required for other anchor types.

10. Pile capacity goes to zero when its capacity as an anchor is exceeded and pullout occurs (is a "non-yielding" anchor).

Table 1.3-3. Features of Direct-Embedment Anchors.

Features of All Direct-Embedment Anchors
1. High capacity (greater than 100,000 lb) achievable.
2. Resists uplift as well as lateral loads, permitting moorings of short scope.
3. Anchor dragging eliminated.
4. Higher holding-capacity-to-weight ratio than other anchor types.
5. Handling is simplified due to relatively light weight.
6. Accurate anchor placement is possible; no horizontal setting distance necessary.
7. Does not protrude above the seafloor.
8. Possibly susceptible to strength reduction accompanying cyclic loading when used in taut moorings in loose sand and coarse silt seafloors.
9. For critical moorings, soil engineering properties required.
10. Anchor typically not recoverable.
11. Anchor may be susceptible to abrasion or fatigue.
Features Unique to Screw-In, Vibrated-In, and Hammer-Driven Plate Anchors
1. Can better accommodate layered seafloors (seafloors with variable resistance) because of continuous power expenditure during penetration.
2. Penetration is controlled and can be monitored.
3. Surface vessel must maintain position during installation.
4. Operational water depth is limited by power and line strength when using surface-powered equipment.
5. Operation limited to sediment seafloors.

Table 1.3-4 lists features of drag-embedment anchors. Although these anchors can develop high capacities, the load on a drag anchor is usually limited to one direction, and the mooring line angle at the seafloor must be virtually horizontal. The holding capacity of drag anchors decreases very quickly as mooring line angles exceeds approximately 6°.

To assist in understanding the advantages and disadvantages of the various anchor types, Table 1.3-5 compares how well they function under different conditions. Judgments of expected performance have been made primarily on the basis of holding capacity and relative cost. It should be noted that Table 1.3-5 is an expeditious guide for general use, and special circumstances can shift the performance ratings.

Table 1.3-4. Features of Drag-Embedment Anchor Systems

Features of Drag-Embedment Anchors
1. Wide range of anchor types and sizes available.
2. High capacity (greater than 100,000 lb) achievable
3. Most anchors are standard off-the-shelf equipment.
4. Broad experience with use.
5. Can provide continuous resistance even though maximum capacity has been exceeded.
6. Is recoverable.
7. Does not function well in rock seafloors.
8. Behavior is erratic in layered seafloors.
9. Low resistance to uplift loads; therefore, large line scope required to cause near horizontal loading at seafloor.
10. If dragging is not acceptable, anchor must be pulled horizontally at high loads (higher than expected service load) to properly penetrate and set.
11. Dragging of anchor to achieve penetration can damage pipelines, cables, etc.
12. Loading must be limited to one direction for most anchor types and applications.
13. Exact anchor placement limited by ability to estimate setting distance.

Table 1.3-5. Performance of Foundation and Anchor Types as a Function of Seafloor and Loading Conditions

Item	Performance[a] for Following Types:			
	Deadweight	Pile	Direct Embedment	Drag
Seafloor Material Type				
Soft clay, mud	++	+	++	++
Soft clay layer (0-20 ft thick), over hard layer	++	++	o	+
Stiff clay	++	++	++	++
Sand	++	++	++	++
Hard glacial till	++	++	++	+
Boulders	++	o	o	o
Soft rock or coral	++	++	++	o
Hard, monolithic rock	++	+	+	o
Seafloor Topography				
Moderate slopes, <10 deg	++	++	++	++
Steep slopes, >10 deg	o	++	++	o
Loading Direction				
Downward load component (foundations)	++	++	o	o
Omni-directional (not down)	++	++	++	o
Uni-directional (not-down)	++	++	++	++
Large uplift component	++	++	++	o
Lateral Load Range				
To 100,000 lb	++	+	++	++
100,000 to 1,000,000 lb	+	++	+	++
Over 1,000,000 lb	o	++	o	o

[a] ++ =   functions well

   +   =   normally is not the preferred choice

   o   =   does not function well

1.4 REFERENCES

1-1.   Unified Facilities Criteria (UFC): Military Harbors and Coastal Facilities, Naval Facilities Engineering Command (Preparing Activity), UFC 4-150-06. Washington, DC, Dec 2001.

1-2.   API Recommended Practice for the Analysis of Spread Mooring Systems for Floating Drilling Units, 2nd Edition, American Petroleum Institute, API RP 2P. Dallas, TX, May 1987.

1-3.   Rules for Building and Classing Mobile Offshore Drilling Units, American Bureau of Shipping, ABS 6-2008. New York, NY, Oct 2008.

[This page intentionally left blank]

2 SITE SURVEY AND IN-SITU TESTING

2.1 INTRODUCTION

2.1.1 Purpose

This chapter summarizes the considerations and methods used to aid in the selection and characterization of a site for bottom-resting or moored platforms in the deep ocean.

2.1.2 Factors Influencing the Site Survey

2.1.2.1 Constraints

There are many factors that influence site surveys, including the survey constraints. In general, the type and detail of site data sought will be a function of the following constraints:

- Value and replacement cost of platform
- Impact of platform failure (primarily) on human life or project risk
- Purpose of the platform
- Topography and seafloor material type
- Any pre-survey requirements for an exact geographical location
- Types of man-induced and environmental loadings
- Type and size of the foundations or anchors
- Availability of personnel, equipment, and survey support platforms

2.1.2.2 Minimum Required Data

Table 2.1-1 is a summary of site data requirements for various geotechnical engineering applications. The level of importance or need for each site data element is also identified in the table. A "High" need indicates the information is mandatory, while "N/A" indicates the data is not needed for design. "Low" in Table 2.1-1 indicates a low requirement level, which may result from either: (1) a low impact of this data element on the system design and performance, as in the low impact of micro-topography on drag anchor performance or (2) a technical inability to use this data element in design because analysis techniques are not developed, as in the inability to use dynamic soil properties in drag anchor design due to an absence of a performance-related model. Table 2.1-2 lists site data required for each of the geotechnical engineering applications.

Table 2.1-1. Site Data Requirements for Categories of Geotechnical Engineering Applications

Geotechnical Engineering Application	Requirements for Following Site Data:								
	Bathymetry		Material Thickness	Sediment				Rock	
	Macro (>3ft)	Micro (<3ft)		Index Properties	In-Situ Strength	Laboratory Strength	Dynamic Response	Index Properties	Laboratory Strength
Shallow Foundations/ Deadweight Anchors	High	High	Low	High	High	Low	Low	Low	High
Deep Foundations/ Pile Anchors	High	Low	High	High	High	High	High	High	High
Direct-Embedment Anchors	Low	N/A	High	High	High	High	High	High	Low
Drag Anchors	High	Low	High	High	Low	Low	Low	N/A	High
Penetration	N/A	N/A	High	High	High	Low	High	High	N/A
Breakout	Low	Low	N/A	High	Low	High	N/A	N/A	Low
Scour	High	High	N/A	High	Low	N/A	N/A	N/A	High
Slope Stability	High	High	High	High	High	High	Low	Low	High

Table 2.1-2. Soil Parameters Normally Required for Categories of Geotechnical Engineering Applications

Geotechnical Engineering Applications	Soil Classi-fication	Grain Size	Atterberg Limits	Strength Properties		Compression Properties			Depth of Survey
				Clay	Sand	Clay		Sand	
				s_u, S_t	$\bar{\phi}$	$c_{sv} k$	C_c	C_c	
Shallow Foundations	Yes	Yes	Yes	Yes	Yes	Yes	Yes	Yes	1.5 to 2 x foundation width
Deadweight Anchors	Yes	No	No	Yes	Yes	No	No	No	1.5 to 2 x anchor width
Deep Pile Foundations	Yes	Yes	Yes	Yes	Yes	Yes	Yes	No	1 to 1.5 x pile group width, below individual pile tips
Pile Anchors	Yes	Yes	Yes	Yes	Yes	No	No	No	To depth of pile anchor
Direct-Embedment Anchors	Yes	Yes	No	Yes	Yes	Yes	No	No	To expected penetration of anchor; maximum 33 to 50 ft clay; 13 to 33 ft sand
Drag Anchors	Yes	Yes	No	Yes	No	No	No	No	33 to 50 ft clay; 10 to 16-½ ft sand for large anchors
Penetration	Yes	Yes	No	Yes	Yes	No	No	No	33 to 50 ft clay; 13 to 33 ft sand
Breakout	Yes	Yes	Yes	Yes	Yes	No	No	No	1 x object width plus embedment depth
Scour	Yes	Yes	No	Yes	No	No	No	No	3.3 to 16-½ ft; related to object size and water motion
Slope Stability	Yes	Yes	Yes	Yes	Yes	No	No	No	33 to 100 ft; more on rare occasions

2.1.2.3 Regional Versus Site-Specific Surveys

Some projects or project phases require general information from a large region, whereas others require more accurate data from a smaller geographic area. For example, a manned habitat installation may require low-precision data resulting from a regional survey over a large area to determine an adequate or a best location for its placement; whereas, design for the habitat's foundation needs high-precision data from the selected site.

Since regional surveys compare sites or cover large distances, detailed information is generally neither possible nor needed. Site-specific studies on the other hand require more detailed information that is used in design. Regional surveys typically include geophysical data collection with limited soil sampling, such as gravity coring. Deep soil borings and in-situ tests are not typically used for site selection. Once a site has been selected however, site-specific data are usually collected by from by soil sampling and by in-situ testing. Geophysical and geological information beyond that collected during the regional survey may also be needed, depending on the project complexity.

2.1.2.4 Hazardous Conditions

The scope of the site investigation will be influenced by an acceptable level of risk of the project to geotechnical hazards; including earthquake loading, faulting, liquefaction, submarine landslides, gas hydrates, erosion, and the presence of underconsolidated sediments. These conditions must be assessed for sites being evaluated.

Table 2.1-3 outlines historical environmental information needed for assessment of geotechnical hazards from earthquakes, winds, waves, and currents. Investigation of environmental factors and hazardous features can be pursued initially by an examination of existing maps, charts, and bottom environmental data (Section 2.2.1).

Table 2.1-3. Historical Environmental Information Needed to Assess Geotechnical Hazards

Hazard	Information Needed
Earthquakes	Frequency, Magnitude, Peak Accelerations, Response Spectra
Wind	Velocity Distribution, Direction Distribution, Maximum Wind Velocity
Waves	Wave Height Distribution, Maximum Wave Height, Direction Distribution
Currents	Vertical Velocity Profile, Distribution of Current Velocity

2.1.2.5 Positioning Capability

The ability to reference a site survey and position a platform on the seafloor may dictate the scope of the site investigation. A precise determination of horizontal and vertical position is a critical aspect of both geophysical and geotechnical investigations for a specific installation.

Positioning an object on the seafloor usually requires location of the object with respect to the surface vessel and location of the vessel with respect to geographical coordinates. Using a state-of-the-art Global Positioning Systems (GPS), the vessel position can be measured to an accuracy of 3 to 5 meters. Positional accuracy can be improved to 1 meter if a Differential GPS (DGPS) system is used. DGPS systems use a network of ground based reference stations to enhance the accuracy of the GPS derived location. The U.S. Coast Guard runs a DGPS system that is broadcast at major waterways and harbors.

Relative seafloor-to-surface positions are measured by sonar transponders to accuracies on the order of 0.1% of the distance being measured.

2.2 DESK TOP STUDY

2.2.1 Information Sources

In a preliminary survey of a site, one important step, generally referred to as a "desk top study," is the search for available information from previous investigations near the site. Findings from the desk top study can provide area information, as well as site-specific data, and aid in planning for a more detailed survey.

Information can be obtained from a variety of governmental, industrial, and educational institutions. Sources of information on geological and geotechnical properties of ocean sediments and on earthquake and earthquake effects are given in the following lists. Many of the sources listed below have online data repositories that can be accessed online by searching for the source name through any standard search engine.

Universities and Government Organizations

- Naval Facilities Engineering Service Center, Oceans Department

- Digital Bathymetric Data Base Variable Resolution (DBDB-V) from the U.S. Naval Oceanographic Office (NAVOCEANO)

- USGS Woods Hole Coastal and Marine Science Center

- USGS Pacific Coastal and Marine Science Center

- National Ocean Service, National Oceanic and Atmosphere Administration

- Lamont-Doherty Earth Observatory, The Earth Institute at Columbia University

- National Environmental Satellite, Data, and Information Service (NESDIS), National Oceanic and Atmospheric Administration Satellite and Information Service

- Naval Oceanographic Office, NAVO

- Scripps Institution of Oceanography

- Woods Hole Oceanographic Institution

- Texas A&M University, Ocean Drilling Program Janus Database

Journals and Conference Proceedings

- U.S. Exclusive Economic Zone (EEZ) GLORIA Mapping Program

- Canadian Geotechnical Journal, National Research Council of Canada

- Civil Engineering in the Oceans (I through VI), American Society of Civil Engineers

- Geotechnique, The Institution of Civil Engineers, London, England

- Journal of Geotechnical and Geoenvironmental Engineering, American Society of Civil Engineers

- Marine Georesources & Geotechnology, Taylor & Francis

- Ocean Engineering, An International Journal of Research and Development, Elsevier

- Offshore Technology Conference, Houston

Sources of Data on Earthquakes and Earthquake Effects

- Incorporated Research Institutions for Seismology, IRIS

- National Earthquake Information Center, NEIC, United States Geological Survey

- Consortium of Organizations for Strong Motion Observations Systems, COSMOS

- United States National Strong-Motion Project, NSMP, United States Geological Survey

- Internet Site for European Strong Motion Data, ISED

- American Meteorological Society

- Bulletin of the Seismological Society of America

- Journal of Geophysical Research, American Geophysical Union

- Seismological Laboratory, California Institute of Technology, Division of Geological and Planetary Sciences

- University of California, Berkeley Seismoligical Laboratory

- University of Hawaii at Manoa, Hawaii Institute of Geophysics and Planetology

- University of Tokyo, Hongo, Bunkyo-Ku, Tokyo, Japan

2.2.2 Typical Ocean Sediments

2.2.2.1 Sediment Types

Seafloor sediments are referred to by origin as either terrigenous (land-derived) or pelagic (ocean derived). Figure 2.2-1 shows the distribution of surface sediments over the world's oceans.

A majority of terrigenious sediments are located on the continental shelves and slopes and are also known as neritic sediments. Terrigenous sediment may also be found beyond the continental slope, on the continental rise and abyssal plain, as a result of transport by slope failures and turbidity currents.

Terrigenous soils, described below, include gravels, sands, silts, and clays. These soils are formed on or adjacent to land; are transported by currents, wind, or iceberg rafting to the deep sea; and contain >30% silt <u>and</u> sand-sized particles of land origin.

- <u>Terrigenous Silty Clays, or Muds</u> – bordering continents

- <u>Turbidites</u> – sand, silt, and clay deposits transported great distances into deep water areas by turbidity currents; characterized by graded beds—sands at bottom grading to clays at top

- <u>Slide Deposits and Volcanic Ash</u> – derived from slumps on marine slopes or from volcanoes

- <u>Glacial Marine Soils</u> – coarse-grained sediments produced by glacial scouring of land features

Pelagic sediments, described below, include abyssal clays, siliceous oozes, and calcareous oozes. These soils are formed in the sea, are composed of clays or their alteration products or skeletal material from plants or animals, and cover 75% of the seafloor.

- Abyssal Clays – contain <30% biogenous material, silty clays of very high plasticity

- Authigenic Deposits – minerals precipitated and crystallized in seawater, predominantly manganese nodules and phillipsite

- Biogenous Oozes – derived from marine organisms and plants

  - Calcareous Oozes – contain >30% biogenous calcium carbonate material, includes coralline deposits, calcareous sands (oolithes) and shells, and fine-grained remains of microscopic animals (oozes)

  - Siliceous Oozes – contain >30% siliceous fine-grained remains of plants and animals

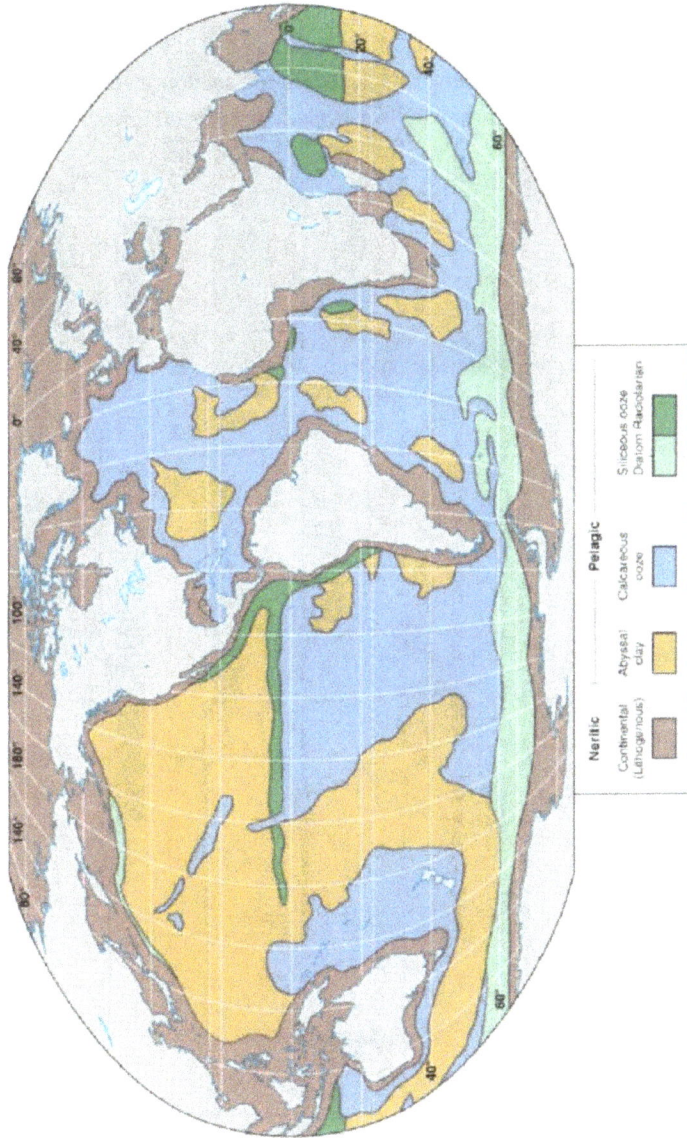

Neritic

Continental
(Lithogenous)

Pelagic

Abyssal
clay

Calcareous
ooze

Siliceous ooze

Diatom Radiolarian

Figure 2.2-1. Ocean Sediment Distribution (Ref. 2-1).

2-9

The pelagic sediments are composed primarily of wind-blown dust (abyssal clays) or calcareous or siliceous biogenous materials. The abyssal clays are usually found at water depths deeper than the carbonate compensation depth (CCD) (Figure 2.2-2). The CCD is defined as the depth above which the calcium carbonate dissolution is less than the carbonate supply. This depth is nominally 4,500 meters but can vary with temperature and salinity. Calcareous ooze (sediment composed of at least 30% by weight of remains of organisms whose hard parts are calcium carbonate) is usually found at depths shallower than the CCD. Siliceous oozes are found in ocean areas of high surface productivity, usually where the seafloor depth is below the CCD. A world wide distrubtuion of calcium carbonate in surface sediments is shown in Figure 2.2-3.

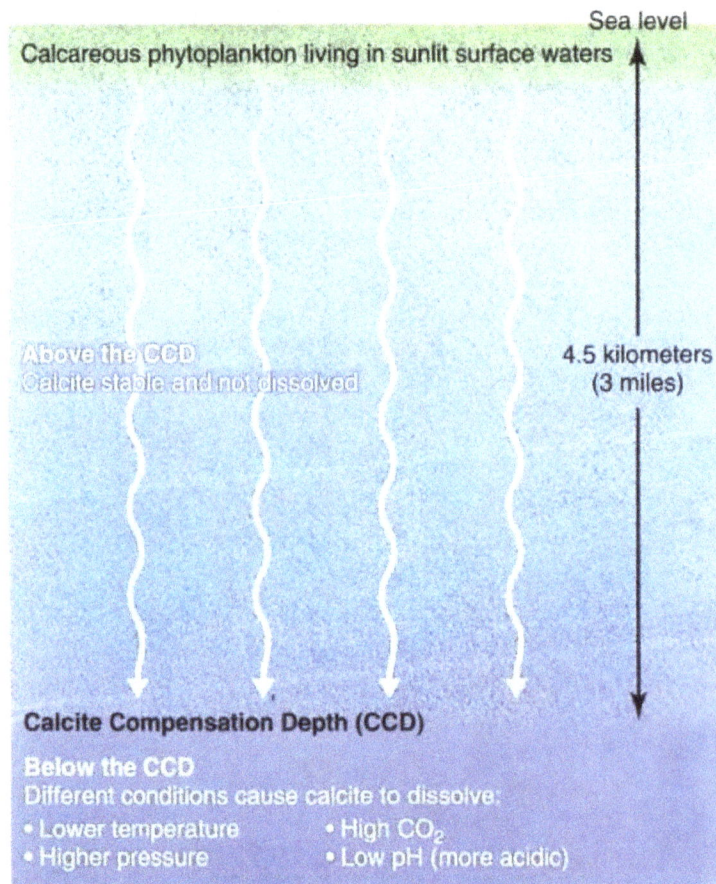

Figure 2.2-2. Characteristics of water above and below the calcite compensation depth (CCD) (Ref. 2-1).

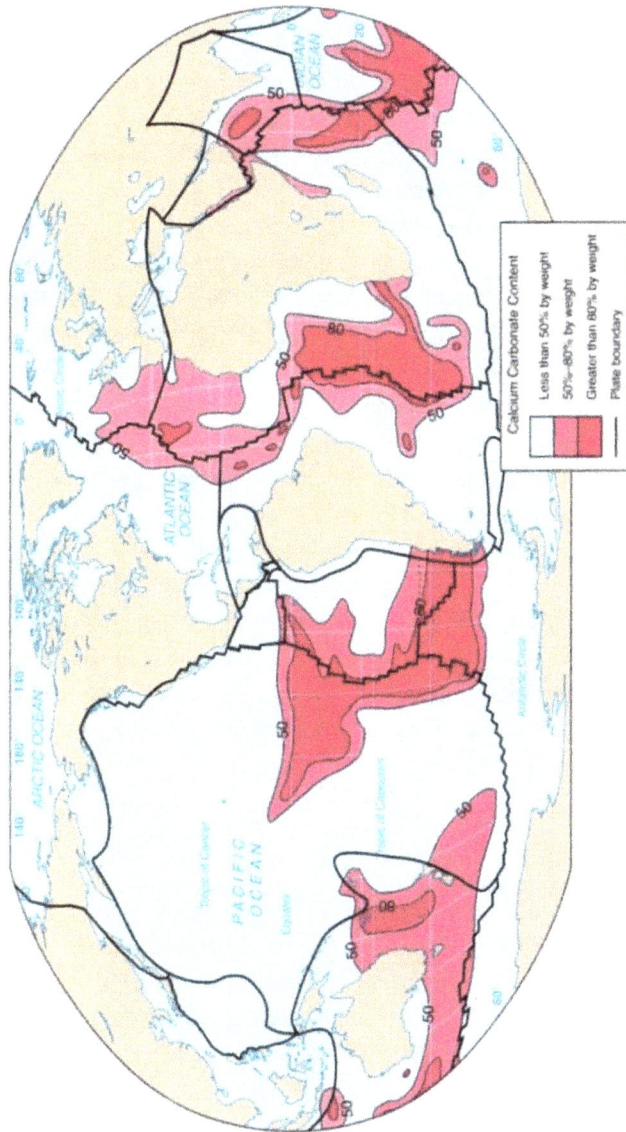

Figure 2.2-3. Distribution of calcium carbonate in modern surface sediments (Ref. 2-1).

2-11

2.2.2.2 Sediment Consolidation State

Three terms are used to describe the existing state of a soil: overconsolidated, normally consolidated, and underconsolidated. Overconsolidated sediments have been subjected to a greater load (overburden) in the past than exists at present. They have been compressed and become stronger. The overconsolidation (OC) phenomenon can also result from many chemical or physical processes. Normally consolidated (NC) materials have never been loaded by overlying material more than they are now. Underconsolidated sediments are "young." That is, they have not come to equilibrium with the weight of overlying materials and are weaker than they will be when this equilibrium is reached. Permeability characteristics generally limit underconsolidation to the fine-grained cohesive soils.

The degree of consolidation is important to a site investigation because it dictates the existing state and, therefore, the strength of the material. In a normally consolidated soil, strength generally increases with depth in proportion to the weight of soils which lie above. For a particular soil, strength at equal depths below the seafloor will be greatest for overconsolidated soils and least for underconsolidated soils. Soil compressibility will vary inversely with the degree of consolidation, being least for overconsolidated and most for underconsolidated.

For pelagic sediments, it is usually correct and conservative to assume that they are normally consolidated. In contrast, terrigenous (neritic) sediments are often overconsolidated, particularly those sediments that were exposed when the sea level was significantly lower than at present. No consistent rule exists for locating overconsolidated sediments except that exposed locations (such as tops of rises or passages) are more likely to be overconsolidated than are protected areas (such as basins). Underconsolidated sediments are almost always found where fine-grained soils are being deposited at a very high rate. In active river deltas, such as near the Mississippi River Delta, there may be little-to-no increase in soil strength with increasing depth below the sea bottom.

2.2.2.3 Estimating Soil Properties

For planning a geotechnical survey or design of a nonsensitive, bottom-resting device, estimates of the soil engineering properties can be developed when the marine geological province is known. First, the probable soil type for that province is identified from Figure 2.2-4; then the soil shear strength and buoyant unit weight parameters are estimated from data extrapolations presented in Figure 2.2-5 through Figure 2.2-9. Additional discussion of sediment types and properties can be found in other chapters where they apply to specific chapter subjects.

CONTINENTAL SHELF AND SLOPE | CONTINENTAL RISE | ABYSSAL PLAIN | MID-OCEAN RISE AND RIDGE | ABYSSAL HILLS

BELOW CCD** | ABOVE CCD** | ABOVE CCD** | BELOW CCD**

TERRIGENEOUS | PELAGIC CLAY OR TURBIDITE (CLAY-SILT-SAND) (OR SILACEOUS OOZE) | CALCAREOUS OOZE | PELAGIC CLAY (OR SILACEOUS OOZE)

ABOVE 10,000 FT | BELOW 10,000 FT

COARSE | FINE

** CCD IS THE CARBONATE COMPENSATION DEPTH (THE DEPTH BELOW WHICH CARBONATE MATERIALS WILL DISSOLVE)

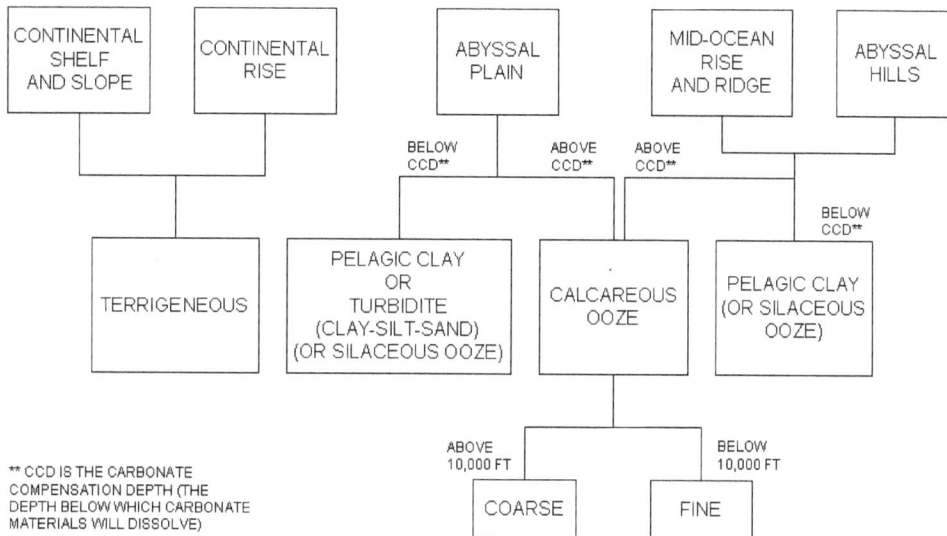

Figure 2.2-4. Marine geological provinces and probable soil types.

When the site is on the continental shelf or slope, the sediment is assumed terrigenous. Available National Ocean Survey charts should be consulted to determine whether the sediment is primarily cohesionless (sandy) or cohesive (mud or clay). If the sediment is cohesive, Figure 2.2-5 is used for the strength distribution of normally consolidated sediment. A literature search is made for strong indications of overconsolidation (e.g., recorded outcrops of older sediments or an exposed location such as rise top, high recorded bottom currents). If sufficient evidence exists that overconsolidated soils are suspected, it would be prudent to drop some penetrometers or short gravity corers.

Nonpenetration or slight penetration with attainment of minimal sample length suggests that overconsolidated sediment does indeed exist. Typical sand properties are given in Figure 2.2-6. If the location is near a large active river delta, the site must be surveyed directly.

When the site location is beyond the continental margins, the probable sediment type can be identified from Figure 2.2-1. If the sediment is classed as a turbidite, Figure 2.2-6 gives typical parameters for proximal and distal turbidites. The distinction is made based on the distance from a source of sand (e.g., the edge of the continental shelf) as follows: if the distance is greater than about 30 miles, the sediment is probably a distal turbidite.

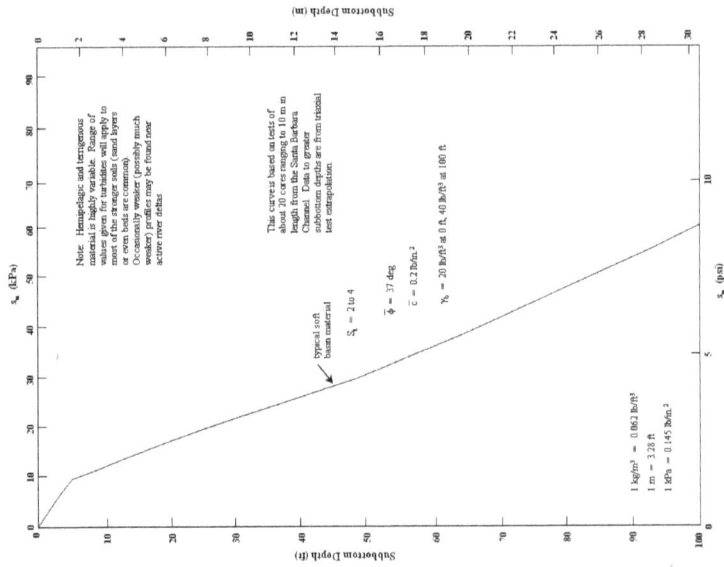

Figure 2.2-6. Typical strength profiles for proximal and distal turbidites.

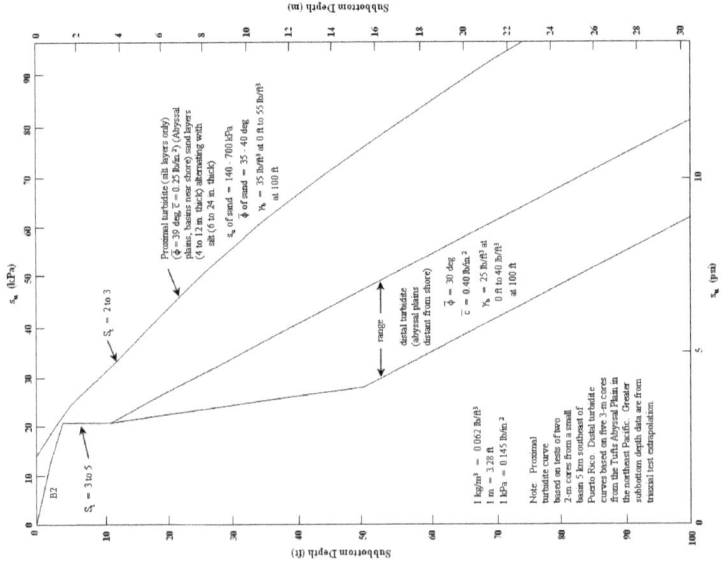

Figure 2.2-5. Typical strength profile for hemipelagic terrigenous silty clay.

2-14

Figure 2.2-8. Typical strength profiles for abyssal clay.

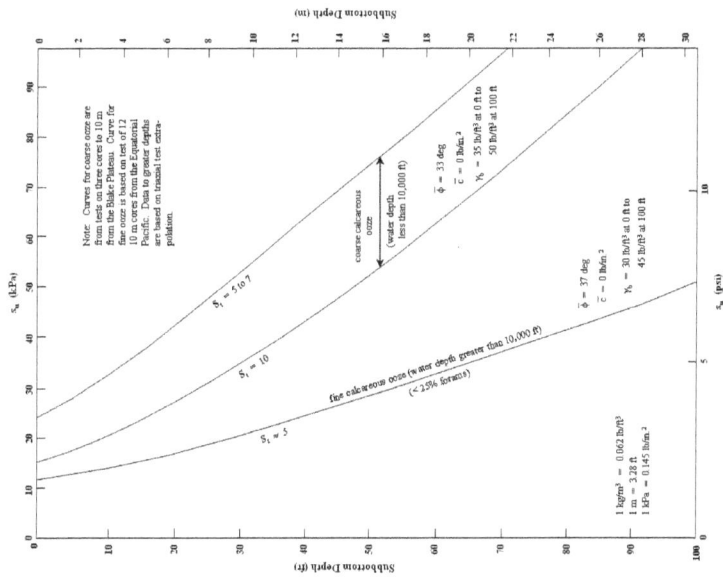

Figure 2.2-7. Typical strength profiles for calcareous ooze.

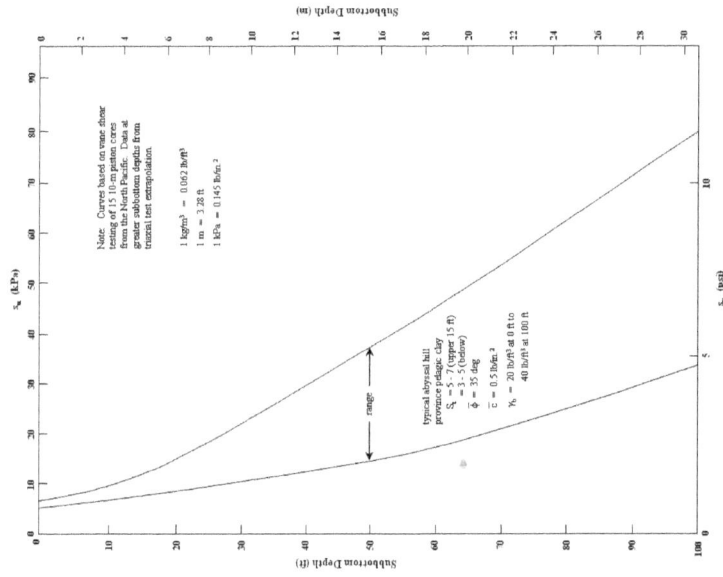

2-15

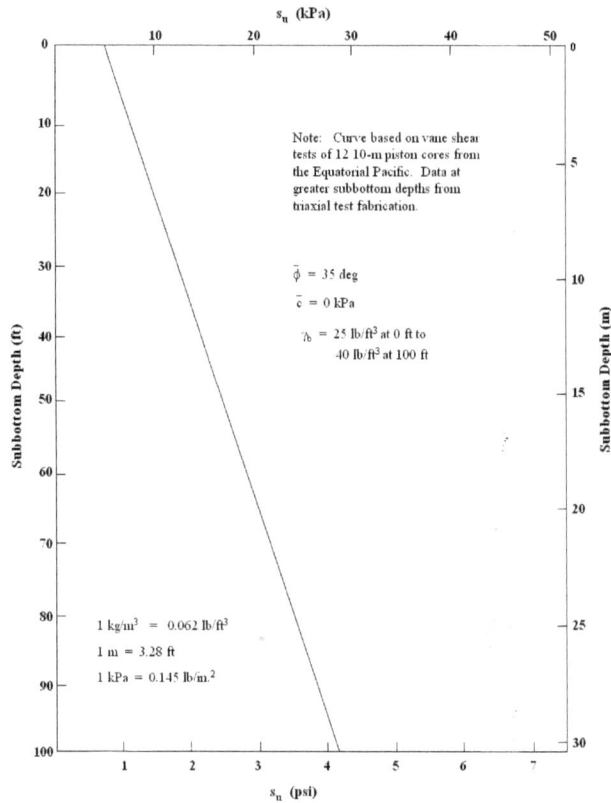

Figure 2.2-9. Typical strength profile for siliceous ooze.

If the sediment is classed as siliceous ooze (diatom or radiolarian ooze, Figure 2.2-1), the typical properties can be found in Figure 2.2-9. If the site is in the deep ocean and not an abyssal plain, it must be determined whether its water depth lies above or below the CCD (Figure 2.2-3).

1. If the site is above the CCD, the sediment is probably calcareous ooze. Figure 2.2-7 gives the typical properties; it should be noted that a further subdivision between coarse and fine ooze is made at the 10,000-foot level.

2. If the site is below the CCD, the sediment is probably abyssal clay or siliceous ooze. Figure 2.2-8 presents estimates of properties for abyssal clay.

Whenever possible, specialists in seafloor soils behavior should be consulted as they can provide information difficult to glean from the open literature. Many parts of the seafloor have been mapped for sediment distributions, and much more detailed information than is given in this discussion may be available. In addition, many core sample descriptions are available. Sources for experts, maps, and core descriptions are listed in Section 2.2.1.

2-16

2.3 REGIONAL SURVEYS

2.3.1 General

Regional survey techniques include subbottom profiling, limited soil sampling, sidescan sonar surveys, and direct (visual) observation. Data obtained from the regional survey are often qualitative only from the standpoint of physical soil behavior. For example, subbottom profiling can be used to delineate major soil layers but gives little if any information about soil properties. Table 2.3-1 summarizes the steps in a typical regional survey.

Table 2.3-1. Steps in a Typical Regional Survey

Steps	Procedure
1	Perform literature and data bank search
2	Identify facility to be installed
3	Identify parameters that impact the siting and design of facility

4	Plan acoustic reconnaissance program for identifying relevant geological hazards
	    a. Acoustic subbottom profiler
	    b. Survey line spacing, 500 ft

5	Plan shallow sampling program consistent with geotechnical hazard identification and soil parameter evaluation
	    a. Gravity cores
	    b. One core per 1/4 mi^2

6	Perform acoustic and shallow sampling

7	Assign soil tests, test soil, and select design soil parameters; for each 6-1/2 ft of core:
	    a. Two bulk unit weight tests
	    b. Two water content tests
	    c. Two grain size analyses (for sand and silt)
	    d. One Atterberg limit test (cohesive soils only)
	    e. One shear strength test (cohesive soils only)

8	Conduct post survey analysis to evaluate cost versus risk for the proposed sites; select and rank the sites

2.3.2 Subbottom Profiling

Subbottom profiling techniques use reflected sound signals to develop a profile of the seafloor bathymetry and subbottom layering. Compressive waves formed by a controlled sound source propagate outward through the seawater at a certain velocity. When the compressive wave encounters another medium (i.e., surficial sediments or deeper, denser sediment layers) with different acoustic properties, a portion of the energy is reflected. The times of arrival of the reflected waves are recorded, producing a profile of acoustic interfaces in the sediments.

High-resolution, continuous, subbottom profiling is commonly used for studying the upper 300 feet of soil. These devices are often referred to as sediment sounders, boomers, or sparkers. They are characterized by their transmission frequency and, consequently, the penetration of the signal and its resolving power. In general, lower frequencies produce greater penetration with lower resolution, while high-frequency systems yield less penetration but have a greater resolution. Typically, geotechnical engineering needs are best served using a 12-kHz sounder system to develop profiles of the seafloor surface and a chirp subbottom profiler system to delineate the sediment strata, near-surface rock contact, and surface faulting. An exmple of data from a 2-16 kHz system is shown in Figure 2.3-1.

2.3.3   Limited Sampling

Regional surveys should include some bottom sampling in order to provide examples of surficial sediment types and consistencies. In areas of outcropping sediment layers, acoustic profiling and surficial sampling can be used together to provide information on the sediment type and projected properties of subbottom layers of an area. Samplers used for such surveys would generally be of the less sophisticated variety – the grab samplers, short gravity corers, and rock dredges. A discussion of sampling equipment is included in Section 2.4.

2.3.4   Sidescan Sonar

Sidescan sonar systems provide graphic records that show two-dimensional plan views of the seafloor topography. Seafloor objects as well as gas bubbles are detected and displayed as in an aerial photograph.

The sidescan sonar operates by emitting high-frequency sound waves in narrow beam pulses from a transducer "fish" that is towed off the stern of a ship (Figure 2.3-2). The fish is towed above the seafloor at a height dictated by the chosen range. The returning acoustic signals are received by the same fish and transmitted by electrical or fiber optic cable to the ship. On deck, data acquisition systems transform these reflected signals into an acoustic water fall image on a monitor. Depending on the scale selected, this image can record a continuous path of the seafloor from 75 to 3,000 feet wide. Individual lines are processed into a mosaic image of the surveyed area.

Sidescan sonar surveys can be used to detect seafloor obstructions, such as sunken ships, pipelines, sediment flows, and rock outcrops (Figure 2.3-3). By studying the results of a sidescan sonar survey, an undesirable site can sometimes be avoided during preliminary site selection.

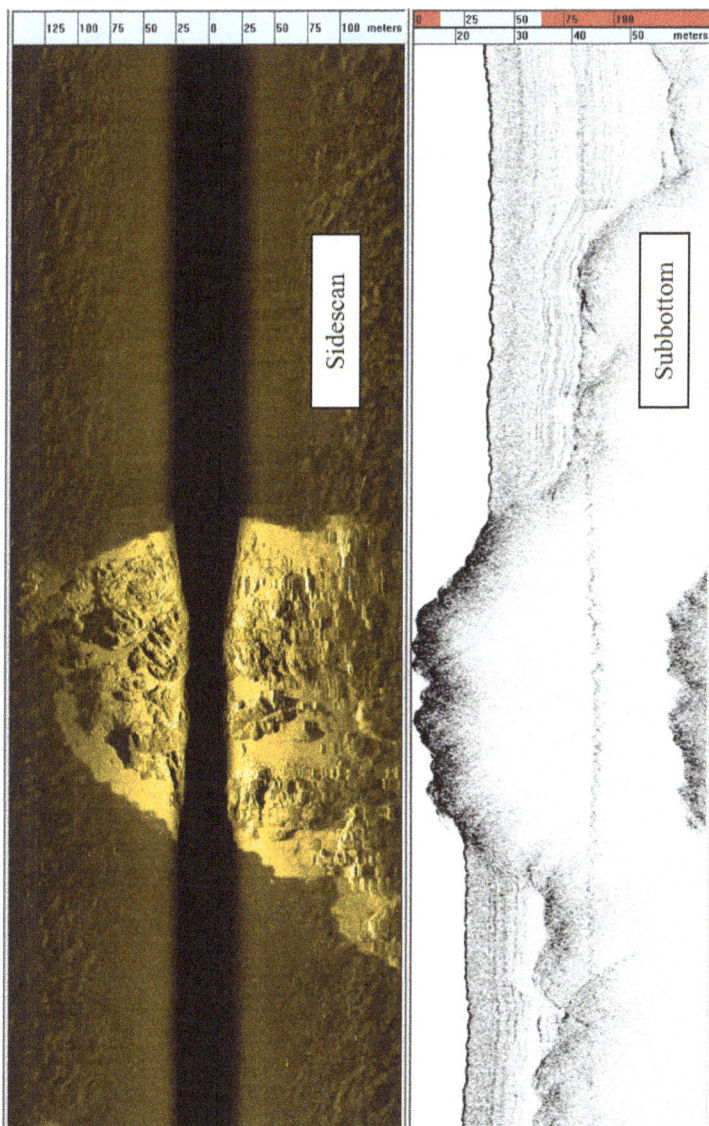

Figure 2.3-1. Example of subbottom profiler data (Ref. 2-2).

2-19

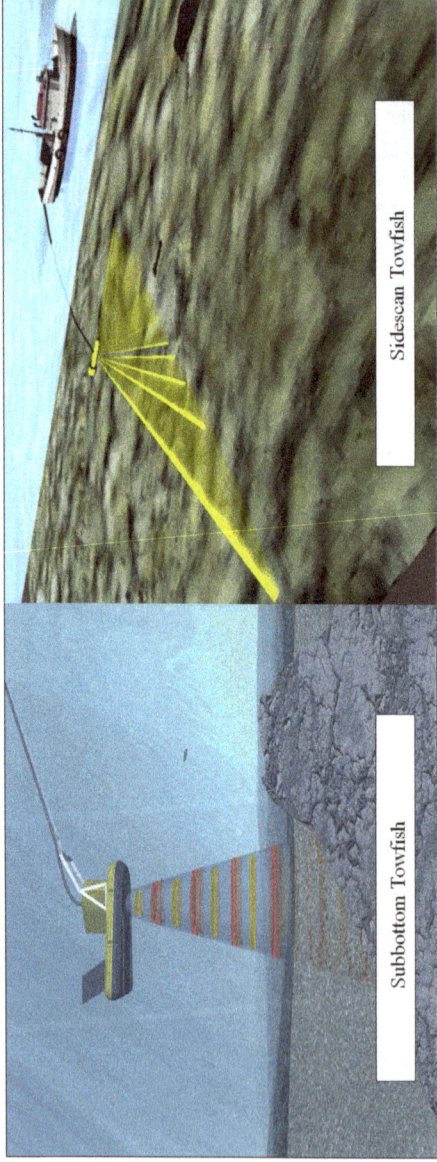

Sidescan Towfish

Subbottom Towfish

Figure 2.3-2. Acoustic data collection operations (Ref 2-2).

WWII Submarine

Pipeline

Figure 2.3-3. Example of sidescan acoustic data (Ref. 2-2).

2.3.5 Visual Observation

Visual observations of the seafloor are made by three general methods: (1) direct observation, (2) use of a remotely controlled still camera, and (3) underwater video cameras. Divers can make direct observations in shallow water. In deep water, visual observations are generally made from a Remotely Operated Vehicle (ROV). Visual observations become necessary when other survey techniques cannot provide necessary site data. For example, remote survey techniques may not be able to delineate the extent of a talus deposit at the base of a steep rock slope. A survey by an ROV, possibly with some waterjetting to remove sediment infilling, could provide such information.

2.3.6 Survey Line Spacing

For small and relatively low risk (unmanned, low cost) installations, a regional survey may not be required if general information on the site bathymetry and stratigraphy is available.

Major structures (such as manned gravity platforms or pile-supported platforms), however, may require a full-scale regional survey. The following guidelines are suggested for mesh spacing of regional surveys for such structures where the overall survey area is to cover a 1-by 1-mile area.

- For bathymetric surveys, a mesh spacing of about 100 feet is recommended at locations where significant bottom change is occurring, and a spacing of 500 feet is recommended at other locations.

- The mesh spacing for subbottom profiling is essentially the same as for bathymetric survey. If the soil is generally homogeneous, the profile spacing can be increased.

2.4 SITE-SPECIFIC SURVEY

2.4.1 General

Soil properties at a site can be established by sampling the soil and returning it to a laboratory for testing. Generally, in shallow waters the techniques of drilling and sampling used on land can be adapted, utilizing a jackup barge or a fixed platform. In the deep ocean, because of the great water depth, the sampling of soil sediments often involves more complicated equipment and techniques than sampling on land or near shore. In deeper waters, sampling must be done from a floating vessel. Gravity corers and vibracorers are usually used to obtain samples in the upper 10 to 20 feet. Below this soil depth, drilling rigs and wireline sampling techniques are normally used. The performance of these sampling techniques in the deep ocean is limited by the handling capability of the supporting vessel and the weather conditions. Table 2.4-1 summarizes the steps of a typical site-specific survey.

Table 2.4-1. Steps in a Typical Site-Specific Survey

Steps	Procedure
1	Check data bank and literature search for completeness
2	Identify facility to be installed
3	Identify soil parameters and geotechnical hazards impacting on design of facility
4	Identify types of information needed to complement existing data (from preliminary studies and regional survey) [see Tables 2.1-1 and 2.1-2]
5	Plan geophysical data collection as necessary to complete data collection a. Close survey line spacing to 100 ft b. Make additional or more accurate seafloor profiling (deep tow) and subbottom profiling
6	Complete shallow sampling. <u>Note</u>: shallow sampling may be sufficient depending on the platform type and size a. Gravity corer generally acceptable b. Spacing of coring locations, 300 ft c. Core to a depth of 1.5 times expected width of foundation or to maximum expected penetration of anchor, if possible
7	Perform deep sampling where necessary; use in-situ tests for high-risk platforms: a. Core to a depth of 1.5 to 2.0 times expected width of foundation or to maximum expected penetration of anchor b. Sampling frequency (within single boring): (1) One sample/5 ft, up to 30 ft (2) One sample/10 ft, between 30- and 200-ft interval (3) One sample/20 ft, over 200 ft c. In-situ tests: (1) Use vane for clay undrained properties (2) Use cone penetrometer for sand or clay and to define strata boundaries (3) Use pressuremeter for soil compression properties (4) Define strata boundaries by borehole logging
8	Assign soil tests, test soil samples, and select design parameters. For every core sample: a. One water content test b. One bulk unit weight test c. One grain size analyses d. One Atterberg limit test every other sample (cohesive soils only) e. Shear strength tests (cohesive soils only)

2.4.2 Shallow Sampling

Shallow soil sampling is usually conducted as part of the regional survey. For small seafloor installations, shallow-penetration samples may provide the soil parameters required for foundation design. The equipment used for shallow sampling in the deep ocean includes grab or dredge samplers, box corers, gravity corers, and i.e., vibracorers. A summary of shallow sampling tools and their application is given in Table 2.4-2.

Table 2.4-2. Shallow Soil Sampler Types and Applications

Sampler Type	Sample Quality	Maximum Sample Length (ft)	Application	Comments
Grab Sampler	low	2	Soil classification (USCS) Index property tests	Inexpensive, no water depth limitation
Box Corer	very high	2	Soil classification (USCS) Sample for: - Strength test - Index test	No water depth limitation; pretripping causes delays in deep water; best for seas below 7 ft
Gravity Corer Free Fall Short Corer Long Corer	high	4 10 30	Soil classification (USCS) Sample for: - Strength test - Index test	Free-fall is limited to 20,000 ft and may be difficult to find upon surfacing; others can use piston for higher sample quality; no water depth limitation; specialty piston corers can sample deeper soils
Vibracorer	moderate	20-40	Soil classification (USCS) Sample for: - Index test	Used primarily in sands; water depth limited by power umbilicals

2.4.2.1 Grab or Dredge Samplers

Grab or dredge samplers (Figure 2.4-1) offer the simplest method for obtaining seafloor soil samples. Because samples obtained by this method suffer from significant disturbance, grab or dredge samples have little value in evaluating soil strength characteristics. Large grab samplers are, however, often the only means for taking samples of gravels and pebbles for surficial sediment identification. The washing out of fines during sample recovery can be a problem with grab samplers. This can be minimized by use of samplers designed to minimize such sample loss (Ref. 2-3).

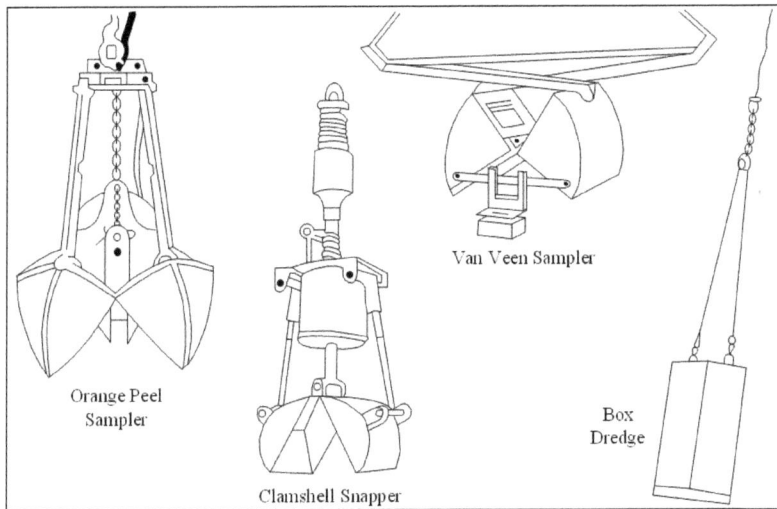

Figure 2.4-1. Grab samplers and dredges (Ref. 2-4).

2.4.2.2 Box Corers

Box corers obtain large volume of relatively undisturbed surficial soil. Sample sizes range from 1 to 2 feet in height and 0.25 to 3 ft^2 in area. A box corer typically consists of a weight column, sample box, spade and spade lever arm, and a tripod support frame (Figure 2.4-2).

2.4.2.3 Gravity Corers

Gravity corers are tube samplers that are driven into the soil by the kinetic energy of their falling mass. Most gravity corers are lowered by winch through the water column and free-fall through only the last 10 to 20 feet of the water column after being released by bottom contact of a trigger weight (Figure 2.4-3) or, when a trigger weight is not used, by free-wheeling of the winch.

Gravity corers may be of open-barrel or piston type. Open-barrel corers are relatively short corers and their use is limited to approximately the upper 10 feet of the seafloor. The piston corer incorporates a piston fixed at the mudline during penetration to improve soil recovery and recovered soil quality. Most gravity corers incorporate plastic barrel liners to promote sample quality and post-recovery soil handling, and use core catchers to limit core loss during retrieval. A comprehensive discussion of corer performance and coring techniques can be found in Reference 2-5. Section 2.2.1 listed many universities and research organizations that make and maintain bottom sampling equipment which may be available for rental. Mooring Systems Inc. sells gravity and piston corers (http://www.mooringsystems.com).

2-25

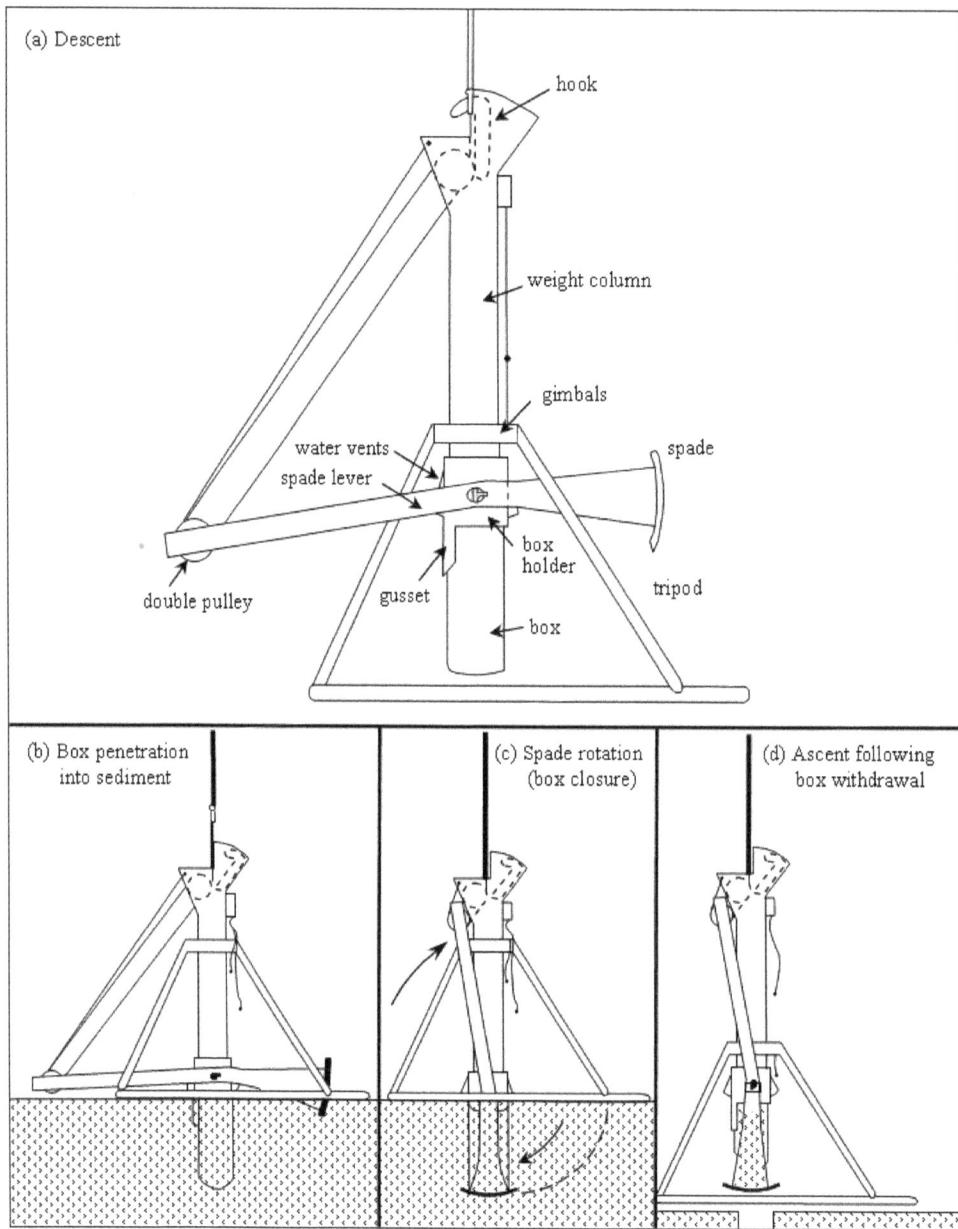

(a) Descent

hook

weight column

gimbals

water vents
spade lever

spade

box
holder

double pulley

gusset

tripod

box

(b) Box penetration
into sediment

(c) Spade rotation
(box closure)

(d) Ascent following
box withdrawal

Figure 2.4-2. Box corer and its operation sequence.

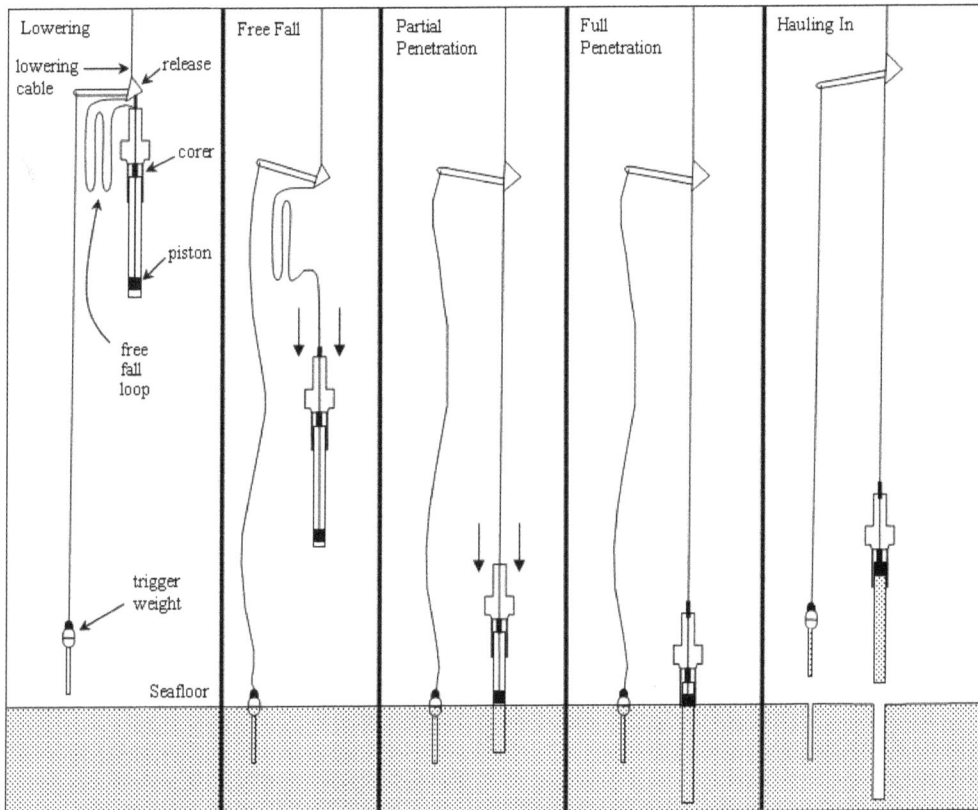

Figure 2.4-3. Long piston corer operation sequence with a short corer used as a trigger weight.

2.4.2.4 Vibracorers

Vibracorers are used in cohesionless sediments, where gravity corers often fail to retain the soil or are capable of only limited penetration. The typical vibracorer consists of a core barrel and driver-vibrator unit supported by a stand, as shown in Figure 2.4-4. The driver-vibrator may be of a rotating eccentric weight or reciprocating piston variety and it may be powered by a pneumatic, hydraulic, or electrohydraulic source. Most government and commercial organizations interested in marine geotechnical investigations maintain vibratory corers – some with only slight differences to accommodate company preferences. Most are limited to operation in less than 600-foot water depths because of the necessity to supply power from a vessel down to the corer. Ability to penetrate varies with the strength of material being cored. Unless very soft sediment is encountered, maximum coring length is limited to 40 to 50 feet.

Figure 2.4-4. Alpine vibracore sampler (from Ref. 2-6).

2.4.3 Deep Sampling

Offshore site investigations for major engineering structures, especially those placed on piles, require expensive deep soil sampling and recovery of representative samples. Deep boring techniques require a higher level of sophistication in equipment and in the drill vessel or platform than is needed to perform shallow sampling. Drilling and sampling operations are generally carried out from a fixed platform, a jackup platform, an anchored barge or ship, or a dynamically positioned ship.

The most versatile and economical approach today to obtain deep samples at sea is wireline sampling, which involves use of an anchored ship with a centerwell. Regular oil platform supply boats, which are about 150 to 200 feet long, are generally suitable for this type of operation. Drilling is performed with a rotary rig mounted on the ship deck over the centerwell. The drill bit is advanced into the soil by rotating a drill pipe while pumping drilling fluid down the pipe. The fluid carries the soil cuttings to the seafloor surface and is not returned to the ship (open circulation). The drill bit and casing are advanced to the elevation where a sample is desired and then the sample is taken.

One of two sampler types is used: wireline hammer samplers and wireline push samplers. The wireline hammer sampler uses successive vertical blows to advance the sampler and provides somewhat disturbed samples. However, it is economical and provides adequate information for pile design. Wireline push samplers advance the sampler with a continuous motion, which produces less soil sample disturbance. However, the operation of a push sampler requires a fairly calm sea since the motion of the ship is transferred to the sampler through the drill pipe. A limited number of drill ships do compensate for motion with specialty equipment, but this is an expensive complication to the basic drilling operation.

A discussion of borehole logging techniques for additional information about soil stratigraphy is given in Section 2.5.6.

2.4.4 Location, Number, and Depth of Sampling

The recommended location and number of borings are based on structure purpose and size, seafloor bathymetry, slope angle of the soil strata, and uniformity of the acoustic profiles. For most structures, if the acoustic profiles are practically uniform over an extensive area, one deep boring at the center of the structure location along with other, more shallow, data is often sufficient. If the subbottom acoustic profiles display a number of irregularities, more deep borings should be drilled. For small, low-risk installations, a few gravity cores taken at the site may provide sufficient information for foundation design.

The depth to which soil sampling is necessary depends on the type and size of the structure. For most pile foundations, the borehole should be at least as deep as the anticipated pile penetration plus three pile diameters. For gravity structures the borehole depths are usually between one and two times the diameter of the foundation, depending on soil conditions. In general, the softer the subsurface soils, the deeper the borehole required. The sampling frequency for a borehole normally varies with depth as approximated in Table 2.4-1.

2.4.5 Sample Handling

2.4.5.1 Preparing and Packaging

To minimize sample disturbance, cores must be prepared and packaged for shipment as soon as possible after recovery.

For gravity cores, liners should be pulled out of the barrel and cut into sections 3 to 6 feet long. The top and bottom of each section and the position of each section in the core should be marked on the liner. The ends of the liner tube should be sealed with plastic caps and electrical tape and preferably sealed again with an appropriate wax.

With box corers, the metal corer box is not watertight. It is very difficult to maintain box corer samples at their natural water content. For quality samples, shipboard subsampling of the recovered sample should be done as soon as possible. Many large-diameter liners with sharpened ends are pushed from top to bottom of the box sample with a rapid and continuous motion. When all liners have been pushed into the box, the sediment around the liners is removed and can be saved in jars or plastic bags for tests that do not require undisturbed samples. The subsamples in the liners are then sealed and marked.

Wireline samples are usually extruded onboard the ship. Clay sediments are normally cut into sections about three diameters long. Each section is then wrapped with wax paper and aluminum foil and sealed with wax in a plastic container. If liners are used, the liners are extruded and separated using a wire saw. Each liner is then sealed with plastic caps, tape, and wax. Cohesionless sediments are usually stored loose in bags or jars because they are quite disturbed in the coring process.

2.4.5.2 Storage

Proper sample storage prior to testing is somewhat controversial. It is usually best that samples are stored vertically, when possible, to maintain their natural orientation and to limit mixing or changes in stress conditions. Cushioning should be used to protect the samples from vibration. Samples should be stored at 5 ±2°C in near 100% relative humidity and away from direct sunlight to prevent biological growth and other physical alterations that might otherwise occur.

2.4.5.3 Transit

Samples should be shipped to the laboratory as soon as possible after the vessel arrives in port. Undisturbed samples are best "hand-carried," either as carry-on luggage in an airplane or in a private vehicle rather than by commercial carrier. For larger shipments, where personal control is not possible, air freight is recommended since it minimizes the time in transit and also reduces vibration and shocks that might further disturb the sample.

Samples should be tested as soon as possible because even proper storage will only slow down and not stop sample property changes that occur with time.

2.5 IN-SITU TESTING

2.5.1 General

The complexity of equipment-seafloor interactions has escalated with the progression of naval operations from shallower (< 200 m) to deeper (> 2000 m) water. With greater depths come new load sources and greater foundation loads that require more reliable and economic design. This puts a premium on the importance of more accurate and detailed investigations to determine seabed soils properties. Deepwater has made traditional investigations of drilling and sampling exponentially more difficult (and costly) and there has been a shift within the ocean community to increased reliance on in-situ testing to determine appropriate strength and deformation parameters.

In-situ testing procedures involve making measurements within the soil with specialized tools. The in-situ tools are advanced into the seafloor by a support vessel that pushes the tool into the soil from the bottom of a drilled borehole (termed "downhole" mode). Alternatively, and now more common, the in-situ tools may be advanced directly into the seabottom from a rig placed on the seabed (termed "seabed" mode).

The most commonly used in-situ testing tools for offshore investigations are the vane and cone penetrometer. In recent years, there has been increased use of "flow" type penetrometers (T-bar and ball) [Ref. 2-7 and 2-8]. Other less common tools include gas probes, piezoprobe, pressuremeter and free-fall penetrometers. The measurements obtained from these in-situ tools are used to identify soil stratigraphy and/or estimate soil parameters for foundation design (e.g., undrained shear strength, friction angle, modulus of elasticity) or to identify possible geohazards.

The attraction of the in-situ test approach is that soils are not removed from their native environment during property evaluation. Marine soil samples are normally subjected to appreciable disturbance and a decrease in hydrostatic and confining soil pressure when retrieved from the seabed. This decrease results in further disturbance to the internal structure of the soil. When combined with disturbance introduced by sampling and subsequent sample handling, the total disturbance can mask the actual in-situ properties of the soil. This is particularly true in soft cohesive soils containing appreciable dissolved gases, such as those found in the Gulf of Mexico. For these soils, decreases in hydrostatic stress often result in gases coming out of solution, which completely disturbs and remolds the soil.

Table 2.5-1 summarizes types of in-situ tests and equipment capabilities. The types of tests, soil information derived from each, and data evaluation will be discussed in subsequent sections. A more detailed discussion of in-situ testing and equipment is given in References 2-9, 2-10, and 2-11.

Table 2.5-1. In-Situ Tests, Applications, and Some Equipment Characteristics

Test Performed	Sediment Type	Parameter Obtained	Apparatus Type	Company	Device Name	Water Depth	Seafloor Penetration
						Operational Limits (feet)	
Vane Shear	Cohesive	s_u	Wireline	Fugro	Dolphin	10,000	5
			Platform	Fugro	Halibut	-	20
				Fugro	Seacalf	19,500	65
Cone Penetrometer	Cohesionless or Cohesive	ϕ or s_u	Wireline	A. P. Vandenberg	Wison-APB	-	10
			Platform	Fugro	Seacalf	19,500	160
				Fugro	SeaRobin	6,600	6
				A. P. Vandenberg	ROSON	6,600	16
				Datem	Neptune 5000	9,900	65
				Gregg Drilling	Seabed CPT	9,900	160
Mini CPT	Cohesionless or Cohesive	ϕ or s_u	Platform	Fugro	Seascout	-	16
				Datem	Neptune 3000	9,900	33
				Gregg Drilling	Mini CPT	9,900	35
Dynamic Penetrometer	Cohesionless or Cohesive	ϕ and D_r or s_u	Freefall	NFESC	eXpendable Doppler Penetrometer (XDP)	4,500	15
Tbar or Ball	Cohesive	s_u	Platform	Fugro	Seacalf	19,500	130
				Gregg Drilling	Seabed CPT	9,900	
Borehole Logging	Cohesionless or Cohesive	γ_b, sediment boundary layers	Wireline	Limited to cored and cased boring holes.			

2.5.2   Vane Shear Tests

The vane shear test is used to determine the peak and remolded undrained shear strength of soft to medium stiff clays and silts (< 200 kPa). In the vane shear test, the torque required to rotate a four-bladed vane embedded (Figure 2.5-1) in the soil is measured and converted into a measure of the shear strength of the soil. Selection of the vane size is dictated by the anticipated soil strength and by the torque measuring limits of the equipment. Vane geometry and rotation rate are standardized (Ref. 2-12). However, offshore operations typically vary somewhat from the standardized rotation rate of 6 deg/min, due to the operator's experience and preference.

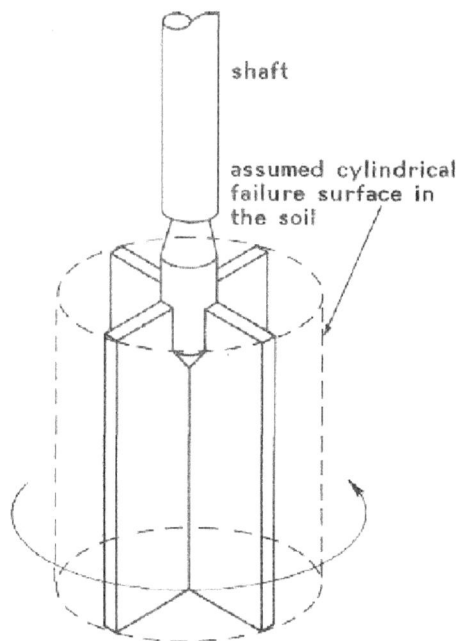

Figure 2.5-1. Vane shear device.

In-situ vane shear test devices are available for either wireline operation through a drill string at the bottom of a borehole or from a seafloor resting platform (Table 2.5-1). The maximum depth of test for the seafloor resting systems is limited by the available reaction force (the underwater weight of the system) to about 20 feet in soft clays for existing devices.

The shear strength can be calculated from measured torque values using:

$$s_{uv} = \frac{2T}{\pi d^2 [H + (d/3)]} \qquad (2\text{-}1)$$

where:

s_{uv} = vane shear strength [F/L^2]

T = applied torque [L·F]

d = diameter of vane blade [L]

H = height of vane blade [L]

To obtain the undrained shear strength, s_u, a correction factor, accounting for effects of anisotropy and strain rate, should be applied to the vane shear strength determined from Equation 2-1 such that:

$$s_u = \mu s_{uv}$$ (2-2)

where:

μ = correction factor from Figure 2.5-2 based on the soil plasticity index (PI)

Many factors can affect the measurement of shear strength. Values obtained in sensitive, overconsolidated, or cemented clays may not be correct due to disturbance of the clay during insertion of the vane. Erratic results are obtained in soils containing shells, gravel, or wood fragments. Effects of anisotropy, strain rate, and sensitivity can often affect vane results but only to the extent that they can be used as a rough measurement of strength variability.

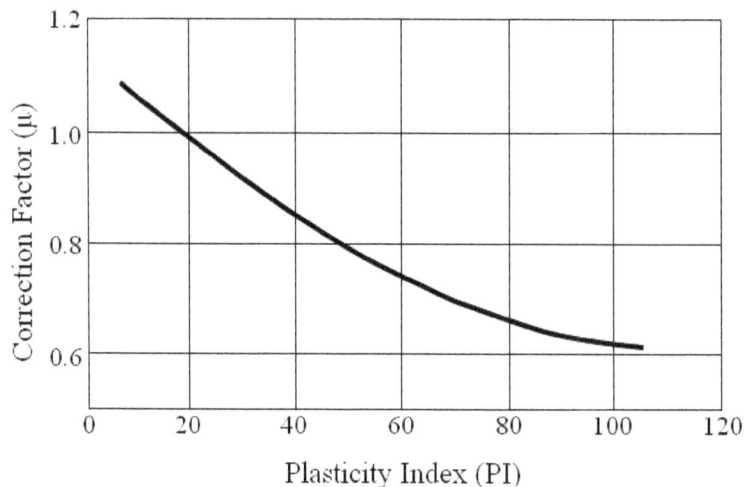

Figure 2.5-2. Correction factor for vane determined shear strength (Ref. 2-13).

2.5.3   Cone Penetration Tests

The cone penetration test (CPT) or cone penetration test with pore pressure measurement (CPTU) is the most widely used in-situ tool for offshore investigations (Ref. 2-9). The penetrometer is pushed into the soil at a constant rate while tip resistance, side friction, and porewater pressure (in the case of CPTU) are measured. The data is then correlated with soil type and estimates of strength parameters are derived for foundation design. The cone penetrometer obtains continuous in-situ information with depth in both coarse-grained cohesionless soils and fine-grained cohesive soils. Fully-drained conditions are assumed in cohesionless soils, which means the excess porewater pressure is zero and the measured porewater pressure is the static in-situ pressure. Undrained conditions are assumed in fine-grained cohesive soils, which means excess porewater pressures will be developed. The measurement of these porewater pressures is useful in interpreting the cone test results.

The standard electronic cone has a 60° apex angle, a base diameter of 3.57cm (1.4 inches), a projected area of 10cm² (1.55 in²) and a friction sleeve area of 150 cm² (23.3 in²). This penetrometer has been accepted as the reference cone according to national and international standards (Ref. 2-14, 2-15, and 2-16).   Figure 2.5-3 illustrates a diagram of a typical cone penetrometer.

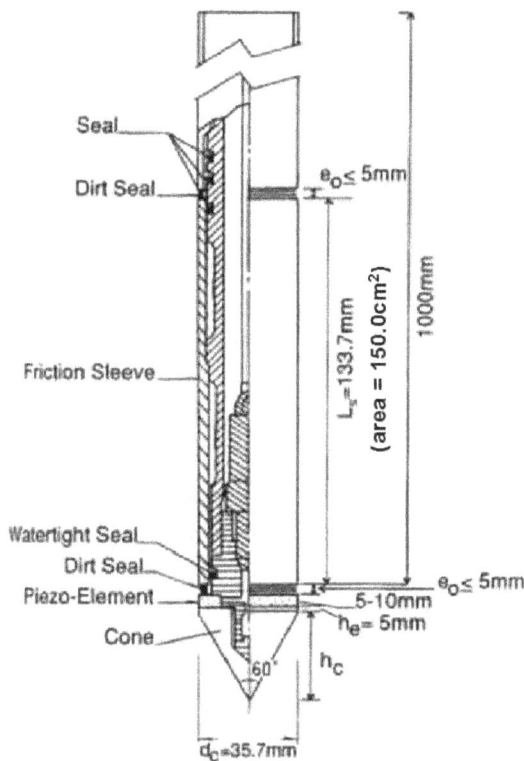

Figure 2.5-3. Cross-section of a typical cone penetrometer (Ref. 2-14).

2.5.3.1 Cone Measurements

Cone tip resistance and sleeve friction measurements are obtained from electrical strain gauge load cells located behind the cone tip. The total cone tip resistance (q_c) is determined by dividing the total cone force Q_c by the projected area of the cone, A_c (10cm^2 in case of the standard cone), as shown in Equation 2-3.

$$q_c = \frac{Q_c}{A_c}$$ (2-3)

The sleeve friction (f_s) is determined by dividing the total force acting on the friction sleeve F_s by the surface area of the friction sleeve, A_s (150 cm^2 in case of the standard cone), as shown in Equation 2-4.

$$f_s = \frac{F_s}{A_s}$$ (2-4)

Porewater pressure measurements are normally recorded by a filter and pressure sensor located just behind the cone tip, although other locations have been used. The advent of pore pressure measurements in the 1980s marked a significant advancement in cone penetration testing because they allowed more accurate soil characterization and strength profiling especially in soft fine-grained sediments. Additionally, cone penetrometers can be outfitted with sensors to measure temperatature, resisitivity, and shear wave velocity.

Cone penetrometers are used both from a wireline system in downhole mode, or now more commonly from a seafloor-resting platform in seabed mode. In offshore testing, it is common practice to zero-out cone resistance and porewater pressure at the seabed prior to testing. Reference 2-17 contains comprehensive information on the use and interpretation of CPT/CPTU equipment.

As mentioned previously, the 10 cm^2 cone is the reference standard. However, larger (15 cm^2) and smaller (1 to 2 cm^2) cones have also been used in offshore practice. Larger cones are capable of increased sensitivity and can provide enhanced accuracy when investigating softer sediments, while smaller diameter "minicones" have the ability to detect thinner soil lenses. The ability to detect thinner lenses is especially important when investigating the upper 1 to 2 meters of the seafloor. The minicone also requires less downward thrust (and thus lighter seabed equipment) to advance the penetrometer into the seafloor, which is advantageous in seabed mode in deepwater environments.

2.5.3.2 Cone Penetrometer Correction

Due to cone design, soil parameters derived from CPTU results should be corrected for porewater pressure effects. This was first noted during early offshore cone testing in deepwater where it was observed that the measured total cone resistance (q_c) was not equal to the hydrostatic water pressure at depth. Because of the way cones are designed, there is a small area behind the cone face where the porewater pressure pushes down on the cone

This imbalance of force aids penetration and should be accounted for to calculate a corrected total cone resistance, q_t as shown in the equation below. This effect is known as the "unequal area effect" and influences the cone and friction sleeve resistances.

$$q_t = q_c + u(1 - a)$$

(2-5)

where:

q_c = total cone tip resistance [F/L^2]

u = measured porewater pressure at the shoulder position [F/L^2]

a = cone area ratio

In this equation, the cone area ratio, a, is approximately equal to the ratio of the area of the load cell or shaft, d, divided by the projected area of the cone tip, D (10 cm^2 in the case of the standard cone). The cone area ratio is provided by the manufacturer but also can be determined by calibration in a laboratory pressure vessel. Good cone design will maximize the area ratio as close to unity as possible. In stiff or dense soils the effect of the correction will be small (since the value of q_c will generally be much larger than u). However, the correction can be significant in soft saturated fine-grained soils where the pore pressures are large relative to the cone resistance.

Likewise, the sleeve friction will be influenced by porewater pressure effects, however, this correction requires knowledge of the porewater pressure behind the sleeve which is normally unavailable. Therefore, without this measurement it has been recommended that this correction not be made (Ref. 2-17).

As more investigations are occurring in deep water, cone manufacturers are developing pressure compensated load cells for cone penetrometers which eliminate the need for the correction factor.

2.5.3.3 Soil Classification by Cone Penetrometer

One of the strengths of the CPT and CPTU is the ability to develop a continuous soil stratigraphy profile from the cone tip, sleeve, and porewater pressure measurments. The classification charts shown on Figure 2.5-4 incorporate all three measurements to classify the soil. The chart on the left should be used if only basic CPT data is available (i.e., q_c and f_s). If porewater pressures are measured, the chart on the right should be used to identify soil type. Note that in practice Q_t is referred to as the normalized cone resistance; F_r is the normalized friction ratio; and B_q is the pore pressure parameter. Normalization of these parameters with effective vertical stress better captures the effects of overburden stress with increasing depth.

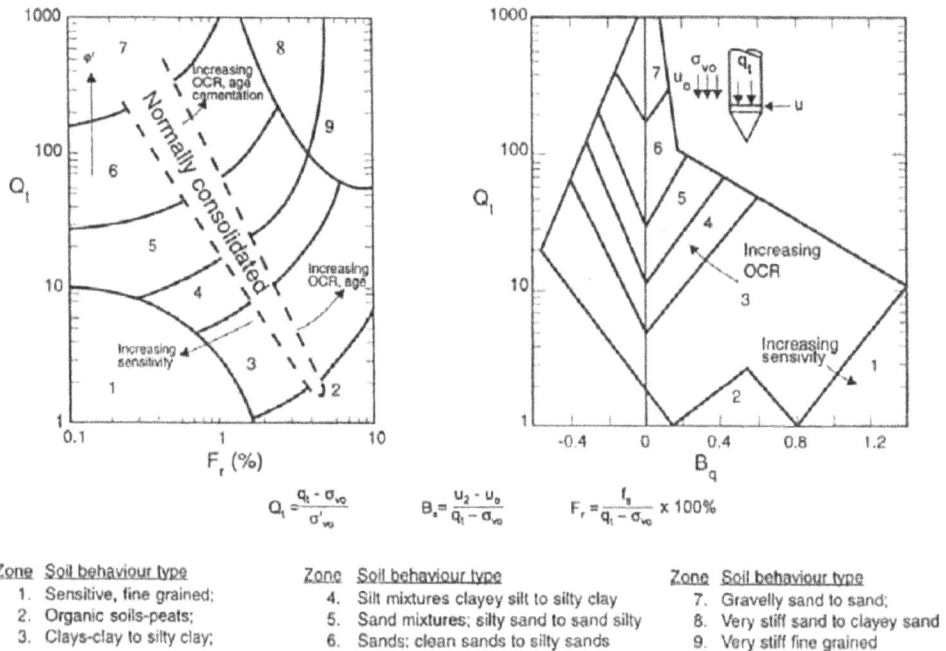

$$Q_t = \frac{q_t - \sigma_{vo}}{\sigma'_{vo}} \qquad B_q = \frac{u_2 - u_0}{q_t - \sigma_{vo}} \qquad F_r = \frac{f_s}{q_t - \sigma_{vo}} \times 100\%$$

Zone	Soil behaviour type	Zone	Soil behaviour type	Zone	Soil behaviour type
1.	Sensitive, fine grained;	4.	Silt mixtures clayey silt to silty clay	7.	Gravelly sand to sand;
2.	Organic soils-peats;	5.	Sand mixtures; silty sand to sand silty	8.	Very stiff sand to clayey sand
3.	Clays-clay to silty clay;	6.	Sands; clean sands to silty sands	9.	Very stiff fine grained

Figure 2.5-4. Soil behavior type classification chart (Ref. 2-18).

2.5.3.4 Estimating Relative Density of Cohesionless Soil using CPT data

Relative density (D_r) is a quantitative parameter typically used to describe the in-situ density state of granular (sandy) soils. It is sometimes used as an intermediate soil parameter in determining the strength of granular soils. Many of the relationships developed for estimating relative density from cone penetration resistance are based on laboratory calibration chamber testing. The most widely accepted correlations are summarized below. It is well known that relative density alone cannot describe the engineering behavior of sand.   Therefore it is suggested that a conservative value be used based on a range of relative densities computed using the following three equations:

1. <u>Kulhawy & Mayne (Ref. 2-19)</u>. The relationship is primarily based on unaged, uncemented sands.

$$D_r = \frac{q_{c1}}{305 \, Q_c \, Q_{OCR}}$$
(2-6)

$$q_{c1} = \frac{\left(\frac{q_c}{p_a} \right)}{\left(\frac{\sigma_v'}{p_a} \right)^{0.5}}$$
(2-7)

where:

q_{c1} = normalized cone resistance (dimensionless)

q_c = measured cone tip resistance [F/L^2]

p_a = atmospheric pressure [F/L^2]

σ_v' = effective vertical stress [F/L^2]

Q_c = compressibility factor ($0.91 < Q_C < 1.09$)

= 0.91 (Low compressibility: quartz sands, rounded grains w/little to no fines)

= 1.0 (Med. compressibility: quartz sands w/some feldspar and/or several % fines)

= 1.09 (High compressibility: high fines content; mica, other compressible minerals)

Q_{OCR} = overconsolidation factor = $OCR^{0.18}$

2. <u>Baldi, Bellotti, Ghionna & Jamiolkowski (Ref. 2-20)</u>. The solution was based on calibration chamber testing of Ticino sand that was subangular to angular, medium to coarse, and primarily quartz.

$$D_r = \frac{1}{C_2} \ln \left(\frac{q_c}{C_0 (\sigma')^{C_1}} \right)$$
(2-8)

where:

C_n = soil contants for normal consolidated sand; C_0 = 157, C_1 = 0.55, C_2 = 2.41

q_c = measured cone tip resistance (kPa)

σ' = effective stress – either mean normal stress, σ_{mean}', or vertical stress, σ_{vo}' (kPa)

3. Lunne & Christoffersen (Ref. 2-21). The solution was derived from two large calibration chamber studies using 5cm² and 10cm² cone penetrometers on pluviated dry sand. The equation is for normally-consolidated, uniform, fine to medium, completely dry or saturated sand. The sand should consist mainly of non-crushable grains or be quartz-based.

$$D_r = \frac{1}{2.91} \ln\left(\frac{q_c}{61(\sigma_v')^{0.71}}\right)$$
(2-9)

where:

q_c = cone-tip penetration resistance (kPa)

σ_v' = effective vertical stress (kPa)

2.5.3.5 Estimating Friction Angle of Cohesionless Soil using CPT data

Numerous correlations relating the cone tip resistance to peak friction angle of sands have been published in the literature (Ref. 2-17). The correlations have been derived using different assumptions and theories but generally can be categorized as empirical methods, bearing capacity methods, or cavity expansion methods.

As with any modeling technique there are strengths and weaknesses of each approach and these have been discussed at great length in the literature (Refs. 2-17, 2-22, and 2-23). Empirical methods are in widespread use but generally lack theoretical background. Cavity expansion models are thought to simulate cone penetration resistance reasonably accurately at deeper penetration and are able to incorporate the effects of soil compressibility and curvature of the strength envelope (Refs. 2-22, 2-24, and 2-25). Bearing capacity models might better simulate the failure mechanism of an advancing cone at shallow penetration (Refs. 2-25 and 2-26) and therefore may be appropriate for estimating friction angles of near surface seafloor soils. However, the bearing capacity methods generally lack the ability to incorporate the effects of soil compressibility and curvature of the strength envelope.

The correlation proposed by Robertson and Campanella (Ref. 2-26) to compute the peak soil friction angle is shown in Equation 2-10. It is empirical in that it is based on the results of calibration chamber tests but it uses the bearing capacity theories of Durgunoglu and Mitchell (Ref. 2-27) and Janbu and Senneset (Ref. 2-28) as upper and lower bounds, respectively, of the data set. The correlation is recommended because of its simplicity and wide acceptance in geotechnical practice. The equation is a function of tip resistance and overburden pressure. Inclusion of overburden pressure in part accounts for the influence of confining pressure on the soil friction angle.

$$\tan\phi' = \frac{1}{2.68}\left[\log\left(\frac{q_c}{\sigma_{vo}'}\right) + 0.29\right]$$ (2-10)

where:

q_c = cone-tip penetration resistance [F/L^2]

σ_{vo}' = effective vertical stress [F/L^2]

The equation provides reasonable estimates of friction angle for normally consolidated, moderately incompressible (F_r of about 0.5%), and predominantly quartz sands.

2.5.3.6 Estimating Young's Modulus of Cohesionless Soil using CPT data

The value of Young's modulus of elasticity (E_s) for sand to be used in settlement computations may be estimated from Equation 2-11:

$$E_s = (1.5 \text{ to } 3.0) \cdot q_c$$ (2-11)

The lower range is applicable to normally consolidated sands, while the higher range applies to overconsolidated sands. Additional information and correlations for Young's modulus and other moduli are provided in Reference 2-17.

2.5.3.7 Estimating Undrained Shear Strength of Cohesive Soil using CPT data

The undrained shear strength of clays may be estimated based on the cone tip resistance and knowledge of the in-situ soil stress and a cone factor. The equation takes the following form:

$$s_u = \frac{q_t - \gamma_t Z}{N_{kt}}$$ (2-12)

where:

q_t = corrected total cone resistance [F/L^2]

γ_t = total unit weight [F/L^3]

Z = depth below seafloor [L]

N_{kt} = cone factor

Laboratory and field studies have shown the values of N_{kt} to vary widely depending upon which methods were used to back-calculate s_u. In general, N_{kt} varies between about 10 and 20 with an average of 15 appropriate for normally consolidated marine clays (Ref. 2-29). The value of N_{kt} tends to increase with increasing soil plasticity and decrease with increasing soil sensitivity. At sites with limited information, use a conservative value of N_{kt} between 15 and 20. At sites with very soft clays where there might be inaccuracies in determining q_t, estimates of s_u can be made using excess pore pressure measurements as given by Equation 2-13:

$$s_u = \frac{\Delta u}{N_{\Delta u}}$$

(2-13)

where:

Δu = pore pressure at shoulder position minus in-situ pore pressure [F/L^2]

$N_{\Delta u}$ = cone factor

Various studies have shown that $N_{\Delta u}$ varies between about 4 and 10 in normally to slightly overconsolidated deposits (Ref. 2-17). In general, a conservative estimate would be between 7 and 10.

In both equations above, the hydrostatic pressure at the mudline should be subtracted from the numerator except when using a compensated cone or where the porewater pressure at the seabed has been zeroed-out prior to penetration. It is recommended that at sites where high-quality field and laboratory data exists, site specific correlations should be used based on reliable values of s_u.

2.5.4   Pressuremeter Tests

The pressuremeter test measures soil deformation as a function of expansion pressure when a membrane is expanded out into the soil from its position down a borehole. Soil parameters that can be estimated from the test are shear strength, deformation modulus (modulus of elasticity), and the horizontal earth pressure at rest. The pressuremeter may be placed either in a predrilled hole or pushed into the soil, Reference 2-30 presents details on the test and on data interpretation.

2.5.5   Dynamic Penetrometer

A dynamic penetrometer is a hydrodynamically shaped cylinder that free falls through the water column and penetrates into the seafloor. Penetrometer velocity is monitored as it is slowed by the soil during penetration. Undrained shear strengths are calculated in a manner similar to the reverse of the dynamic penetration prediction procedure of Chapter 8. In this method, the change in velocity over a short penetration interval is used to calculate the kinetic energy consumed, and then to calculate the soil strength required to consume that energy.

The Navy's expendable Doppler penetrometer (XDP) system is designed to make possible, through data analysis, an accurate determination of the mechanical strength of various types of seafloor soils. The system does this by recording the signal transmitted by a precise-frequency sound source within the probe as it falls through the water and penetrates the seafloor. The system derives the instantaneous velocity of the probe from changes in the frequency of the received signal caused by the Doppler effect. By analyzing the velocity data as the probe impacts the bottom and decelerates, the system provides an estimate of the strength profile and penetrability of the seafloor. The XDP system has the following components:

1. A penetrometer with its constant frequency source

2. A receiving hydrophone with a preamplifier

3. A receiver for processing the incoming data

The Navy system is capable of operating in water depths of up to 4,500 feet and has penetrated up to 15 feet in some soils.

2.5.6   Borehole Logging Techniques

Borehole logging techniques are used to enhance knowledge of soil changes in lieu of more expensive and time-consuming testing. They can be performed in the drill pipe after the drilling and sampling operations. In one type of test, an electric probe is lowered in the drill pipe while continuously measuring natural gamma emissions of the soil. The higher radioactive mineral contents of clays yield higher gamma emissions than sands. Thus, a correlation between natural gamma logs and soil types (stratigraphy) can be established. The importance of this type of testing will be greatly increased where coring problems are encountered and core recovery (percent) is very low.

Information on thermal and magnetic properties of the soil mass around the borehole can be gathered by other logging techniques.

2.6 REFERENCES

2-1.   A. Trujillo and H. Thurman. *Essentials of Oceanography*, 9th edition, Upper Saddle River, NJ, Pearson Education, Inc., 2008.

2-2.   V. Capone. Images provided by Black Laser Inc., 2010.

2-3.   R. McQuillin and D.A. Arders. *Exploring the Geology of Shelf Seas*. London, England, Graham and Trotman, Ltd., 1977.

2-4.   I. Noorany. "Underwater Soil Sampling and Testing - State-of-the-Art Review," in Underwater Soil Sampling, Testing, and Construction Control, American Society for Testing and Materials, ASTM STP 501. Philadelphia, PA, 1972, pp. 3-41.

2-5.   H.J. Lee and J.E. Clausner. Seafloor Soil Sampling and Geotechnical Parameter Determination Handbook, Civil Engineering Laboratory, Technical Report R-873. Port Hueneme, CA, Aug 1979.

2-6.   Alpine staff. "Vibracore 3." Alpine Ocean Seismic Survey, Inc. Website. http://www.alpineocean.com/vibracore.html# (January 11, 2007).

2-7.   T. Lunne, M.F. Randolph, S.F. Chung, K.H. Anderson, and M. Sjursen. "Comparison of Cone and T-bar Factors in Two Onshore and One Offshore Clay Sediments," in Proceedings Frontiers in Offshore Geotechnics: ISFOG 2005 – Gourvenec and Cassidy, eds., pp. 981-989, 2005.

2-8.   M. F. Randolph. "Characterisation of Soft Sediments for Offshore Applications," Proceedings ISC-2 on Geotechnical and Geophysical Site Characterization, Fonseca and Mayne, eds., pp. 209-232, 2004.

2-9   T. Lunne. "In Situ Testing in Offshore Geotechnical Investigations," in Proceedings of the International Conference on Insitu Measurement of Soil Properties, Bali, Indonesia, pp. 61-81, 2001.

2-10.   T. Lunne, S. Lacasse, and N.S. Rad. "General Report/Discussion Session 2: SPT, CPT, Pressuremeter Testing and Recent Developments in In-Situ Testing Part 1: All Tests Except SPT," Proceedings of The Twelfth International Conference On Soil Mechanics and Foundation Engineering, Vol. 4, Rio De Janeiro, August 1989, pp. 2339-2403.

2-11.   H.J. Kolk, and J. Wegerif. "Offshore Site Investigations: New Frontiers," Frontiers in Offshore Geotechnics: ISFOG 2005-Gourvenec and Cassidy, eds., Taylor and Francis Group, London, 2005, pp. 145-161.

2-12. "Standard Test Method for Field Vane Shear Test in Cohesive Soil," from ASTM Volume 04.08 Soil and Rock (I): D420 – D5876, American Society for Testing and Materials, ASTM D2573-08. Philadelphia, PA, 2008.

2-13. L. Bjerrum. "Embankments on Soft Ground," in Performance of Earth and Earth-Supported Structures, Vol. II, American Society of Civil Engineers. New York, NY, 1972, pp. 1-54.

2-14. "Standard Test Method for Performing Electronic Friction Cone and Piezocone Penetration Testing of Soils" from ASTM Volume 04.08 Soil and Rock (I): D 420-D 5876, American Society for Testing and Materials, ASTM D5778-95. Philadelphia, PA, 2008.

2-15. "International Reference Test Procedure for Cone Penetration Test (CPT)." Report of the ISSMGE Technical Committee on Penetration Testing of Soils – TC 16, International Society for Soil Mechanics and Geotechnical Engineering, 1989.

2-16. Svenska Geotekniska Föreningen. "Recommended Standard for Cone Penetration Tests," Swedish Geotechnical Society , SGF Report 1:93 E, 1993.

2-17. T. Lunne, P.K. Robertson, and J.J.M. Powell. *Cone Penetration Testing in Geotechnical Practice*. Blackie Academic and Professional, 1997.

2-18. P.K. Robertson. "Soil Classification using the Cone Penetration Test," Canadian Geotechnical Journal , Vol. 27, No.1, pp. 151-158, 1990.

2-19. F. H. Kulhawy, and P.W Mayne. "Manual on Estimating Soil Properties for Foundation Design". Cornell University, Geotechnical Engineering Group; Electric Power Research Institute, EPRI EL-6800 Project 1493-6, 1990.

2-20. G. Baldi, R. Bellotti, V. Ghionna, and M. Jamiolkowski, M. "Interpretation of CPTs and CPTUs; 2nd part: Drained Penetration of Sands," in Proceedings of the Fourth International Geotechnical Seminar, Singapore, pp. 143-156, 1986.

2-21. T. Lunne, and H.P. Christoffersen. "Interpretation of Cone Penetrometer Data for Offshore Sands" in Proceedings Offshore Technology Conference, OTC 4464, 1983.

2-22. H.S. Yu, and J.K. Mitchell, J.K. "Analysis of Cone Resistance: Review of Methods." Journal of Geotechnical and Geoenvironmental Engineering, Vol. 124, No. 2, pp. 140-149, 1998.

2-23. J.W. Chen, and C.H. Juang. "Determination of Drained Friction Angle of Sands from CPT." Journal of the Geotechnical Engineering, Vol. 122, No. 5, pp.374-381, 1996.

2-24.   A. Puech, and P. Foray. "Refined Model for Interpreting Shallow Penetration CPTs in Sands." Proceedings Offshore Technology Conference, OTC 14275, 2002.

2-25.   J. H. Schmertman. "Guidelines for cone penetration test, performance and design." U.S. Department of Transportation, Federal Highway Administration, FHWA-TS-78-209, 1978.

2-26.   P.K. Robertson, and D. Woeller. "Cone Penetration Testing Geotechnical Applications Guide." Proceedings from a Short Course on Geotechnical Site Investigation using the Cone Penetration Test, Gregg, CA, Sec. 3, pp. 15-16, 2005.

2-27.   H.T. Durgunoglu, and J.K. Mitchell. "Static Penetration Resistance of Soils: I – Analysis, II – Evaluation." Proceedings of the Conference on In Situ Measurement of Soil Properties, ASCE, Vol. 1, pp. 151-189, 1975.

2-28.   N. Janbu, and K. Senneset, K. "Effective Stress Interpretation of In-Situ Static Penetration Tests." Proceedings of the European Symposium on Penetration Testing, ESOPT, Stockholm, Sweden, Vol. 2, pp. 181-193, 1974.

2-29.   T. Lunne, and A. Kleven. "Role of CPT in North Sea Foundation Engineering," Cone Penetration Testing and Experience, American Society of Civil Engineers, NY, October 1981, pp. 76-101.

2-30.   J. Briaud. In Situ Tests to Measure Soil Strength and Soil Deformability for Offshore Engineering, Texas A&M Research Foundation. College Station, TX, Oct 1980.

2.7 SYMBOLS

A_c — Projected area of the cone [L^2]

A_s — Surface area of the friction sleeve [L^2]

a — Cone area ratio

Bq — Pore pressure parameter

C_c — Compression index

CCD — Carbonate compensation depth, below which carbonate materials will dissolve

$\overline{c}$ — Drained cohesion intercept (effective soil cohesion) [F/L^2]

c_v — Coefficient of consolidation [L/T]

$C_{0,1,2}$ — Soil contants for normal consolidated sand

d — Diameter of vane blade [L]; Also, area of the load cell or shaft [L^2]

D — Projected area of the cone tip [L^2]

D_r — Relative density

E_s — Constrained modulus of elasticity for sand [F/L^2]

F_r — Normalized friction ratio

F_s — Total force acting on the friction sleeve [F]

f_s — Sleeve friction [F/L^2]

H — Height of vane blade [L]

k — Permeability [L/T]

N_{kt} — Cone factor

$N_{\Delta u}$ — Cone factor

PI — Cohesive soil plasticity index

p_a — Atmospheric pressure [F/L^2]

Q_c — Total cone force [F]

Q_{OCR} — Overconsolidation factor $= OCR^{0.18}$

Q_t — Normalized cone resistance

q_c — Cone tip resistance [F/L^2]

q_{cl} — Normalized cone resistance

q_t — Corrected total cone resistance [F/L^2]

S_t — Sensitivity

s_u — Undrained shear strength [F/L^2]

s_{uv} — Vane shear strength [F/L^2]

T — Torque applied to vane [$L \cdot F$]

u	Measured porewater pressure at the shoulder position [F/L^2]
Z	Depth below seafloor surface [L]
Δu	Pore pressure at shoulder position minus in-situ pore pressure [F/L^2]
σ'	Effective stress [F/L^2]
σ_v'	Effective vertical stress [F/L^2]
σ_{vo}'	Effective vertical stress [F/L^2]
γ_b	Buoyant unit weight of soil [F/L^3]
γ_t	Total unit weight [F/L^3]
μ	Correction factor for vane shear strengths
$\bar{\phi}$	Drained, or effective, friction angle [deg]

3 LABORATORY DETERMINATION OF SOIL PROPERTIES

3.1 INTRODUCTION

3.1.1 Scope

Table 2.1-2 of Chapter 2 identified engineering properties of soils required for analysis and design for several applications in the deep ocean environment. Chapter 2 outlined elements of preliminary "desk-top" and field surveys, including the acquisition of some engineering properties from in-situ tests. This chapter describes the laboratory phase of soils classification and engineering properties determination.

3.1.2 Special Considerations

Most considerations and concepts developed in conventional onshore geotechnical engineering apply also in the marine environment. Differences in handling, testing, and data evaluation techniques arise because of the very low effective stresses in surficial materials (and resulting very soft or loose physical state); the new soil materials encountered (primarily the biogenous remains and the authigenic precipitates); and, to a lesser extent, the salt content of the pore fluid. This chapter often cites conventional soils testing references (Refs. 3-1 and 3-2), with most of the material devoted to presenting necessary deviations from standard procedures.

3.2 SOIL CLASSIFICATION

3.2.1 Classification Information

Soils may be classified in a number of manners, including by origin, by grain size, and by a combination of grain size and behavior. A brief discussion of those classification methods is presented in Sections 3.2.2 through 3.2.4. Additional information on soil classification can be found in the article "Classification of Marine Sediments" by Iraj Noorany (American Society of Civil Engineers Journal of Geotechnical Engineering, Vol. 115, No. 1, January 1989).

3.2.2 Classification by Origin

Marine geologists classify seafloor soil types by origin. This classification system was introduced in Section 2.2.2, where characteristic soil strength profiles were discussed for each major type, based on origin. A soil sample in this system can often be classified with visual examination of core material by experienced personnel. Classification by inexperienced personnel or where soil does not cleanly fit into one of the major types requires properties testing. This is done with tests to establish grain size distribution (Section 3.3.7), to determine carbonate and organic carbon content (Section 3.3.8), and to identify and sort constituents by visual microscopic examination. With soils found in the deep ocean, classification by origin is often a necessary element in predicting the engineering behavior.

3.2.3 Classification by Grain Size

The marine geologist also classifies sediments strictly by grain size, according to the Wentworth scale (Table 3.2-1a) or according to an American Society for Testing and Materials (ASTM) scale that has slightly different grade limits and subdivides material types (Table 3.2-1b). The portion of the sample below gravel size from each of the size groups (sand, silt, and clay), in percent of sample dry weight, is often reported on a trilineal plot (Figure 3.2-1). This Trilineal Classification System by itself normally does not provide an adequate description of a soil for engineering purposes, but it is a rapid, size classification tool.

Table 3.2-1. Size Range Limits for Two Soil Classification Systems

Material Type	Diameter Range (mm)
(a) Wentworth Scale Grade Limits	
Gravel	64 – 2.0
Sand	2.0 – 0.062
Silt	0.062 – 0.0036
Clay	< 0.0036
(b) ASTM Grade Limits (Ref. 3-1)	
Cobbles	> 76.2
Gravel	
Coarse	76.2 – 19.1
Fine	19.1 – 4.76
Sand	
Coarse	4.76 – 2.00
Medium	2.00 – 0.42
Fine	0.42 – 0.074
Silt	0.074 – 0.005
Clay	< 0.005
Colloids	< 0.001

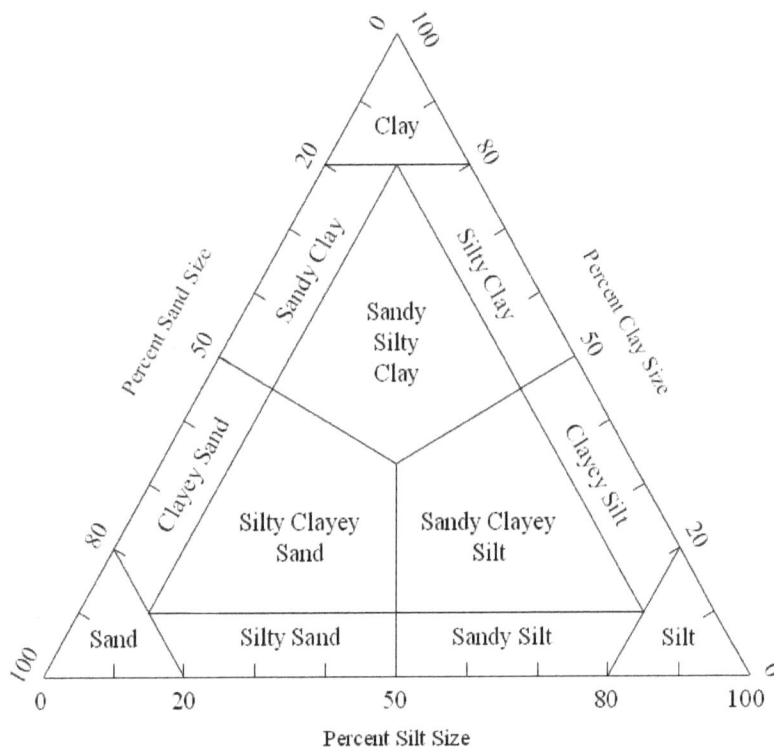

Figure 3.2-1. Trilineal soil classification plot – normally used with Wentworth grade limits.

3.2.4 Classification by Grain Size and Behavior

The Unified Soil Classification System is based on the soil's grain size distribution and its index properties. A sieve analysis for grain size distribution (Section 3.3.7) and simple index tests (Section 3.3.6) are necessary for classification. Data from these tests are input for the Unified Soil Classification Chart (Figure 3.2-2) in developing a soil's classification.

The System first divides soil into three groups: coarse-grained (gravels and sands), fine-grained (silts and clays), and highly organic materials. The classifications indicate that more than 50% of the sample grains, based on dry weight, are larger (coarse-grained) or smaller (fine-grained) than 0.074 mm in diameter (no. 200 sieve). Highly organic soils are identified by their black or dark gray color and by their hydrogen sulfide odor.

The coarse-grained soils are further subdivided by their predominant grain size and by the index properties of their fine fraction. The fine-grained soils are subdivided entirely based on their index properties (see Figure 3.2-2). The Unified System is described in detail in References 3-1 and 3-3.

UNIFIED SOIL CLASSIFICATION
INCLUDING IDENTIFICATION AND DESCRIPTION

			FIELD IDENTIFICATION PROCEDURES (Excluding particles larger than 3" and basing fractions on estimated weights)	GROUP SYMBOLS[a]	TYPICAL NAMES	LABORATORY CLASSIFICATION CRITERIA
COARSE GRAINED SOILS (More than half of material is larger than No. 200 sieve size[b])	GRAVELS (More than half of coarse fraction is larger than No. 4 sieve size) (For visual classification, the ¼" size may be used as equivalent to the No. 4 sieve size)	Clean Gravels (Little or no fines)	Wide range in grain size and substantial amounts of all intermediate particle sizes	GW	Well graded gravels, gravel-sand mixtures, little or no fines	$C_u = \dfrac{D_{60}}{D_{10}}$ Greater than 4; $C_c = \dfrac{(D_{30})^2}{D_{10} \times D_{60}}$ Between one and 3
			Predominantly one size or a range of sizes with some intermediate sizes missing	GP	Poorly graded gravels, gravel-sand mixtures, little or no fines	Not meeting all gradation requirements for GW
		Gravels with Fines (Appreciable amount of fines)	Non-plastic fines (for identification procedures see ML below)	GM	Silty gravels, poorly graded gravel-sand-silt mixtures	Atterberg limits below "A" line, or PI less than 4 / Above "A" line with PI between 4 and 7 are borderline cases requiring use of dual symbols
			Plastic fines (for identification procedures see CL below)	GC	Clayey gravels, poorly graded gravel-sand-clay mixtures	Atterberg limits above "A" line with PI greater than 7
	SANDS (More than half of coarse fraction is smaller than No. 4 sieve size)	Clean Sands (Little or no fines)	Wide range in grain sizes and substantial amounts of all intermediate particle sizes	SW	Well-graded sands, gravelly sands, little or no fines	$C_u = \dfrac{D_{60}}{D_{10}}$ Greater than 6; $C_c = \dfrac{(D_{30})^2}{D_{10} \times D_{60}}$ Between one and 3
			Predominantly one size or a range of sizes with some intermediate sizes missing	SP	Poorly-graded sands, gravelly sands, little or no fines	Not meeting all gradation requirements for GW
		Sands with Fines (Appreciable amount of fines)	Non-plastic fines (for identification procedures see ML below)	SM	Silty sands, poorly graded sand-silt mixtures	Atterberg limits below "A" line, or PI less than 4 / Above "A" line with PI between 4 and 7 are borderline cases requiring use of dual symbols
			Plastic fines (for identification procedures see CL below)	SC	Clayey sands, poorly graded sand-clay mixtures	Atterberg limits above "A" line with PI greater than 7

Determine percentage of gravel and sand from grain size curve. Depending on percentage of fines (fraction smaller than No. 200 sieve size) coarse grained soils are classified as follows:

Less than 5% — GW, GP, SW, SP
More than 12% — GM, GC, SM, SC
5% to 12% — Borderline cases requiring use of dual symbols

Use grain size curve in identifying the fractions as given under field identification

IDENTIFICATION PROCEDURES ON FRACTION SMALLER THAN No. 40 SIEVE SIZE

			DRY STRENGTH (Crushing Characteristics)	DILATENCY (Reaction to Shaking)	TOUGHNESS (Consistency Near Plastic Limit)	GROUP SYMBOLS[a]	TYPICAL NAMES
FINE GRAINED SOILS (More than half of material is smaller than No. 200 sieve size) (The No. 200 sieve size is about the smallest particle visible to the naked eye)	SILTS AND CLAYS	Liquid Limit less than 50	None to slight	Quick to slow	None	ML	Inorganic silts and very fine sands, rock flour, silty or clayey fine sands with slight plasticity
			Medium to high	None to very slow	Medium	CL	Inorganic clays of low to medium plasticity, silty clays, gravelly clays, sandy clays, lean clays
			Slight to medium	Slow	Slight	OL	Organic silts and organic silt-clays of low plasticity
	SILTS AND CLAYS	Liquid Limit greater than 50	Slight to medium	Slow to none	Slight to medium	MH	Inorganic silts, micaceous or diatomaceous fine sandy or silty soils, elastic silts
			High to very high	None	High	CH	Inorganic clays of high plasticity, fat clays
			Medium to high	None to very slow	Slight to medium	OH	Organic clays of medium to high plasticity
HIGHLY ORGANIC SOILS			Readily identified by color, odor, spongy feel and frequently by fibrous texture			Pt	Peat and other highly organic soils

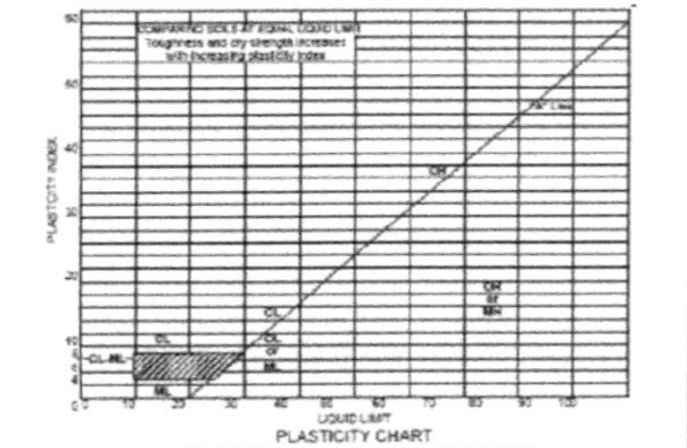

PLASTICITY CHART
FOR LABORATORY CLASSIFICATION OF FINE GRAINED SOILS

[a] Boundary classifications - Soils possessing characteristics of two groups are designated by combinations of group symbols. For example GW-GC, well graded gravel-sand mixture with clay binder.

[b] All sieve sizes on this chart are U.S. standard.

Figure 3.2-2. Unified Soil Classification Chart (Ref. 3-3).

3-4

3.3   INDEX PROPERTY TESTS

3.3.1   General

Index tests provide information on the present condition (water content) and on the physical and chemical composition (grain size distribution, Atterberg limits, and carbonate content) of a soil sample. The index tests can be run quickly and inexpensively, compared to most tests for engineering properties. Empirical relationships have been developed between several index properties and engineering properties (Chapter 2, Table 2.1-2) of soils found on land. Most of these empirical relationships can be expected to apply to terrigenous marine soils because these soils are essentially similar soils moved offshore or submerged due to sea level or land elevation changes. However, when dealing with pelagic soils, the previously established empirical correlations between index and engineering properties may not be applicable.

Table 3.3-1 lists pertinent information on the most widely used index property tests and the standard ASTM references. This section will discuss particularly important aspects of these test procedures as applied to marine soils and will describe modifications to the standard test and data reduction procedures necessary to properly evaluate marine soils. Some index and engineering properties determined from a range of marine soil types are shown in Table 3.3-2.

3.3.2   Sample Preparation

ASTM D421 (Ref. 3-1) describes the standard method for dry preparation of soil samples for grain size and Atterberg limit tests. This dry preparation technique can be used for cohesionless terrigenous soils having no more than a trace of biogenous material. However, for those soils with measurable proportions of biogenous material, dry preparation should not be used. Air drying removes water from intra-particle voids of biogenous material, and these voids are not resaturated later during the Atterberg limit tests. Further, the mechanical disaggregation technique used (grinding with mortar and rubber-covered pestle) is far too abrasive for use with fragile biogenous materials, including coralline sands.

Instead, marine soil samples are normally prepared in a wet state and are only dried when the test is completed to obtain dry sample weight. Disaggregation, if required, is best accomplished using an ultra-sonic bath, with the sample immersed in a deflocculant solution.

Table 3.3-1. Requirements for Index Property Tests (Ref. 3-4)

Index Property Test	Reference for Standard Test Procedure	Variations From Standard Test Procedure	Type of Sample for Test[a]	Size or Weight of Sample for Test[b]
Sample preparation	ASTM D421	None	Disturbed or undisturbed.	As required for subsequent tests.
Water content	ASTM D2216	None	Disturbed or undisturbed with unaltered natural water content.	As large as convenient.
Dry unit weight	ASTM D2937	Determine total dry weight of a sample of measured total volume.	Undisturbed with unaltered natural volume.	As large as convenient.
Specific gravity:				
Material smaller than No. 4 sieve size	ASTM D854	Volumetric flask preferable; vacuum preferable for de-airing.	Disturbed or undisturbed.	25 to 50 gm for fine-grained soils; 150 gm for coarse-grained soil.
Material larger than No. 4 sieve size	ASTM C127	None	Disturbed or undisturbed.	500 gm.
Atterberg limits[c]:				
Liquid limit	ASTM D4318	Harvard liquid limit device and grooving tool acceptable; open wire grooving tool acceptable.	Disturbed or undisturbed, fraction passing No. 40 sieve.	50 to 100 gm.
Plastic limit	ASTM D4318	Ground glass plate preferable for rolling.	Disturbed or undisturbed, fraction passing No. 40 sieve.	15 to 20 gm.
Gradation:				
Sieve analysis	ASTM D422-63	Selection of sieves varies for samples of different gradation.	Disturbed or undisturbed, nonsegregated sample, fraction larger than No. 200 sieve size.	600 gm for finest grain soil; to 4,000 gm for coarse-grained soils.
Hydrometer analysis	ASTM D422-63	Fraction of sample for hydrometer analysis may be that passing No. 200 sieve. Entire sample of fine-grained soil may be used.	Disturbed or undisturbed, nonsegregated sample, fraction smaller than No. 10 sieve size.	65 gm for fine-grained soil; 115 gm for sandy soil.

[a] Disturbed or undisturbed indicates that the source sample may be of either type.
[b] Sample weights for tests on air-dried basis. Dry weight estimated before test and determined after index test is run.
[c] Material for these tests should not be dried before the test is run.

Table 3.3-2. Some Index and Engineering Properties of Ocean Sediments
(Most Data Limited to Upper 2 Meters of Seafloor) (Ref. 3-5).

Core No.		Water Depth (m)	w_n%	w_L	w_p	γ (kN/m³)	CaCO₃%	S.G. Solids	s_u, Vane kPa	S_t Vane	$\bar{c}$ (kPa)	$\bar{\phi}$ (deg)	C_c	$\dfrac{C_c}{1+c_0}$
	TERRIGENOUS													
1	Terrigenous Clayey Silt	180	44-59	35-48	29-32	16.6-17.8	negligible	?	7-25 in-situ 1.4-17 lab	2-3 3-4	-	-	0.22-0.40	0.10-0.17
2	Terrigenous Clayey Silt (basin)	370	73-108	73-88	43-49	14.4-15.4	negligible	?	1.4-17 in-situ 3.4-25 lab	2 4-very high	1.38	37	0.68-0.91	0.20-0.25
3	Terrigenous Clayey Silt	1,700	104-144	109-121	61-89	13.5-14.3	negligible	?	3.4-14 lab	3-5				
4	Calcareous Proximal Turbidite (silt-clayey silt)	2,000	52-79	N.A.	N.A.	-	78	2.71-2.72	1.7-10 ship	4-very high	-	-	-	-
	PELAGIC - Calcareous ooze													
5	Calcareous ooze	3,160	86-116	43	NP[a]	13.8-15.4	88	2.67	11-14	11	-	-	0.91	0.21
6	Calcareous ooze	4,480	84-110	56	NP[a]	13.8-15.4	93	2.76	11-12	?	29	27	0.67	0.19
7	Calcareous ooze	3,500	168-236	-	-	12.4-12.9	72-78	2.54-2.59	1.1-4.9	6	-	-	-	-
8	Calcareous ooze	3,530	156-212	-	-	12.6-13.1	72-77	2.53-2.58	2.9-12.9	8	-	-	-	-
9	Calcareous ooze	3,690	203	-	-	12.9-13.3	73-79	2.55-2.58	2.3-5.8	8	-	-	-	-
10	Calcareous ooze	3,670	150-218	-	-	12.9-13.5	74-82	2.55-2.63	1.6-13.0	4-12	-	-	-	-
11	Calcareous ooze	3,940	127-210	-	-	13.0-13.9	76-84	2.51-2.61	2.0-4.1	4-10	-	-	-	-
12	Calcareous ooze	4,100	112-248	-	-	12.4-14.0	68-81	2.51-2.62	2.6-11.6	4-12	-	-	-	-
13	Calcareous ooze	4,560	128-414	-	-	11.4-13.6	58-81	2.33-2.60	2.1-8.1	5-7	-	-	-	-
14	Calcareous ooze	4,700	143-415	-	-	11.6-13.3	47-79	2.40-2.62	1.7-6.6	4-6	-	-	-	-
15	Calcareous ooze	4,500	172-255	-	-	12.4-13.6	64-85	2.49-2.61	1.6-3.4	2-12	-	-	-	-
16	Calcareous ooze	4,370	120-250	-	-	12.2-13.7	56-86	2.55-2.62	2.3-5.9	4-11	5.9	35	(2.7 kPa, 40°)[b]	
17	Calcareous ooze	4,450	81-215	-	-	12.8-14.3	70-87	2.54-2.61	2.6-7.5	5-9	3.7	40	(1.4 kPa, 44°)[b]	
18	Calcareous ooze	4,300	102-196	-	-	13.4-14.7	79-88	2.57-2.65	2.6-3.2	5	5.7	35	(1.4 kPa, 42°)[b]	
19	Calcareous ooze	4,300	172	-	-	14.8	90	2.57	5.0	8	-	-	-	-
20	Calcareous ooze	4,520	100-164	-	-	14.0-14.9	80-86	2.62-2.66	1.9-4.3	6	-	-	-	-
21	Calcareous ooze	4,850	296-324	-	-	11.8-12.3	34-47	2.53-2.56	1.0-5.0	5	-	-	*at stress < 32 kPa	-
22	Nanno foram chalk ooze	1,950	115	112	45			2.73	14		-	-	-	-
23	Nanno ooze	4,721	123	146	55	-	-	2.60	14		-	-	-	-
24	Clay rich nanno ooze	4,431	60	59	16	-	-	-	-	-	-	-	-	-
25	Calcareous ooze	3,930	98-110	66-70	42-57	14.2-14.3	56-75	2.65-2.69	4.8-8.9 lab	3-6	11	28	0.64-0.89 (3.4 kPa, 34°)[c]	0.17-0.23
26	Calcareous ooze	1,100	51-71	N.A.	N.A.	-	80-86	2.68-2.72	14-27 in-situ 0.4-1.8 ship	5-10 in-situ	0	32	-	
	Siliceous ooze													
27	Radiolarian and ash rich Diatom Ooze	5,518	219	198	71				19	-	-	-	-	-
28	Clayey Diatom Ooze (1.2-10m)	2,649	151	71	47	13-14	small	-	1.6	-	-	-	-	-
	Diatom ooze (170m)	2,649	185	NP[a]	NP[a]	12-13	small	-	18-44	-	-	-	2.64	0.49
	Silt rich Diatom ooze (230m)	2,649	119	NP[a]	NP[a]	13.0-13.5	small	-	83	-	-	-	2.78	0.67
29	Diatom ooze (118m)	2,414	87-106	-	-	13.8-13.9	1	2.38	43	-	0	36	-	-
	Diatom ooze (155m)	2,414	83-171	-	-	13.1	1	2.30	40	-	0	41	-	-
	Pelagic Clay													
30	Pelagic clay	5,460	94-112	73-111	35-48	-	negligible	?	2.2-4.4 ship	2-3	3.1	27	-	-
31	Pelagic clay	4,600	144-162	115-124	55-56	-	very low	2.72	0.68-1.35	3-12	0	37	1.82	0.35
32	Pelagic clay	4,600	135-162	105-109	56-61	-	very low	2.69-2.73	0.73-1.37	3-12	(3.1 kPa, 30°)[d]	37	1.70	0.34
33	Pelagic clay (Abyssal hills)	5,421	112-138	91-103	35-38	14.1	0.1	2.69-2.79	4.3-6.0	1.8-4	1.7	37	-	-
34	Pelagic clay (Abyssal hills)	5,768	115-122	74-86	36-37	14.3-14.4	0	2.74-2.81	1.3-2.8	2.4	3.4	35	0.72	0.20
35	Iron oxide (Abyssal hills)	5,644	186-284	157-223	79-102	12.2	0.1	2.70	9.3-13.1	2-3	3.5	38	-	-
36	Iron oxide (Abyssal hills)	5,163	202-235	225-229	97-101	13.0	0	2.84-3.30	12.2-13.5	1.8-2	4.3	38	2.06	0.29

a: NP indicates Non-Plastic

b: Data in parentheses are $(\bar{c}, \bar{\phi})$ at normal effective stress < 32 kPa. Data in $\bar{c}$ and $\bar{\phi}$ columns at normal effective stress > 32 kPa.

c: Data in parentheses are $(\bar{c}, \bar{\phi})$ at normal effective stress < 80 kPa. Data in $\bar{c}$ and $\bar{\phi}$ columns at normal effective stress > 80 kPa.

d: Data in parentheses are $(\bar{c}, \bar{\phi})$ for core 31 at normal effective stress <25kPa. Data in $\bar{c}$ and $\bar{\phi}$ columns at normal effective stress >25kPa.

3.3.3 Water Content

ASTM D2216 (Ref. 3-1) describes the standard method for the laboratory determination of water content for terrestrial soils. In marine soils, salt comprises a small portion of the fluid phase in the natural state. For highly accurate computations, a correction should be applied to the equation for calculating the water content. This correction, however, is most often not made.

The water content of the soil sample, corrected for dissolved salts in the pore fluid, is obtained from:

$$\overline{w} = \frac{(W_1 - W_2) + r(W_1 - W_2)}{(W_2 - W_c) - r(W_1 - W_2)} \cdot 100 \tag{3-1}$$

where:

$\overline{w}$ = water content corrected for salt content [%]

r = salinity of pore fluid defined as the ratio between the weight of dissolved salt and the weight of seawater; for engineering, assume a value of 0.035

W_1 = weight of container and moist soil [F]

W_2 = weight of container and oven-dried soil [F]

W_c = weight of container [F]

3.3.4 Unit Weight

The wet and dry unit weights are determined from relatively undisturbed soil samples obtained directly from core tube or liner sections of known length and diameter (ASTM D2937, Ref. 3-1) or from a carved sample of known volume, such as a consolidation sample or a triaxial cylindrical sample. The wet, or bulk, unit weight of the soil sample is:

$$\gamma = \frac{W}{V} \tag{3-2}$$

where:

W = wet weight of the soil sample [F]

V = volume of the soil sample [L^3]

The adjusted dry density, corrected for salt content, is:

$$\overline{\gamma}_d = \frac{(W_2 - W_c) - r(W_1 - W_2)}{V} \tag{3-3}$$

3.3.5 Specific Gravity

Two methods are used for determining the specific gravity, or grain density, of materials that make up soil samples. For that portion of a sample finer than the no. 4 sieve [4.76 mm], the pycnometer method is used (ASTM D854, Ref. 3-1). For the coarser portion of a sample, a technique better suited to the larger grain sizes is used (ASTM C127, Ref. 3-6). Most marine soils are finer than 4.76 mm, and a discussion of the pycnometer method will suffice here. ASTM D854 provides guidance on computing the weighted average specific gravity for those samples containing both coarse and fine materials.

The soil sample is first leached of soluble salts by placing the sample on filter paper in a Büchner funnel and washing the sample with distilled water. This sample is then washed into the pycnometer, and the test is run as described in the ASTM standard.

The specific gravity of soil grains is calculated from:

$$G_s = \frac{G_w W_d}{W_d - (W_b - W_a)}$$ (3-4)

where:

G_w = specific gravity of distilled water at the temperature t of the pycnometer and contents (see ASTM D854)

W_d = weight of oven-dry soil [F]

W_a = weight of pycnometer filled with distilled water at temperature t [F]

W_b = weight of pycnometer filled with distilled water and soil sample at temperature t [F]

For most seafloor soils (except for siliceous oozes and pelagic clays of high iron oxide content), the specific gravity can be estimated as 2.7 without incurring significant error. Table 3.3-2 lists some measured values for specific gravity of ocean sediments (Ref. 3-5).

3.3.6 Liquid Limit, Plastic Limit, and Plasticity Index

The liquid limit and plastic limit are water contents of soils at borderlines used to describe significant changes in physical properties. They are known as Atterberg limits, after the man who designed the test. Although the test specifics were somewhat arbitrary, they are now a primary standard for indexing behavior of fine-grained soils. ASTM D4318 (Ref. 3-1) describes the standard test methods for determining the liquid limit, plastic limit, and plasticity index of soils.

3.3.6.1 Liquid Limit

The liquid limit (w_L) is the water content at the transition between the liquid and plastic states of a soil. This is determined from testing with a cupped device into which remolded soil is placed. A groove is scoured into the soil, separating it into two halves. The liquid limit is arbitrarily defined as the water content at which the two halves will flow together when the finely calibrated cup containing the soil halves is dropped a specific distance 25 times at a specified rate.

The liquid limit test is intended to be performed only on that portion of a soil sample passing a no. 40 sieve (less than 0.42 mm in diameter). It is very important that preparation of the test sample does not change the sample characteristics and cause this boundary to shift. Shifting of the boundary is likely with marked change of: (1) the soil's pore water salt content and (2) the sediment grain characteristics.

To minimize the impact of salt concentration change in the pore water, the liquid limit test on marine soils should be run on material taken directly from the stored sample tubes. Distilled or deionized water should be added to the sample to raise the water content, or the sample should be allowed to lose moisture by air drying with air blower or heat-lamp assistance until the liquid limit has been defined by running the test at several water contents. Although this procedure will result in some change in water-salt concentration, the effect of this small change on the liquid-plastic boundary is minimal.

To minimize degradation of sediment grains, especially for the pelagic oozes, the mechanical agitation and remolding of the soil must be minimized. Dry preparation should not be used for specimens containing measurable percentages of biogenous or organic materials (Section 3.3.2).

For those cohesive soil samples containing significant amounts of coarser materials, separation of coarse and fine material may be possible but is a laborious process. The sample can be soaked in distilled water and gently pushed through a no. 40 sieve. The empty sieve is first placed in a pan; and then the soaked sample is poured into the sieve. Distilled water is added to the pan, bringing the water level to 1 cm over the screen. The soil is then gently stirred with the fingers while the sieve is agitated up and down. Remains of the original sample are worked until all the fine material has passed through the sieve into the pan. Most water from the pan sample is removed by passing through filter paper in a funnel. The moist soil fines retained on the filter paper are then warmed by heat lamp or air blower until the soil reaches a puttylike consistency suitable for the liquid limit test. However, because this method of sample preparation removes most of the pore water salt from the sample, it may influence test results. If at all possible, direct use of marine soils from the stored sample tubes and physical separation of coarse materials is preferred for Atterberg limit determinations.

Except for the recommended changes in sample preparation and mixing, the liquid limit test is conducted as specified in ASTM D4318 (Ref. 3-1). It is most desirable to run the test at water contents just above and below that which would require 25 drops to cause sample closure. Two tests should be run at each of these levels, and interpolation should be used to determine the 25-drop water content. The one-point method described in the standard should not be used for marine soils.

3.3.6.2 Plastic Limit

The plastic limit (w_p) of a soil is the water content at the transition between the plastic and semi solid states. The plastic limit is arbitrarily defined as the lowest water content at which the soil can be rolled into threads 1/8 inch in diameter without the threads breaking into pieces (ASTM D4318, Ref. 3-1).

All comments pertaining to the preparation and handling of the sample for the liquid limit test also apply to the plastic limit test.

3.3.6.3 Plasticity Index

The plasticity index (*PI*) is calculated as the difference between the liquid and plastic limits; i.e.,

$$PI = w_L - w_p \tag{3-5}$$

Values of the liquid limit and plastic limit for samples of seafloor soil types are reported in Table 3.3-2.

3.3.7   Grain Size Analysis

The determination of the grain size distribution of marine soils is performed in the same way as for terrestrial soils (ASTM D422-63, Ref. 3-1), with some variation in sample preparation to limit grain particle degradation.[1] The distribution of particle sizes larger than 0.075 mm (no. 200 sieve) is determined by sieving. The distribution of sizes finer than 0.075 mm is determined by hydrometer test. Grain sizes up to 2.0 mm (no. 10 sieve) can be included in the hydrometer test sample to provide an overlapping of grain size distribution curves from the two methods. Since most marine soils, including the brown clays and oozes, are finer than 2.0 mm, separation of the sample on the no. 10 sieve is not necessary; the hydrometer test can be performed directly on the sample.

---

[1] Salt content of the water plays no part in this test.

Samples of marine soils for grain size sieving are prepared by the wet method. The soil is not oven-dried prior to the test because this would remove the water from within the biogenous structures and could alter the structure of clay-sized particles or could cause particle bonding into larger particles. These effects are not reversed upon rewetting. When washing these sediments on the sieves, the agitation must be kept to a minimum to limit particle degradation.

Samples of marine soils for the hydrometer test are prepared by rough mixing the sample with water to promote separation of the very small particles. Some pelagic clay samples, notably those having a high iron oxide content, are very difficult to separate into individual particles. One technique places the sample in a solution of dispersing agent, sodium hexametaphosphate. The soil and dispersant are then mixed in a blender, followed by centrifuging of the mixture to separate the solid particles. This process is repeated perhaps three times to reduce the natural flocculated structure to a dispersed structure. Particle degradation for clay-sized material is not a significant problem, as the material is primarily minute plate-shaped particles that do not break down.

The grain size curves from the sieve analysis for larger particles and from the hydrometer analysis for smaller particles may not agree exactly where the test data overlap. Part of this deviation arises because the theory on which the hydrometer grain size analysis is based (Stoke's theory) assumes a spherical-shaped particle. Clay particles and much of the foram fragments are plate shaped and do not conform to the theory. Further, the whole biogenous shells, especially the foraminifera, are hollow spheres. Thus, their effective specific gravity is lower than that measured (Section 3.3.5). The specific gravity error causes the percentage of the sediment classified as fine-grained to be larger than it really is.

3.3.8   Carbonate and Organic Carbon Content

The organic and carbonate carbon contents of the marine sample should be measured for those soils with suspected high carbonate content (>30%) and for those giving off hydrogen sulfide gas. ASTM D4373-02 (Ref. 3-1) describes the standard test methods for determining the carbonate content of soils. The test is a gasometric method that utilizes a simple portable apparatus. Note that this method does not distinguish between the carbonate species and such determination must be made using quantitative chemical analysis methods such as atomic absorption.

The organic carbon content may be determined by wet combustion using an elemental chemical analysis. Details of this test method are presented in the ASTM Special Procedures for Testing, STP38516S.

3.4 ENGINEERING PROPERTY TESTS

3.4.1 General

Engineering property tests define properties of soil or soil samples at specific states of stress. The most important is the undisturbed state – or as the soil exists in its natural environment. Most testing attempts to establish properties for this condition.

The undrained shear strength (s_u) of samples of cohesive marine soils can be measured either by a vane shear test or by an unconsolidated-undrained (UU) triaxial test. The unconfined compression test, a special case of the UU triaxial test where confining pressure is zero, is often run because of its simplicity. Soil strength is measured differently in each of the tests, but the value will be approximately the same as undrained shear strength.

Laboratory vane shear testing is uniquely suited to very soft sediments that cannot stand under their own weight outside the core liner (a prerequisite for sample preparation in most UU testing). In addition, the vane shear test is the only laboratory test used to date for determining the sensitivity of soft marine cohesive soils. The vane shear test can be used only on cohesive soils. Tests should never be taken in granular soils as any such measurement is meaningless and misleading.

Other, generally more complex, testing is done for soil parameters useful in predicting soil behavior under conditions different than those existing (for example, the different stress conditions created by placement of a structure on or in the soil). The effective stress parameters, $\bar{c}$ and $\bar{\phi}$, define a generalized soil failure criterion (Mohr-Coulomb failure envelope) and are usually determined in a consolidated-drained (CD) or consolidated-undrained $(\overline{CU})$ triaxial test with pore pressure measurements.

The compression index, C_c, and coefficient of consolidation, c_v, are determined from the one-dimensional consolidation test. The permeability, k, can be calculated either from consolidated triaxial tests or from one-dimensional consolidation tests.

A summary of test requirements is given in Table 3.4-1. Care should be taken that the testing conditions represent stress states for the soil being investigated.

Table 3.4-1. Requirements for Engineering Property Tests (Ref. 3-4)

Test	Reference for Suggested Test Procedure	Applicability and Variations From Suggested Test Procedure	Size or Weight of Sample for Test (Undisturbed or Remolded)
Vane Shear	ASTM D4648	Applicable to very soft to stiff saturated fine-grained (cohesive) soils.	Test usually run on exposed sample at end of core liner tube. Specimen height ≥ 3 3 vane height. Specimen diameter ≥ 3 3 vane diameter.
Unconfined Compression (UC)	ASTM D2166	Pelagic clays from depths greater than 15ft; biogenous oozes, and nearshore terrigenous silts normally too soft to be properly prepared for this test.	Minimum cross-sectional area 10 cm^2. Length = 2 to 3 3 vane diameter.
Triaxial Compression Unconsolidated Undrained (UU)	ASTM D2850	Information similar to that for UC test; can be used with softer and more pervious sediments than UC.	Same as UC test.
Consolidated-Drained (CD)	Ref. 3-2	Rate of shear limited to allow complete drainage.	Same as UC test.
Consolidated-Undrained (CU)	ASTM D4767 and Ref. 3-2	More common and generally less time-consuming than CD. Pressure lines leading to sample should be seawater filled. Loading rod friction should be minimized by using air bushing or equivalent.	Same as UC test.
One-Dimensional Consolidation	ASTM D2435	Must provide for very low initial load increments. Sample should be submerged in seawater.	Sample diameter ≥ 50 mm or ≥ 2 to 5 3 height. Sample height ≥ 13 mm.
Direct Shear	ASTM D3080	Limited to consolidated shear tests on fine-grained soils.	Sample diameter ≥ 50 mm, or ≥ 2 3 height. Sample height ≥ 12.5 mm.

3.4.2 Vane Shear Test

ASTM D4648 (Ref. 3-1) standardized the laboratory vane shear tests. The following is based this standard.

1. Scope. The laboratory vane shear test is applicable to very soft to stiff saturated fine-grained (cohesive) soils. The laboratory vane shear test should not be used in soils with very high undrained shear strengths (s_u greater than 14 psi) because vane failure conditions in these higher strength soils may deviate from the assumed cylindrical failure surface and cause significant error in the measured strength. This method includes use of either conventional calibrated spring units or electrical torque transducer units with a motorized miniature vane.

2. Description of the Test. The miniature vane shear test consists of inserting a four-bladed vane in an undisturbed tube sample or remolded sample and rotating it at a constant rate to determine the torsional force required to cause a cylindrical surface to be sheared by the vane. Force is measured by a calibrated torque spring or torque transducer directly attached to the vane. This force is then converted to a shearing resistance primarily on the cylindrical surface.

3. Apparatus. The vane should consist of a rectangular four-bladed vane as illustrated in Figure 3.4-1. It is recommended that the height of the vane be twice the diameter, although vanes with other ratios can be used, including a height equal to the diameter. Vane blade diameter typically varies from 0.5 to 1.0 inch. Variations from recommended values would be made where sample size presents constraints or where other special conditions exist. The vanes should be "thin" so as to displace no more than 15% of the soil when inserted into the soil.

    Torque is applied to the vane by manual or motorized power. The shaft should be rotated at a constant rate of 60 to 90 deg/min. Another, slower standard rate (6 deg/min) is sometimes used. Torque is measured through a spring or an electrical transducer rotating with the shaft.

4. Preparation of Samples. Soil samples into which the vane is inserted should be large enough to minimize influence of container sides on the test results. The test should not be centered closer to the edge of the container than 1.5 times the vane diameter. Tests run in the same container should be at least 2.0 times the vane diameter apart from each other.

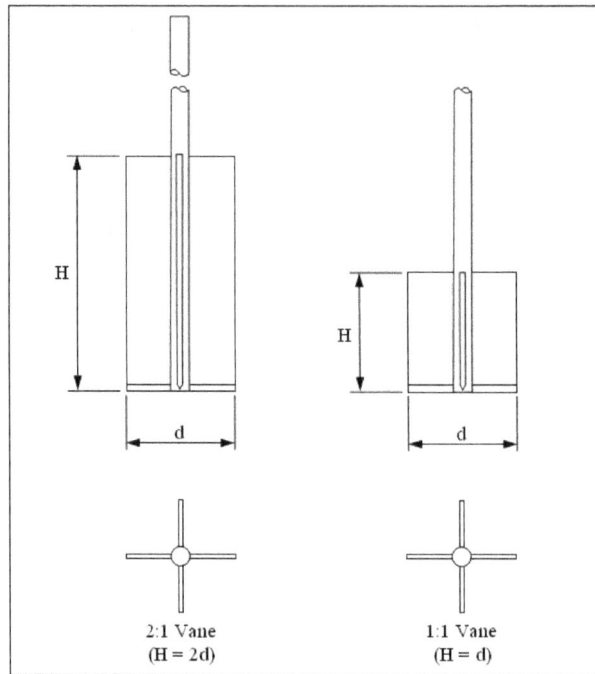

Figure 3.4-1. Miniature vane blade geometry.

5. <u>Test Procedure</u>. The vane shear unit should be securely fastened to a table or frame to prevent movement during a test. The vane is inserted and fixed at an elevation in the sample so that the vane top is embedded by an amount at least equal to the vane height. The sample should be held firmly to prevent rotation. Torque readings should be recorded at a frequency that will allow good definition of the torque-rotation curve (approximately every 5° of rotation) or until a maximum of 180° of rotation is obtained. The vane blade is removed and cleaned, and a representative sample of the specimen is taken from the vicinity of the test to determine the water content. The soil is inspected for sand, gravel, and other inclusions that may have influenced test results. Care should be taken to make notes of all sample or test peculiarities observed.

The remolded vane strength, if desired, is obtained following the test on the undisturbed sample and prior to removal of the vane and soil sampling. Following the initial test, the vane rotation should be continued (at a more rapid rate) until two complete revolutions have been completed from the original position of the vane when it was inserted. Determination of the remolded strength should be started immediately after completion of rapid rotation. The procedure, outlined above, is followed with vane removal and soil sampling done at its conclusion.

6. <u>Calculations</u>. A graph is prepared showing the applied torque versus the rotation angle. The vane shear strength (s_{uv}) is computed from the maximum torque value using the same equation previously introduced for computing in-situ vane shear strength [Equation 2-1]. The remolded shear strength is computed the same way. Soil sensitivity (S_t) is derived from the ratio of undisturbed to remolded shear strength as follows:

$$S_t = \frac{s_{uv}(undisturbed)}{s_{uv}(remolded)} \tag{3-6}$$

The shear strengths measured by the laboratory vane shear test are influenced by effects of anisotropy and strain rate, as described in Section 2.5.2 for in-situ vane tests. For comparison with undrained shear strengths determined by triaxial testing, a correction factor should be applied as done for the in-situ vane data through use of Equation 2-2 and the correlation with plasticity index from Figure 2.5-1.

3.4.3 Unconfined Compression Test

The standard test method for determining the unconfined compressive strength of marine soils is defined in ASTM D 2166 (Ref. 3-1).

The unconfined compression test should only be performed on samples that are cohesive, are relatively impervious, and have sufficient strength to stand under their own weight. Pelagic clays from subbottom depths beyond 15 feet would usually meet these criteria; biogenous oozes and nearshore silty sediments generally would not.

3.4.4 Unconsolidated, Undrained Triaxial Compression Test

The unconsolidated, undrained (UU) triaxial comprerssion test for marine soils is run as described in ASTM D2850 (Ref. 3-1). Although the UU test determines the same type of information as the unconfined compression test, it can be run on somewhat softer and more pervious sediments, but requires very careful sample preparation and handling.

3.4.5 Consolidated-Undrained and Consolidated-Drained Triaxial Compression Tests

The consolidated-undrained ($\overline{CU}$) and consolidated-drained (CD) triaxial tests are used to measure the effective strength parameters, $\overline{c}$ and $\overline{\phi}$, of cohesive marine soils. These parameters can also be measured from the drained direct shear test (Section 3.4.6). The $\overline{CU}$ triaxial test is standardized by ASTM D4767 (Ref. 3-1) but the CD test is not. Comprehensive test descriptions and procedures for both tests are given on pages 122-137 of Reference 3-2 and in Reference 3-7.

3.4.6 Consolidated-Drained Direct Shear Test

The consolidated-drained direct shear test is another alternative for determining the effective strength parameters, $\bar{c}$ and $\bar{\phi}$, for marine cohesive soils. This test is standardized as ASTM D3080 (Ref. 3-1). The direct shear test is well-suited to a consolidated-drained condition because the drainage paths through the test specimen are short, thus allowing consolidation to take place fairly rapidly. However, the test is not suited to the development of exact stress-strain relationships within the test specimen because of the nonuniform distribution of shearing stresses and displacements. The slow rate of displacement provides for dissipation of excess pore pressures, but it also permits plastic flow of soft cohesive soils.

3.4.7 Consideration for Triaxial Testing of Marine Soils

Some special considerations must be made for triaxial testing of marine sediments. First, since marine sediments have seawater as a pore fluid, the pressure lines leading to the sample should be filled with seawater rather than freshwater, if possible, when this water may enter the sample. Although data have not been published to show that freshwater changes soil behavior, in theory, changes in pore water salt content accompanying water entry to marine samples could significantly alter behavior. Using saltwater in the pressure lines may, however, introduce a corrosion problem requiring use of stainless steel fittings at critical points. Secondly, many marine sediments are considerably softer than those usually found on land. Load and pressure transducers used to make measurements need to be sensitive or accurate at abnormally low readings: accuracy to within 0.02 pound for load and 0.02 psi for pressure. Devices such as air bushings must be used to reduce friction between loading rod and cell considerably below what is normally acceptable for soils testing.

3.4.8 One-Dimensional Consolidation Test

Procedures for the one-dimensional consolidation test have been standardized in ASTM D2435 (Ref. 3-1). Engineering properties determined by this test include the Compression Index, C_c; the Recompression Index, C_s; the coefficient of consolidation, c_v; the coefficient of permeability, k; and the coefficient of secondary compression, c_α. These data are used for estimating the amount and time rate of settlement under applied loads.

The degree of overconsolidation is also typically determined from this test. It is an engineering property of high value because of the high impact that soil stress history has on shear strength and other soil behavior. A soil that has undergone consolidation under a higher effective vertical overburden pressure than presently exists is overconsolidated. The ratio of this past effective pressure to the present effective pressure is called the overconsolidation ratio (OCR). OCR may also be determined through triaxial testing, which measures consolidation.

The only major procedural difference in adapting the standardized consolidation test to marine soils is a need for applying very low loads, as low as 8 psf. This is normally achieved by placing small weights on the loading cap for the first few load increments. In addition, the soil sample should be submerged in seawater rather than freshwater.

3.5 PROPERTY CORRELATIONS

3.5.1 General

This section presents correlations of soil engineering properties with index properties more easily measured. The correlations can be used as a rough guide for estimating properties when only limited site survey information is available. They should not be used for design without evaluation of actual material properties through laboratory or in-situ tests.

3.5.2 Nearshore Sediments

The nearshore terrigenous sediments are highly variable in composition. A considerable amount of information in geotechnical engineering literature can be applied to these sediments. Figure 3.5-1 shows a correlation between $s_u / \bar{p}_{vo}$ versus plasticity index for normally consolidated (NC) glacial clays on land and in coastal regions (Ref. 3-8). In this figure, s_u is the undrained shear strength, and $\bar{p}_{vo}$ is the effective vertical overburden pressure. In Figure 3.5-1, "young" refers to normally consolidated recent sediments, and "aged" refers to clays that have developed higher strength due to higher inter-particle bonding that has occurred with aging. The effects on strength are similar to a mild overconsolidation.

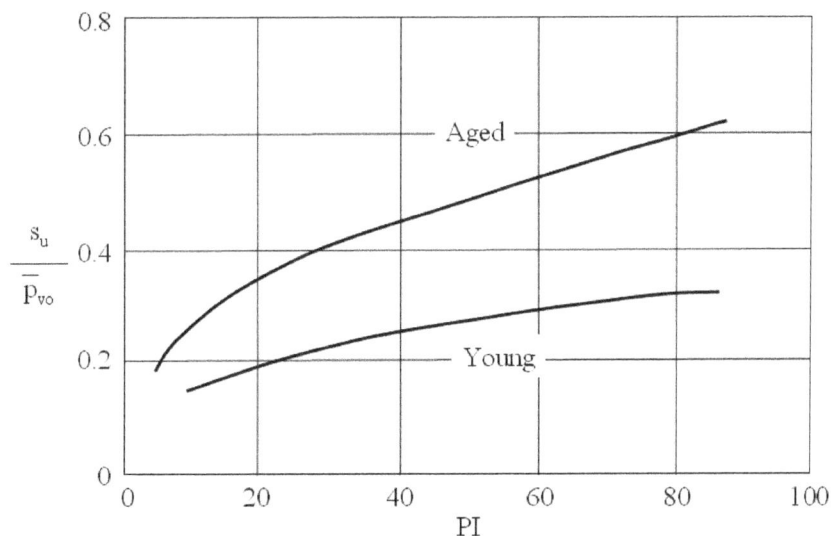

Figure 3.5-1. Relationship between $s_u/\bar{p}_{vo}$ and PI for normally consolidated late glacial clay (Ref. 3-8).

The $s_u/\bar{p}_{vo}$ ratio for overconsolidated (OC) soils will be higher than the range shown in Figure 3.5-1. Compared with normally consolidated (NC) soils, the ratio $s_u / \bar{p}_{vo}$ has been observed to increase as a function of the over-consolidation ratio (OCR) as follows:

$$\frac{s_u / p_{vo} \ \ for \ OC \ soil}{s_u / p_{vo} \ \ for \ NC \ soil} = (OCR)^{0.8} \qquad (3\text{-}7)$$

This relationship and the data in Figure 3.5-1 can be used to make a rough estimate of the in-situ strength of overconsolidated nearshore marine clays.

Figure 3.5-2 shows a correlation between friction angle (angle of shearing resistance, $\bar{\phi}$) and plasticity index for normally consolidated, fine-grained soils. Although this was not developed with data from marine soils, it can be used to make a rough estimate of $\bar{\phi}$ for near-shore, terrigenous, fine-grained soils.

Figure 3.5-3 shows a correlation between coefficient of consolidation (c_v) and liquid limit for fine-grained soils. Although this was not developed with data from marine soils, it can be used to make a rough estimate of c_v for nearshore, terrigenous, fine-grained soils.

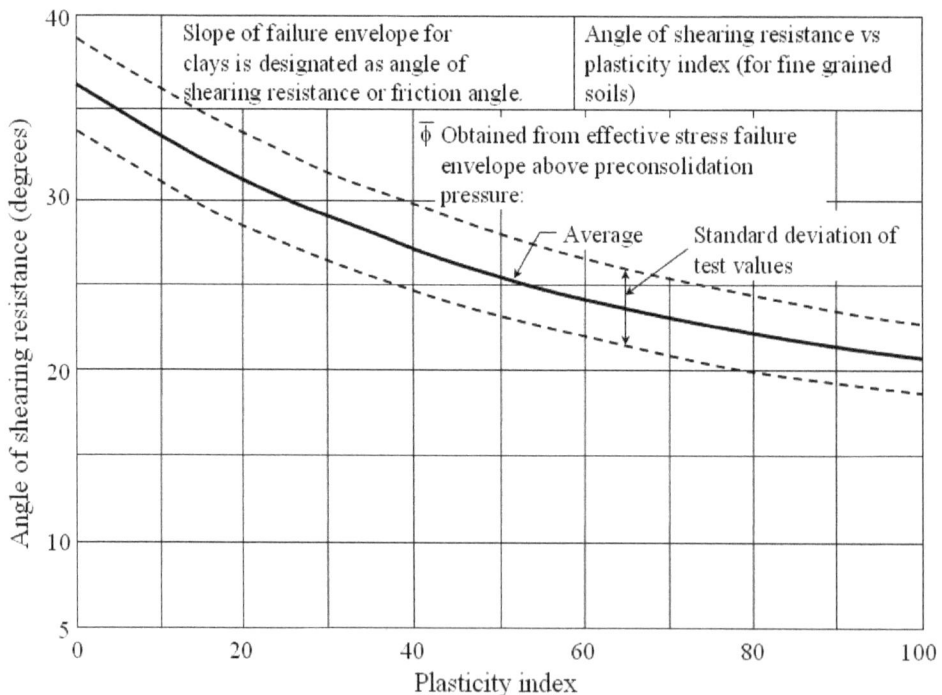

Figure 3.5-2. Relationship between friction angle and PI for normally consolidated fine-grained soils (Ref. 2-5).

3-20

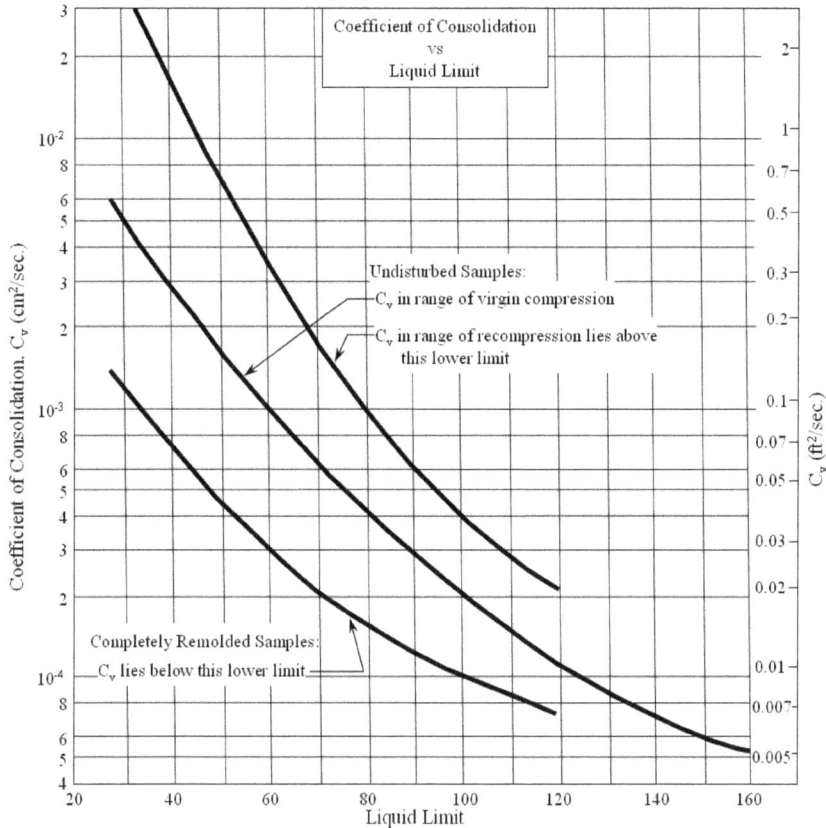

Figure 3.5-3. Correlation between coefficient of consolidation and liquid limit (Ref. 3-4).

The soil deposits in many river delta front regions of the continental shelf are under consolidated (have not fully consolidated under their present effective overburden pressure) and have strengths lower than would be expected for their existing stress condition. References 3-9, 3-10, and 3-11 contain information on the properties of these special types of sediments.

The properties of nearshore calcareous sediments, particularly their interaction with pile foundations, have been found to be different from those of terrigenous sediments. Calcareous sands are highly variable in character and behavior due to mode of deposition and alterations that take place after deposition. For this reason, typical properties cannot be suggested for nearshore calcareous sands. Reference 3-12 gives a summary of available information.

3.5.3 Deep Sea Sediments

Although the deep sea region is a relatively calm depositional environment, other processes, such as turbidity currents, can cause deposition in a considerably different manner. Still, vast regions of the seafloor are covered with sediments relatively uniform in profile (compared to adjacent sediments at the same depth) with corresponding relatively uniform engineering properties at equivalent depths. Chapter 2 presented estimated strength profiles for soils found in deep ocean regions to assist in planning for site surveys. Most data were extrapolated from shallow soil samples by consolidating them to the state of stress found at deeper elevations. The following sections present additional data on deep sea sediments, most of which were obtained in connection with shallow (upper 60 cm) exploration for manganese nodule deposits.

3.5.3.1 Pelagic Clays

The pelagic brown or red clays are very fine-grained silty clays typically with more than 60% particles finer than 0.002 mm. Liquid limits range from 75 to 275, and plasticity indices range from 40 to 180. As shown in Figure 3.5-4, the plasticity data for pelagic clays indicate they behave like highly compressible clayey silts and silty clays (this is indicated by data plotting close to the "A" line and to the right of the "B" line).

The water content for pelagic clays is usually higher than the liquid limit. The average undrained shear strength within the upper 60 cm of the soil profile is in the range of 0.5 to 1.5 psi. The laboratory-measured sensitivity is in the range of 2 to 10. Effective cohesion values, $\bar{c}$, in the range of 0.14 to 0.45 psi and a friction angle, $\bar{\phi}$, in the range of 27° to 37° are indicated (from very limited test data).

The range of compression index (C_c) for pelagic clays is shown in Figure 3.5-5. In consolidation tests, pelagic clays from shallow embedment depths exhibit mild overconsolidation behavior. In the figure, $\bar{e}_0$ refers to the void ratio, which is equal to the volume of voids divided by the volume of solids for the soil sample.

3.5.3.2 Pelagic Oozes

Pelagic oozes are calcareous or siliceous remains of tiny marine organisms or plants and have properties based on the type of sediment and the amount of clay in the sediment. Only a limited amount of data is available on the engineering properties of pelagic oozes in the upper few feet of the soil profile. These indicate that oozes have water contents in the range of 50 to 100% (or up to 300% for siliceous ooze) and shear strengths in the range of 0.5 to 1.5 psi.

Very limited test data indicate effective cohesion values, $\bar{c}$, in the range of 0 (not measurable) to 4 psi and a friction angle, $\bar{\phi}$, in the range of 27° to 37°. The range of compression index (C_c) for calcareous oozes is illustrated in Figure 3.5-5.

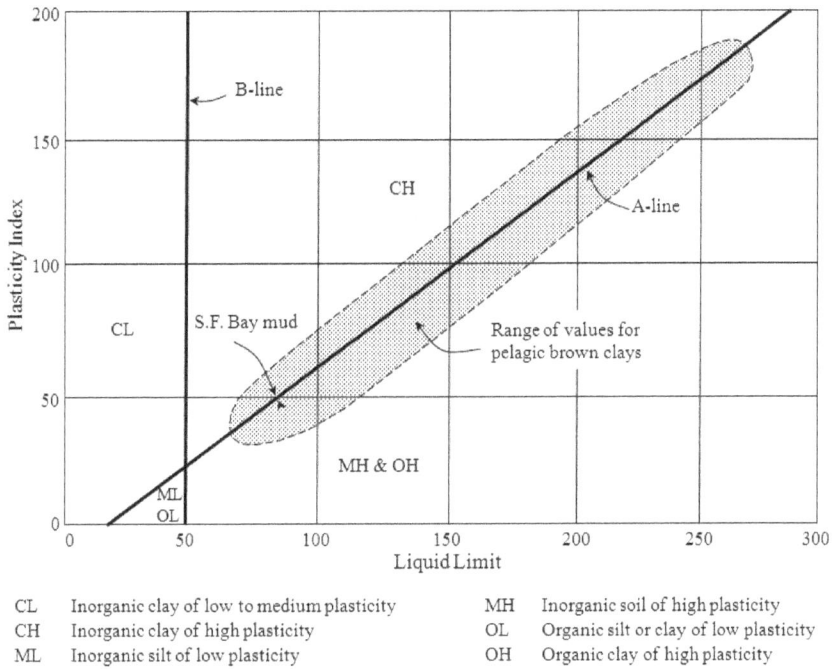

Figure 3.5-4. Range of PI values for pelagic clay.

CL	Inorganic clay of low to medium plasticity	MH	Inorganic soil of high plasticity
CH	Inorganic clay of high plasticity	OL	Organic silt or clay of low plasticity
ML	Inorganic silt of low plasticity	OH	Organic clay of high plasticity

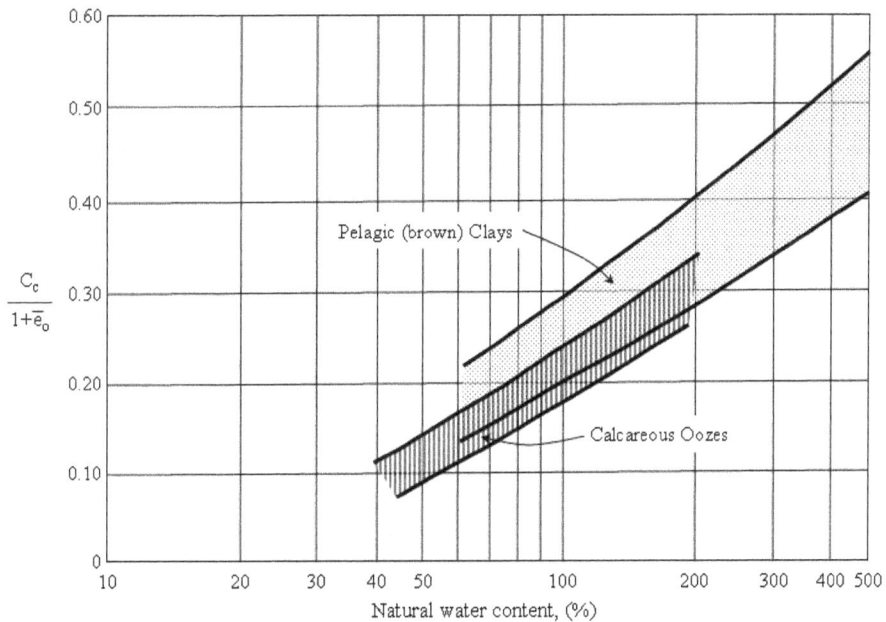

Figure 3.5-5. Correlation between water content and $C_c / (1 + \bar{e}_0)$ for pelagic clay and calcareous ooze.

3.6 REFERENCES

3-1.   "04.08 Soil and Rock (I): D 420 – D 5876," in Annual Book of ASTM Standards, American Society for Testing and Materials. Philadelphia, PA, Mar 2010.

3-2.   T.W. Lambe. Soil Testing for Engineers. New York, NY, John Wiley and Sons, 1951.

3-3.   Earth Manual, 3rd Edition, Bureau of Reclamation, U.S. Department of the Interior. Denver, CO, 1998.

3-4.   Unified Facilities Criteria (UFC): Soil Mevhanics, Naval Facilities Engineering Command (Preparing Activity), UFC 3-220-01N. Washington, DC, Jun 2005.

3-5.   P.J. Valent. Engineering Behavior of Two Deep Ocean Calcareous Sediments, Including Influence on the Performance of the Propellant Driven Anchor, Ph.D. Thesis, Purdue University. Lafayette, IN, Aug 1979, pp 19-20.

3-6.   "04.02 Concrete and Aggregates," in Annual Book of ASTM Standards, American Society for Testing and Materials. Philadelphia, PA, Oct 2009.

3-7.   A.W. Bishop and D.J. Henkel. The Measurement of Soil Properties in the Triaxial Test, 2nd Edition. London, England, Edward Arnold Ltd., 1962.

3-8.   L. Bjerrum. "Embankments on Soft Ground," in Proceedings of ASCE Specialty Conference on Performance of Earth and Earth-Supported Structures, Vol. 2, Purdue University. Lafayette, IN, 1972.

3-9.   I. Noorany and S.F. Gizienski. "Engineering Properties of Submarine Soils, State-of-the-Art Review," in Journal of Soil Mechanics and Foundations Division, ASCE, Vol. 96, No. SM5, Sep 1970, pp 1735-1762.

3-10.   A.F. Richards. Marine Geotechnique. University of Illinois Press, 1967.

3-11.   K. Terzaghi. "Varieties of Submarine Slope Failures," in Proceedings of Eighth Texas Conference on Soil Mechanics and Foundation Engineering, University of Texas, 1956.

3-12.   K. R. Demars, editor. Geotechnical Properties, Behavior, and Performance of Calcareous Soils, American Society for Testing and Materials, STP 777. Philadelphia, PA, 1982.

3.7 SYMBOLS

$\bar{c}$	Effective soil cohesion (drained cohesion intercept) [F/L^2]
C_c	Compression index, also grain size coefficient
CD	Consolidated-drained triaxial compression test
CH	Inorganic clay, high plasticity
CL	Inorganic clay, low to medium plasticity
C_s	Recompression index
$\overline{CU}$	Consolidated-undrained triaxial compression test (with pore pressure measurement)
C_u	Coefficient of (grain size) uniformity
c_v	Coefficient of consolidation [L^2/T]
c_α	Coefficient of secondary compression
d	Diameter of vane blade [L]
D_{10}	Sample grain diameter, below which 10% of material falls [L]
D_{30}	Sample grain diameter, below which 30% of material falls [L]
D_{60}	Sample grain diameter, below which 60% of material falls [L]
$e_0, \bar{e}_0$	Void ratio
GC	Clayey gravel
GM	Silty gravel
GP	Poorly graded gravel
G_s	Specific gravity (see also *S.G.*)
G_w	Specific gravity of distilled water
GW	Well-graded gravel
H	Height of vane [L]
k	Permeability [L/T]
LI	Liquidity index
MH	Inorganic elastic silt
ML	Inorganic silt, low plasticity
NC	Normally consolidated
OC	Overconsolidated
OCR	Overconsolidation ratio
OH	Organic clays, medium to high plasticity
OL	Organic silts and clays, low plasticity
PI	Plasticity index

Pt	Peat and other highly organic soils
$\overline{p}_{vo}$	Vertical effective stress, or soil overburden pressure [F/L^2]
r	Salt content
SC	Clayey sand
$S.G.$	Specific gravity (see also G$_s$)
SM	Silty sand
SP	Poorly graded sand
S_t	Sensitivity
s_u	Undrained shear strength [F/L^2]
s_{uv}	Vane shear strength [F/L^2]
SW	Well-graded sand
t	Reference temperature for pycnometer test weights
UC	Unconfined compression test
UU	Unconsolidated-undrained condition for triaxial testing
V	Volume [L^3]
W	Wet weight of soil sample [F]
W_a	Weight of pycnometer filled with distilled water [F]
W_b	Weight of pycnometer filled with distilled water and soil sample [F]
W_c	Weight of container in water content determination [F]
W_d	Weight of oven-dry soil [F]
W_1	Weight of container, and moist soil [F]
W_2	Weight of container and oven-dried soil [F]
$\overline{w}$	Water content corrected for salt content
w_L	Liquid limit
w_P	Plastic limit
w_n	Natural water content
γ	Wet, or bulk, unit weight [F/L^3]
$\overline{\gamma}_d$	Dry unit weight [F/L^3]
$\overline{\phi}$	Effective, or drained, friction angle [deg]

4   SHALLOW FOUNDATIONS AND DEADWEIGHT ANCHORS

4.1   INTRODUCTION

4.1.1   General

Shallow foundations and deadweight anchors are typically similar structures, their main element being a footing that interacts with the soil. Shallow foundations primarily resist downward-bearing and sideward forces, while deadweight anchors resist upward and sideward forces.

The design methods in this chapter are applicable to shallow foundations and deadweight anchors located in the deep and shallow ocean areas and follow an iterative or trial-and-error process. The process starts with an estimation of reasonable or "convenient" foundation or anchor dimensions, and then an analysis is made to predict performance. If the proposed foundation or anchor is found to be inadequate or to be excessively overdesigned, the dimensions are changed and the analysis process is repeated. In some cases the selected shallow foundation or deadweight anchor for the given soil conditions may be found impractical or too costly. Other foundation types (such as piles) must then be considered.

4.1.2   Definitions/Descriptions

4.1.2.1 Shallow Foundations

Generally, to be considered "shallow" a foundation would have a depth of embedment, D_f, less than the minimum lateral dimension (width) of the foundation, B. The horizontal base dimensions of a shallow foundation are generally large relative to the foundation thickness. Figure 4.1-1 is a sketch of a simple foundation. In the figure, z_s is the shear key height, and H is the foundation base height. Other types of shallow foundations are shown in Figure 4.1-2. Some shallow foundations for use in soft ocean soils are constructed with shear keys or skirts that extend below the foundation base to improve the lateral load resistance of the foundation.

The loading on a shallow foundation will be the combination of structure weight, environmental loading from current and wave forces (and possibly from wind and earthquake forces), and other externally applied forces. Loadings may include overturning moments, which create uplift (tensile) as well as downward (compressive) pressures. Virtually all loadings (except gravitational loadings on a horizontal seafloor) develop some load component parallel to the seafloor (lateral loads).

The type of loading will determine the methodology used in design. If the foundation loading is compressive (downward), resistance is derived from the bearing capacity of the soil. If a portion of the foundation is loaded in tension (upward), the uplift resistance will depend on the submerged weight of the foundation, the soil friction on the embedded surfaces, and "suction" beneath the foundation.

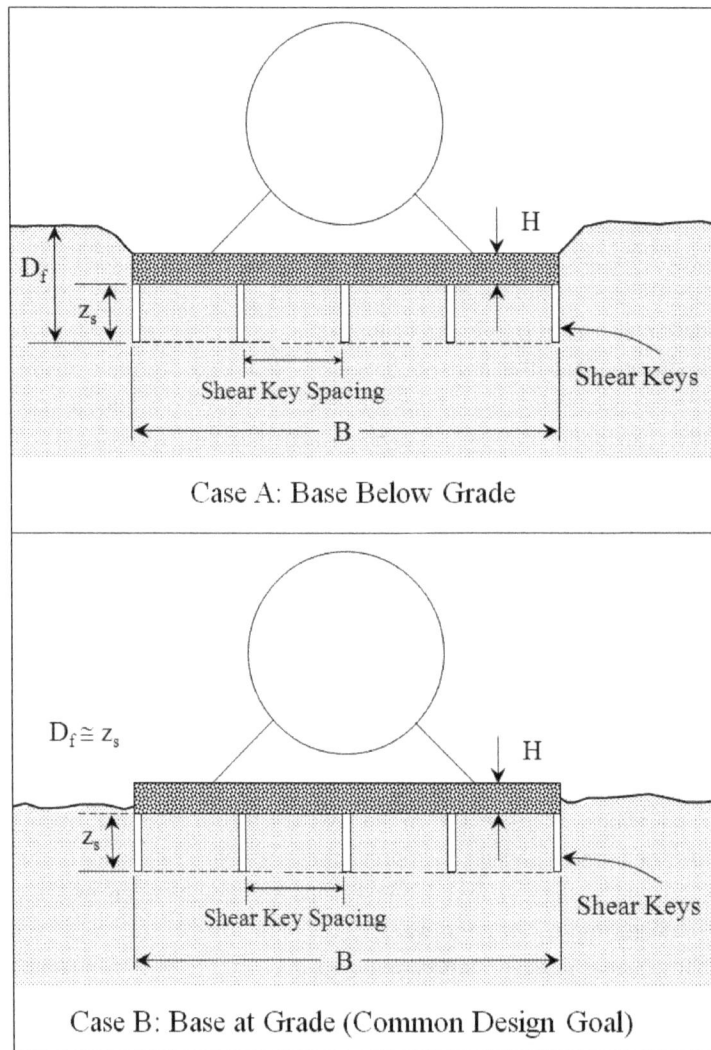

Figure 4.1-1. Features of a simple shallow foundation.

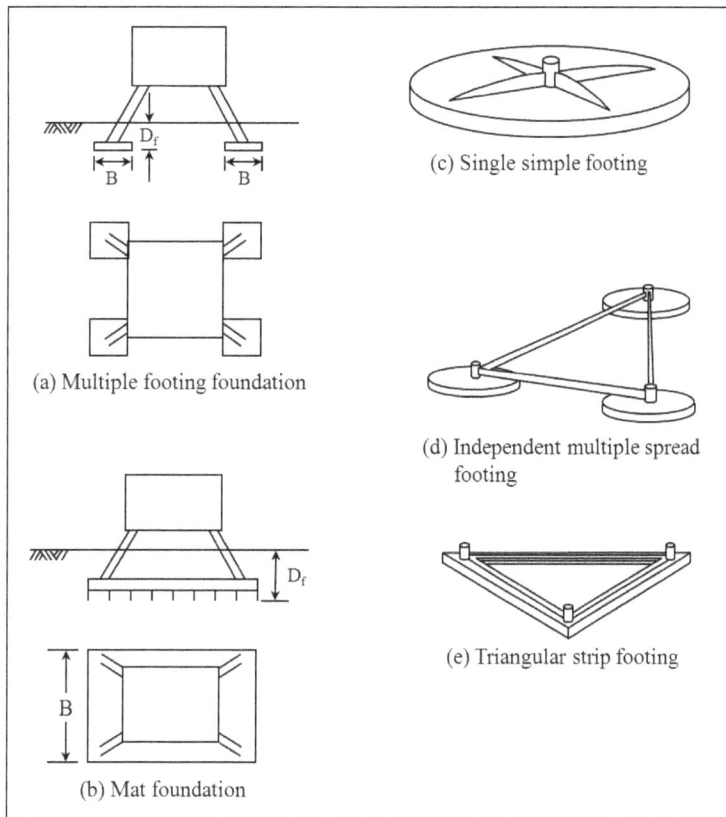

Figure 4.1-2. Types of shallow foundations.

4.1.2.2 Deadweight Anchors

A deadweight anchor can be any heavy object that is placed on the seafloor. Deadweight anchors can rest on the seafloor or be partially or even completely buried within it. The primary purpose of a deadweight anchor is to resist uplift and lateral forces from a mooring line connected to a buoyant object. With the exception of a few specially shaped deadweight anchors (designed to dig into the soil to a limited extent as the anchor is dragged), the behavior of a deadweight anchor is practically the same as the behavior of a shallow foundation subjected to an uplift load. The uplift resistance is provided primarily by the net submerged weight of the anchor. Often, these specially shaped anchors provide little additional uplift resistance above that provided by their submerged weight.

Ten types of deadweight anchor are shown in Figure 4.1-3. They range from relatively sophisticated anchors with shear keys to simple concrete clumps or clumps of heavy scrap materials. Anchors with shear keys provide greater lateral load resistance than do those without shear keys. However, the additional capacity of the more sophisticated deadweight anchors may be offset by increased costs for fabrication and installation.

4-3

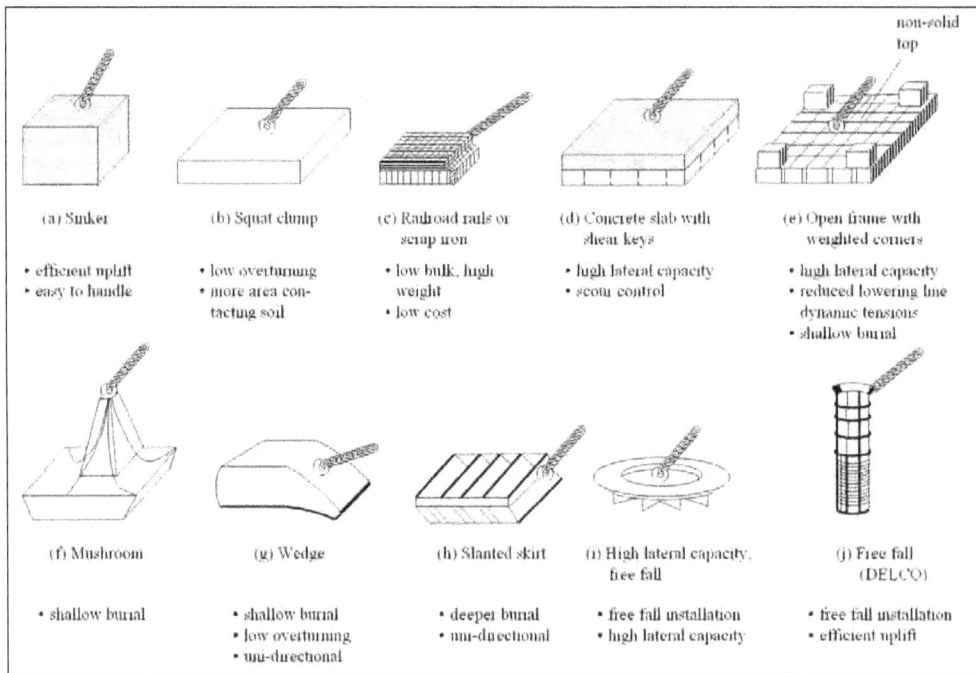

Figure 4.1-3. Types and significant characteristics of deadweight anchors.

4.2 DESIGN CONSIDERATIONS

4.2.1 General

First, information about potential sites and the loading characteristics for the foundation or anchor must be determined. This information is used for the following purposes:

- To define the appropriate types of geotechnical information needed which will allow foundation or anchor design.

- To select sites attractive from geotechnical considerations and to avoid sites containing hazards.

- To obtain the needed specific soil parameters for design.

Factors influencing a foundation or deadweight anchor design are then weighed as the selection and design process is carried out. These factors include:

- Knowledge of site characteristics (water depth, bathymetry and slope, stratigraphy, environmental loading conditions, potentially hazardous features, among others).

- Structure or moored platform characteristics (and the nature and relative importance of the structure or platform).

- Soil characteristics (vertical and lateral extent of the soil investigation to determine existing conditions and parameters for design).

- Extent of knowledge and local experience on the behavior of similar foundation or anchor types.

- Cost and level of risk of failure.

The foundation design process is interactive and involves all these considerations in varying degrees. The listed considerations should influence the choice of safety factors for design calculations which contain uncertainties.

4.2.2 Site

Site characteristics important to shallow foundations and deadweight anchor design include water depth, topographic features, data on environmental conditions, stratigraphic profiles, sediment characteristics, and potentially hazardous seafloor features. At the earliest opportunity, attention should be given to identifying seafloor features that suggest steep slopes, erosion, existing slumps, or under-consolidated sediment. These indicate that excessive settlement, overstressing, or large foundation movement can occur. If a more suitable site cannot be located, minimization of the effects of these problems should be considered during design.

4.2.3 Structure

Characteristics of the structure or moored platform drive the foundation or anchor design. Weight, configuration, stiffness, purpose, design life, and cost, as well as other information, are relevant. These dictate the loading conditions and other design considerations and relevant factors of safety which determine the type and size of the foundation or anchor.

4.2.4 Loading

The following loading conditions should be determined:

1. Static long-term loading (i.e., relatively constant loads applied for a long time period) for cohesionless soils only. For example: the deadweight loading from the structure and foundation.

2. Static short-term loading (i.e., relatively constant loads applied for a short time period). For example: a downward force applied during installation to insure penetration of shear keys, or a load increase on the foundation or anchor during infrequent replacement of a subsurface buoy which had been applying a significant buoyant force.

3. Rapid cyclic loading (i.e., significant repetitive forces occurring over a relatively short period, so that excess pore pressures generated by the loading do not have time to dissipate). For example: mooring line loads created from storm waves, or earthquake loading of the structure or of the soil mass itself.

4. Slow cyclic loading (i.e., cyclic forces that occur over a sufficiently long time period so that excess pore pressures generated by the loading have sufficient time to dissipate between loads). For example: mooring line load variations created by tide-related changes in current.

4.2.5 Geotechnical

The design of shallow foundations and deadweight anchors requires that the following items affecting geotechnical aspects of design be considered:

- Foundation instability: bearing capacity failure and other failures due to uplifting, overturning, horizontal sliding, or combinations of these.

- Slow foundation displacements – primarily excessive consolidation settlement.

- Installation problems associated with the use of shear keys.

- Recovery problems associated with high resistance to breakout (Chapter 9).

The soil data required include soil type, index properties, density, strength under the conditions of the applied loads, and deformation characteristics under static and cyclic loading conditions. Table 4.2-1 lists the property values necessary to evaluate the loading conditions discussed in Section 4.2.4 for cohesive and cohesionless soils.

At the site where the foundation or deadweight anchor is to be placed, the depth to which soils data are required equals approximately the foundation width or diameter. The soil characteristics and design parameters should be obtained through on-site and laboratory testing. For unmanned or other noncritical installations, and for small structures and low loads, where overdesign is not costly, soil information can often be estimated from available literature (Chapters 2 and 3 include properties for typical soil types and engineering property correlations with more easily obtained index properties). However, this lack of high quality soils data must be reflected by use of a high factor of safety.

4.2.6 Factor of Safety

A safety factor must be applied during the design of foundations and deadweight anchors to account for uncertainties in loading, soils data accuracy, and analytical procedure accuracy. Table 4.2-1 lists recommended factors of safety to be applied to the loading conditions discussed in Section 4.2.4 when soils properties are accurately known (by field and laboratory testing). When these data are not well known, and for a critical installation, the safety factors should be increased by multiplying the table value by 1.5.

Table 4.2-1. Soil Properties Required for Analysis and Recommended Factors of Safety

Loading Condition	Recommended Factor of Safety (F_s) for Stability[a]	Soil Properties for Cohesive Soil[b]	Soil Properties for Cohesionless Soil[b]
Long-Term Static Loading	2.0	Drained Parameters[c] $\bar{c}, \bar{\phi}, \gamma_b, C_c, E, v, e_o$	Drained Parameters $\bar{\phi}, \gamma_b, E, v$
Short-Term Static Loading	1.5	Undrained Parameters $s_u, S_t, \gamma_b, C_c, E, v, e_o$	Drained Parameters $\bar{\phi}, \gamma_b, E, v$
Rapid Cyclic Loading[d]	1.5	Undrained Parameters $s_u, S_t, \gamma_b, C_c, E, v, e_o$	Undrained Parameters ϕ_u, γ_b, E, v
Slow Cyclic Loading[d]	2.0	Undrained Parameters $s_u, S_t, \gamma_b, C_c, E, v, e_o$	Drained Parameters $\bar{\phi}, \gamma_b, E, v$

[a] These factors are recommended for the cases where properties data are accurately known. These factors should be increased by multiplying the listed value by 1.5 if geotechnical data are not accurately known or if the installation is particularly critical (see above).

[b] In the absence of site-specific data on E and v, the following are recommended:
For cohesive soil: E as given by Equation 4-47, with $PI = 45\%$, and $v = 0.45$.
For cohesionless and mixed soils: E as given by Equations 4-48 and 4-47 (with $PI = 45\%$, and $v = 0.45$), respectively.

[c] Long-term static loading in cohesive soils is rarely the limiting design case compared to short-term static loading as cohesive soil tends to gain strength over time due to consolidation, and is not covered in this chapter. Also, determination of drained cohesive parameters is expensive and time prohibitive.

[d] Usually treated as equivalent to short-term static loading.

4.3 DESIGN METHODOLOGY AND PROCEDURE

4.3.1 General

The design procedure for a shallow foundation or deadweight anchor is an iterative process. A foundation or anchor trial size is selected, and then checked for adequacy. When that size is found to be inadequate, it is modified and checked again until a satisfactory design results. A flow chart for design of shallow foundations and deadweight anchors is given in Figure 4.3-1. The individual steps in the design process are summarized in Table 4.3-1. The design must consider all applicable factors discussed in Section 4.2.

Table 4.3-1. Summary of Steps in the Design of Shallow Foundations and Deadweight Anchors

Cohesionless Soils	Cohesive Soils
Shallow Foundation	
1. Assume regular configuration.	1. Same.
2. Assume z_s.	2. Same.
3. Bearing capacity consideration (Equations 4-1 through 4-31). Determine W_{bf}, B, and A.	3. Same.
4. Lateral load resistance consideration (Equation 4-33). Determine W_{bf}, B, and A.	4. Consider lateral load resistance (Equation 4-32). Determine W_{bf}, B, and A.
5. Preliminary sizing. Optimize W_{bf}, B, A and H based on results of steps 3 and 4.	5. Same.
6. Determine number of shear keys (Equations 4-40, 4-42 and 4-43).	6. Determine number of shear keys (Equations 4-40 and 4-41).
7. Determine thickness of shear keys from structural considerations.	7. Same.
8. Check penetration of shear keys (Section 4.3.5.2).	8. Same.
Deadweight Anchors	
1. Assume configuration of deadweight anchor.	1. Same.
2. Assume z_s.	2. Same.
3. Consider lateral load resistance (Equation 4-33). Determine W_{bf}, B, and A.	3. Consider lateral load resistance (Equation 4-32 or 4-37). Determine W_{bf}, B, and A.
4. Check bearing capacity with eccentricity (Equations 4-1 through 4-31). If no good, adjust W_{bf}, B, and A, and go to step 3. If okay go to step 5.	4. Same.
5. Determine preliminary sizing. Optimize W_{bf}, B, A, and H based on steps 3 and 4.	5. Same.
6. Determine number of shear keys (Equations 4-40, 4-42 and 4-43).	6. Determine number of shear keys (Equations 4-40 and 4-41).
7. Determine thickness of shear keys from structural considerations.	7. Same.
8. Check penetration of shear keys (Section 4.3.5.2).	8. Same.

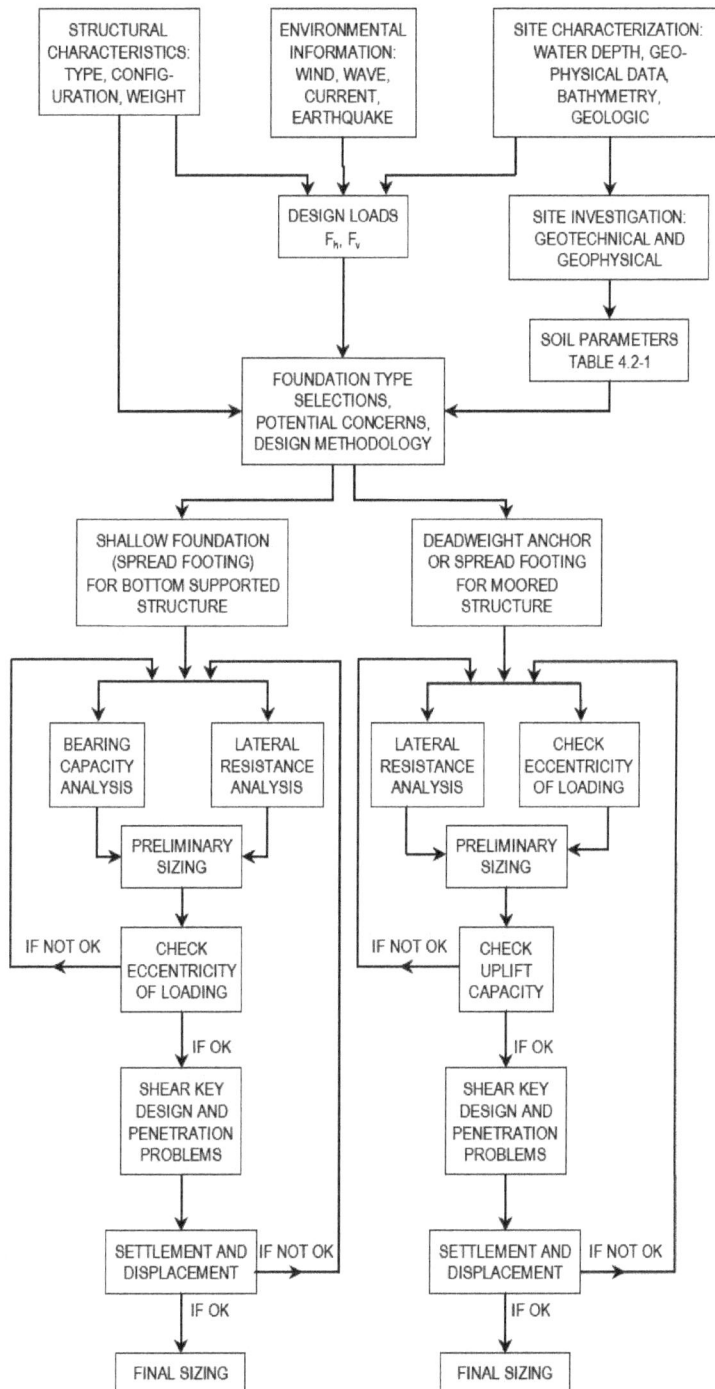

Figure 4.3-1. Flow chart for the design of shallow foundations and deadweight anchor.

4.3.2   Bearing Capacity

4.3.2.1 General Relationship

The bearing capacity of a seafloor soil is dependent upon the following factors:

- Engineering properties of the soil profile
- Type and size of foundation
- Depth of embedment
- Load direction
- Inclination of the ground surface

The maximum bearing capacity, Q_u, for the trial size foundation is calculated using the general formulation presented in Equation 4-1, below. This general equation is based on Reference 4-1, with the addition of side traction, represented as an equivalent base stress due to side adhesion and friction. The maximum bearing capacity is compared to the sum of all forces acting normal to the seafloor surface, with an appropriate safety factor applied to these normal forces. The effects if eccentric loading are addressed in Sections 4.3.2.2 and 4.3.2.3. Specific equations for the maximum bearing capacity under various loading conditions and soil types are presented in Sections 4.3.2.6 through 4.3.2.10.

$$Q_u = A'(q_c + q_q + q_\gamma) + P\,H_s\left(\frac{s_{ua}}{S_t} + \gamma_b\,z_{avg}\,\tan\delta\right) \tag{4-1}$$

where:

A' = effective base area of foundation depending on the load eccentricity $[L^2]$ (Section 4.3.2.3)

q_c = bearing capacity stress for cohesion = $s_{uz}\,N_c\,K_c$ $[F/L^2]$

q_q = bearing capacity stress for overburden = $\gamma_b\,D_f\,[1 + (N_q\,K_q - 1)f_z]$ $[F/L^2]$

q_γ = bearing capacity stress for friction = $\gamma_b\,(B'/2)\,N_\gamma\,K_\gamma f_z$ $[F/L^2]$

f_z = depth attenuation factor for the frictional portion of bearing capacity stress, to extend the formulation to any footing depth, as described in Section 4.3.2.6, Equation 4-24

P = base perimeter = $2B + 2L$ $[L]$

H_s = side soil contact height = min $(D_f, H + z_s)$ $[L]$

s_{ua} = undrained shear strength averaged over the side soil contact zone $[F/L^2]$

S_t = soil sensitivity = ratio of undisturbed to remolded strength

4-10

γ_b = buoyant unit weight of soil above the foundation base [F/L^3]

z_{avg} = average depth over side soil contact zone = ½ [D_f + max(0, D_f–H– z_s)] [L]

δ = effective friction angle between the soil and the side of the foundation [deg]

= ϕ– 5 deg   for rough-sided footings, or

= 0   for smooth-sided footings or where the soil is greatly disturbed

ϕ = soil friction angle ($\phi = \phi_u$ for undrained case; $\phi = \bar{\phi}$ for drained case)

s_{uz} = undrained shear strength effective for base area projected to depth of shear key tip [F/L^2] = average strength from shear key tip depth to 0.7B' below

D_f = depth of embedment of foundation [L]

z_s = depth of shear key tip below foundation base [L]

B' = effective base width depending on eccentricity [L] (Section 4.3.2.3)

B = base width [L]

L = base length [L]

H = base block height [L]

N_c, N_q, N_γ = bearing capacity factors (Section 4.3.2.4)

K_c, K_q, K_γ = bearing capacity correction factors (Section 4.3.2.5)

4.3.2.2 Nominal Bearing Pressure Distribution from Eccentric Loads

Most shallow foundation or anchors on the seafloor will be under an eccentric load due to waves, currents, residing on a slope, or a horizontal component of the mooring line for anchors. These loads will result in a moment being placed on the foundation or anchor in addition to the normal force (Figure 4.3-2). The nominal non-uniform pressure distribution on the soil is to assume a linear distribution ranging from a maximum to a minimum as:

$$q_{max} = \frac{F_n}{BL} + \frac{6M}{B^2 L} \qquad (4\text{-}2)$$

$$q_{min} = \frac{F_n}{BL} - \frac{6M}{B^2 L} \qquad (4\text{-}3)$$

where:

q_{max} = estimated maximum bearing pressure [F/L^2]

q_{min} = estimated minimum bearing pressure [F/L^2]

F_n = normal bearing load [F]

M = applied moment [F·L]

B = foundation width [L]

L = foundation length [L]

The normal bearing load and the moment can be represented as single offset normal load as shown in Figure 4.3-3. The amount of eccentricity can be calculated by Equation 4-4.

$$e = \frac{M}{F_n} \hspace{10cm} (4\text{-}4)$$

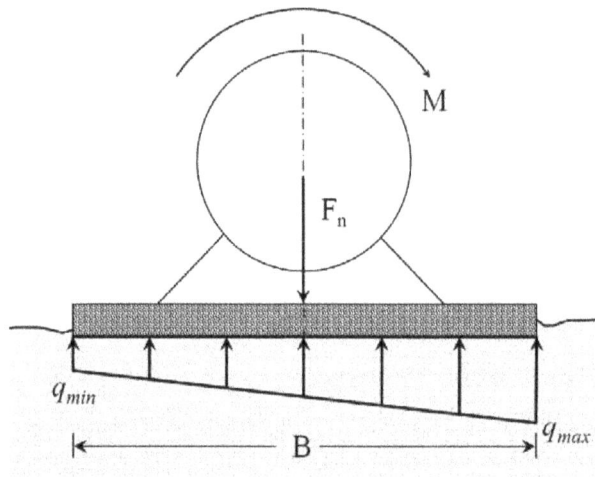

Figure 4.3-2. Linear bearing pressure distribution due to eccentric loading

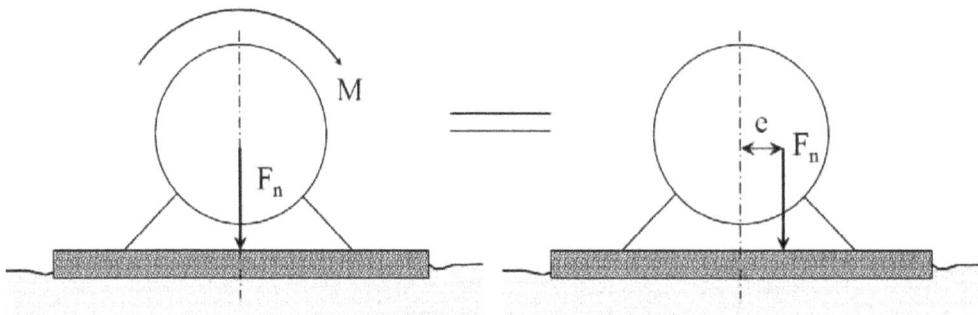

Figure 4.3-3. Normal bearing load and moment depicted as an equivalent offset load.

Substituting Equation 4-4 into Equations 4-2 and 4-3 results in the following equations for q_{max} and q_{min} when $e < B/6$:

$$q_{max} = \frac{F_n}{BL}\left(1 + \frac{6e}{B}\right)$$
(4-5)

$$q_{min} = \frac{F_n}{BL}\left(1 - \frac{6e}{B}\right)$$
(4-6)

When $e > B/6$, Equations 4-7 and 4-8 are used to estimate the maximum and minimum bearing pressure, repectively.

$$q_{max} = \frac{4F_n}{3L(B - 2e)}$$
(4-7)

$$q_{min} = 0$$
(4-8)

As a general design guide it is recommended that the eccentricity be less than or equal to the the foundation width, B, divided by 6. This ensures that the entire foundation will be under pressure. If the eccentricity is greater than $B/6$, some of the soil under the foundation may be under tension or may separate from beneath the footing; this is generally not a desireable condition.

It is noted, however, that for anchoring applications, a design that exceeds this limit may be acceptable, so long as the bearing capacity of the reduced-area footing is adequate. Calculation of the maximum allowable eccentricity for this option is discussed near the end of Section 4.3.2.3.

Examples of estimated pressure distributions for various eccentricities are shown in Figure 4.3-4.

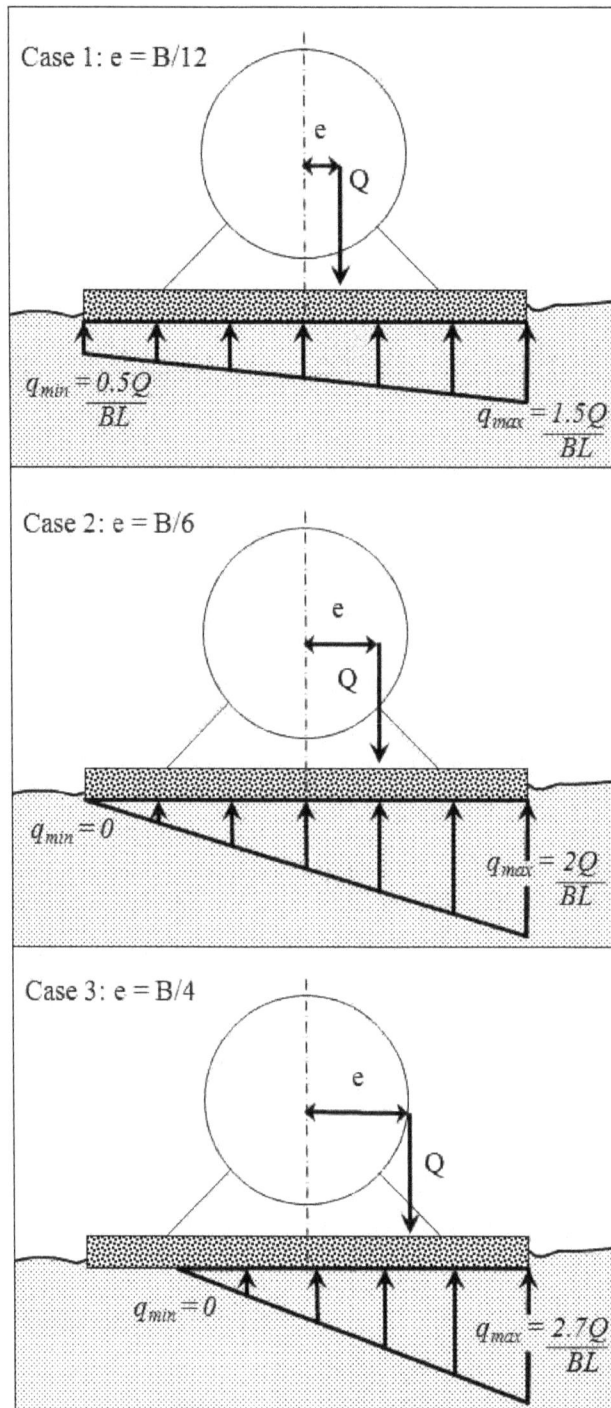

Figure 4.3-4. Examples of estimated pressure distributions for various eccentricities.

4.3.2.3 Bearing Capacity of Foundations with Eccentric Loading

Meyerhoff (Ref. 4-2) developed a method called the effective base dimensions to account for eccentric loading when calculating bearing capacity, a reduced foundation base-to-soil contact area is used to determine bearing capacity. The equivalent or resultant vertical load acts at the center of the reduced area determined as shown in Figure 4.3-5. For a rectangular base area, eccentricity can occur with respect to either, or both, axis. The altered effective length and width of the footing are:

$$L' = \text{larger of} \quad L - 2e_1 \quad \text{or} \quad B - 2e_2 \tag{4-9}$$
$$B' = \text{smaller of} \quad L - 2e_1 \quad \text{or} \quad B - 2e_2$$

The effective area is:

$$A' = B'L' \tag{4-10}$$

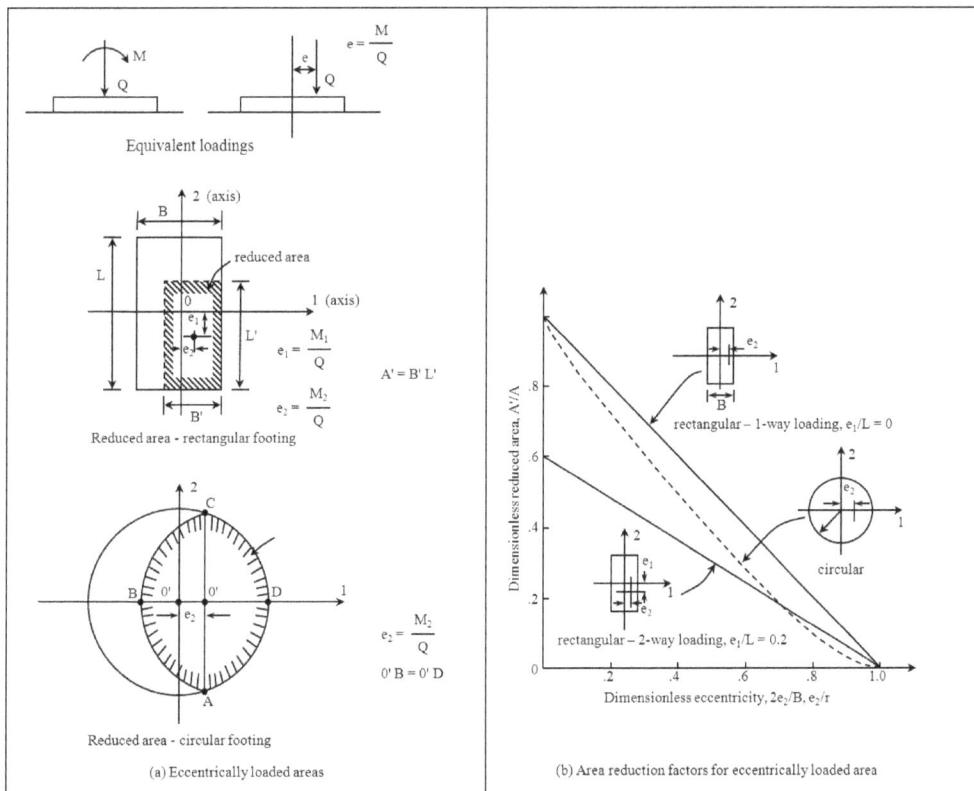

Figure 4.3-5. Area reduction factors for eccentrically loaded foundations (Ref. 4-3).

For shallow foundations in general, the rule of thumb is that eccentricity should not exceed one-sixth the footing width (as measured in the pull direction). However, for anchors, it is acceptable to exceed this limit so long as adequate bearing capacity is maintained over the reduced area as calculated using Equations 4-9 and 4-10. The maximum allowable eccentricity for this option, which is generally greater than $B/6$ or $L/6$, is derived from setting the ultimate bearing capacity equal to the normal force multiplied by a suitable factor of safety, then rearranging Equation 4-1 to obtain the minimum reduced area, and using Equations 4-9 and 4-10 to obtain a maximum value for eccentricity.

Thus, the required bearing capacity is given by Equation 4-11:

$$Q_u = F_n F_s (bce)$$
(4-11)

where:

$F_s(bce)$ = factor of safety for bearing capacity with eccentric loading

Next, the associated minimum allowable reduced area (from Equation 4-1, solving for A') is given by:

$$A' = \frac{F_n F_s (bce) - P H_s \left(\frac{S_{ua}}{S_t} + \gamma_b z_{avg} \tan \delta \right)}{(q_c + q_q + q_\gamma)}$$
(4-12)

If the pull direction is parallel to the length and perpendicular to the width, the maximum allowable eccentricity in the length direction is computed according to Equation 4-13:

$$e_{1max} = \frac{L - A'/B}{2}$$
(4-13)

If the pull direction is parallel to the width and perpendicular to the length, the maximum allowable eccentricity in the width direction is given by Equation 4-14:

$$e_{2max} = \frac{B - A'/L}{2}$$
(4-14)

To minimize the potential for excessive eccentricity leading to massive foundation rotation, the distance $H' + H_s$ (the moment arm of the lateral load component) should be kept as small as possible. This is most easily done by minimizing the foundation base height. It is recommended that H be limited to $0.25B$ where possible.

4.3.2.4 Bearing Capacity Factors

The bearing capacity stresses q_c, q_q, and q_γ contain a number of bearing capacity factors and correction factors in their formulation. The bearing capacity factors N_c, N_q and N_γ may be computed from Equation 4-15 through 4-17 (Ref. 4-1). Figure 4.3-6 graphically illustrates the relationship between the bearing capacity factors and soil friction angle based on the equations.

$$N_q = \exp(\pi \tan \phi) N_\phi \qquad\qquad (4\text{-}15)$$

$$N_c = \frac{N_q - 1}{\tan \phi} \quad \text{for } \phi \neq 0, \qquad N_c = 2 + \pi \quad \text{for } \phi = 0 \qquad\qquad (4\text{-}16)$$

$$N_\gamma = 2 \, (1 + N_q) \tan \phi \, \tan(\pi / 4 + \phi / 5) \qquad\qquad (4\text{-}17)$$

where:

$$N_\phi \;=\; [\tan(\pi/4 + \phi/2)]^2 \quad \text{and} \quad \phi \text{ is in radians}$$

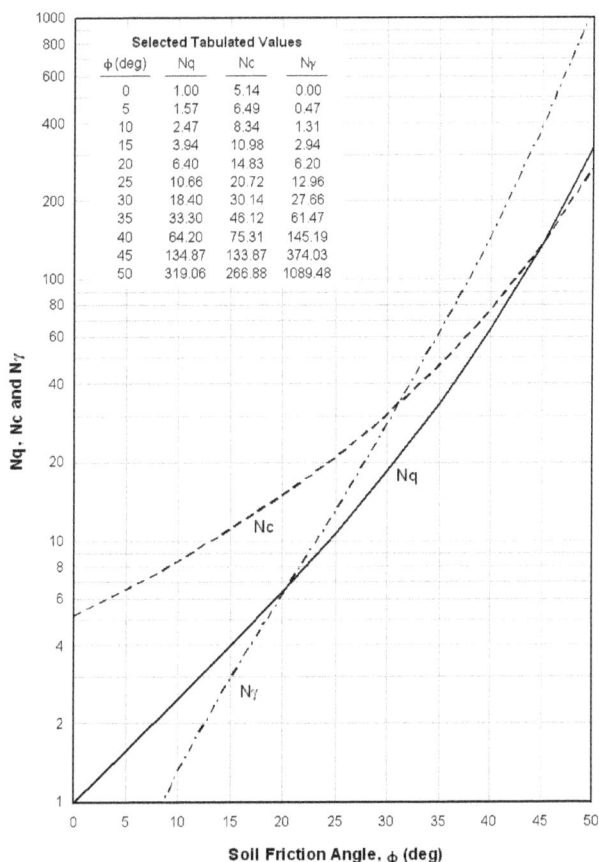

Figure 4.3-6. Bearing capacity factors as a function of soil friction angle.

4-17

4.3.2.5 Correction Factors

The correction factors K_c, K_q, and K_γ used in the calculation of bearing capacity stresses (Equation 4-1, computation of q_c, q_q, and q_γ), each represent a subgroup of factors, which account for the following:

Item	Designation
Load Inclination	i
Foundation Shape	s
Depth of embedment	d
Inclination of foundation base	b
Inclination of ground surface	g

The correction factors for cohesion, overburden, and density (K_c, K_q, and K_γ, respectively) are evaluated from:

$$K_c = i_c s_c d_c b_c g_c$$
$$K_q = i_q s_q d_q b_q g_q \tag{4-18}$$
$$K_\gamma = i_\gamma s_\gamma d_\gamma b_\gamma g_\gamma$$

Load Inclination. Seafloor foundations and deadweight anchors are often subjected to a large lateral load component arising from wave and current loadings and, occasionally, from wind loading on a surface float which is connected by a mooring line. This large lateral load component, combined with the gravity load component of a structure or deadweight, forms a resultant load of substantial inclination to the vertical. This inclination of the resultant load causes a change in the form of the bearing capacity failure surface, permitting failure to take place at a lower load. Subgroup correction factors, which account for inclination of the resultant load, are:

$$i_c = i_q - \frac{1 - i_q}{N_c \tan \overline{\overline{\phi}}}$$

$$i_q = \left(1 - \frac{F_h}{F_v + B'L'\overline{c} \cot \overline{\overline{\phi}}} \right)^m \tag{4-19}$$

$$i_\gamma = \left(1 - \frac{F_h}{F_v + B'L'\overline{c} \cot \overline{\overline{\phi}}} \right)^{m+1}$$

where:

F_h = horizontal component of design load [F]

F_v = vertical downward component of all loads [F]

$\bar{c}$ = effective cohesion, usually obtained from drained triaxial test of undisturbed soil sample [F/L^2]

$$m = \frac{2+(L'/B')}{1+(L'/B')}\cos^2\theta + \frac{2+(B'/L')}{1+(B'/L')}\sin^2\theta$$

θ = angle between the line of action of F_h and the long axis of the foundation in the horizontal plane

For the case of an undrained bearing capacity failure in cohesive soil (ϕ = 0), the correction factor i_c is obtained from:

$$i_c = 1 - \frac{mF_h}{B'L's_{uz}N_c} \tag{4-20}$$

Foundation Shape. The basic bearing capacity factors of Figure 4.3-6 and Equations 4-15 through 4-17 are derived for the two-dimensional failure case or for an infinitely long strip foundation. Corrections to the calculated two-dimensional bearing capacity prediction for the more likely rectangular and circular bearing areas are calculated from:

$$s_c = 1 + (B'/L')(N_q/N_c)$$
$$s_q = 1 + (B'/L')\tan\phi \tag{4-21}$$
$$s_\gamma = 1 - 0.4(B'/L')$$

For circular foundations loaded without eccentricity, the shape correction factors are calculated by setting $B'/L' = 1$.

Depth of Embedment. The basic bearing capacity factors are derived for a footing on the soil surface. Corrections to the calculated bearing capacity prediction for a footing at a shallow depth of embedment (generally less than the footing width for small footings), subgroup d, are given by Equation 4-22.

These correction factors for depth of embedment, subgroup d, are sensitive to soil disturbance along the sides of the embedded base. Therefore, if the depth of the footing is less than the width or if the placement of the footing has disturbed the soil, it is typically wise to discount entirely the beneficial effect of overburden shear strength. This is done by setting $d_c = d_q = d_\gamma = 1$.

Alternately, if the footing has been in place long enough so that the soil has reestablished its strength (by compaction, aging, vibration, etc.) the following equations for the depth of embedment correction factors may be applied. Note that these equations are valid for all footing depths. For extension to any footing depth, a depth attenuation factor based on the critical confining pressure is applied to q_γ and the frictional portion of q_q, as described in the Section 4.3.2.6.

$$d_c = 1 + 2\left(1 - \sin\phi\right)^2 \arctan\left(\frac{D_f}{B'}\right) \cdot \frac{N_q}{N_c}$$

$$d_q = \left[1 + 2\left(1 - \sin\phi\right)^2 \arctan\left(\frac{D_f}{B'}\right)\tan\phi\right]$$ (4-22)

$$d_\gamma = 1$$

Note that for purely cohesive soils, the soil friction angle, ϕ, is zero, so the depth of embedment correction factors presented in Equation 4-22 reduce to those shown in Equation 4-23, below. However, because for purely cohesive soils the bearing capacity factor $N_q = 1$, the value of K_q (and hence, d_q) is inconsequential in Equation 4-1.

$$d_c = 1 + 2\arctan\left(\frac{D_f}{B'}\right) \cdot \frac{1}{N_c}$$

$$d_q = 1$$ (4-23)

$$d_\gamma = 1$$

Inclination of Foundation Base and Ground Surface. The correction factors for inclination of the foundation base, subgroup *b*, and for the inclination of the seafloor, subgroup *g*, can be set equal to 1 where the foundation is placed nearly level on a near-horizontal seafloor. Thus, because nonsloping sites are usually sought for foundations and the foundation is usually placed in a near-horizontal orientation, normally: $b_c = b_q = b_\gamma = 1$, and $g_c = g_q = g_\gamma = 1$.

Note that when a best estimate of the bearing capacity failure load is required, or for those instances where foundations will be installed deeply in the seafloor, at a severe inclination or on a slope, relationships for the correction factor subgroups *d*, *b*, and *g* can be found in References 4-1 and 4-3.

4.3.2.6 Extension to Deep Behavior

The portion of bearing capacity due to intergranular friction becomes attenuated at depth because the high intergranular stresses cause particle crushing during shear failure. This attenuation applies to the q_γ term in Equation 4-1, and to the net q_q term after subtracting the soil buoyancy portion $\gamma_b D_f$.

$$q_q = \gamma_b D_f + q_{qf}$$
$$q_\gamma = \gamma_b (B'/2) N_\gamma K_\gamma f_z \qquad\qquad (4\text{-}24)$$

where:

$$q_{qf} = \gamma_b D_f (N_q K_q - 1) f_z$$

$$f_z = \frac{\arctan\left(D_f / D_t\right)}{\left(D_f / D_t\right)}$$

D_t = transition characteristic depth related to onset of grain crushing behavior (Equation 4-27) [L]

The transition characteristic depth is determined by equating the frictional portion of the bearing capacity stress evaluated at a much greater depth with the maximum bearing stress that can be developed plastically by the soil undergoing grain crushing at the critical confining pressure. This maximum bearing stress is given by Equation 4-25:

$$q_{fmax} = s_{ucr} N_{cclay} K_{cclay} \qquad\qquad (4\text{-}25)$$

where:

s_{ucr} = effective shear strength at critical confining pressure [F/L^2]
   = $\sigma_{cr} \sin(\phi) / \{1\text{-}\sin(\phi)\}$

N_{cclay} = $2 + \pi$

K_{cclay} = $i_{cclay} s_{cclay} d_{cclay} b_{cclay} g_{cclay}$

s_{cclay} = $1 + (B'/L') / N_{cclay} = 1 + (B'/L') / (2 + \pi)$

d_{cclay} = $1 + (2 / N_{cclay}) \arctan(D_f / B') = 1 + \pi /(2 + \pi)$ at great depth

i_{cclay} = $b_{cclay} = g_{cclay} = 1$

σ_{cr} = critical confining pressure [F/L^2]; may be measured in the laboratory or estimated from the empirical relationship $\sigma_{cr} \approx D_r^{1.7} \cdot 20{,}000 \, psf$

4-21

$$D_r = \text{fractional relative density} = \frac{(1/\gamma_{b\min})-(1/\gamma_b)}{(1/\gamma_{b\min})-(1/\gamma_{b\max})}$$

Note: In the computation of the fractional relative density, D_r, the minimum and maximum values of the buoyant unit weight may be measured in the laboratory. Alternatively, so long as the result is not zero or negative, use may be made of the empirical relationship $D_r \approx (\gamma_b - 56.5\ pcf) / 11.5\ pcf$.

At a depth much greater than the transition characteristic depth, the attenuation factor becomes:

$$f_{zdeep} = (\pi/2)\,(D_t/D_f) \tag{4-26}$$

Using this expression for f_z and equating q_{fmax} (Equation 4-25) with the sum of q_{qf} and q_γ (Equation 4-24), and considering that the ratio B'/D_f approaches zero at great depth, gives the transition characteristic depth, D_t, as:

$$D_t = \frac{q_{f\max}}{(\pi/2)\gamma_b\left[(N_q K_q - 1)+\{(B'/2)/D_f\}N_\gamma K_\gamma\right]} \tag{4-27}$$

4.3.2.7 Static Short-Term Loading and Cyclic Loading in COHESIVE SOILS

Under static short-term loadings and all cyclic loadings, failure on cohesive soils will occur before excess pore pressure can dissipate. These are, therefore, undrained failures and the soils properties used in the design are undrained properties (Table 4.2-1).

The maximum downward vertical load, Q_u, that a foundation on or in cohesive soil can support under undrained conditions is calculated by Equation 4-28. This equation results from setting the soil friction angle to zero in the general equation for bearing capacity (Equation 4-1).

$$Q_u = A'\{s_{uz}N_c K_c + \gamma_b D_f\} + (2B + 2L)\big(\min(D_f, H + z_s)\big)\frac{s_{ua}}{S_t} \tag{4-28}$$

where:

A' = effective base area of foundation depending on the load eccentricity [L^2] (Section 4.3.2.3)

s_{uz} = undrained shear strength of cohesive soil-averaged over the distance $0.7B$ below the foundation base [F/L^2]

N_c = bearing capacity factor (Section 4.3.2.4); for undrained failure $N_c = 5.14$

K_c = correction factor that accounts for load inclination, foundation shape, embedment depth, inclination of foundation base, and inclination of ground (Section 4.3.2.5). For a nearly square or round footing on a nearly level seafloor with a vertical load, $K_c = 1.2$

γ_b = buoyant unit weight of soil above the foundation base [F/L³]; for a soft cohesive seafloor, $\gamma_b \approx 20$ lb/ft³

D_f = depth of embedment of foundation [L]

z_s = shear key height [L]

B = base width [L]

L = base length [L]

H = base block height [L]

s_{ua} = undrained strength averaged over the side soil contact zone [F/L²]

S_t = soil sensitivity; for a soft cohesive seafloor, $S_t \approx 3$

For the simple case of a vertical load applied concentrically on a square or circular foundation resting on the seafloor surface, where both the foundation base and seafloor are horizontal, Equation 4-28 reduces to Equation 4-29, where A is the foundation base area.

$$Q_u = (6.17) A\, s_{uz} \qquad\qquad (4\text{-}29)$$

4.3.2.8 Static Long-Term Loading in COHESIVE SOILS

The static long-term loading condition on cohesive soils exists after the excess pore water pressures have dissipated. Static long-term bearing loading in cohesive soils is rarely the limiting design case compared to static short-term loading, as cohesive soil tends to gain strength over time due to consolidation. (Note that this is not necessarily true for lateral loading, as discussed in Section 4.3.3.1.) Also, the determination of drained cohesive parameters is expensive and time prohibitive. Therefore, the effect of static long-term loading in cohesive soils on bearing capacity will not be further addressed in this chapter.

4.3.2.9 Static Short- and Long-Term Loading in COHESIONLESS SOILS

Bearing capacity failure on cohesionless soils normally occurs under drained conditions. Cohesionless soils have a sufficiently high permeability to allow water drainage and rapid dissipation of excess pore pressures. Therefore, both the short- and long-term designs use drained soil properties (Table 4.2-1).

The maximum downward vertical load capacity of a shallow foundation or deadweight anchor on cohesionless soils is calculated by:

$$Q_u = A'\gamma_b \left[D_f \{1 + (N_q K_q - 1)f_z\} + (B'/2)N_\gamma K_\gamma f_z \right]$$
$$+ (2B + 2L)\left(\min(D_f, H + z_s)\right)\gamma_b \tan\delta \left(D_f + \max(0, D_f - H - z_s)\right)/2 \qquad (4\text{-}30)$$

where:

A' = effective base area of foundation depending on the load eccentricity [L^2] (Section 4.3.2.3)

γ_b = buoyant unit weight of soil above the foundation base [F/L^3]

D_f = depth of embedment of foundation [L]

z_s = shear key height [L]

N_q, N_γ = bearing capacity factors obtained from Figure 4.3-6 or from Eqns. 4-15 and 4-17

K_q, K_γ = correction factors dependent on load inclination, foundation shape, embedment depth, inclination of foundation base, and inclination of ground (Section 4.3.2.5). For a nearly square or round footing on a nearly level surface with a vertical load, K_q = 1.0, and K_γ can be assumed as 0.6.

f_z = depth attenuation factor

B' = effective base width of foundation depending on the load eccentricity [L^2] (Section 4.3.2.3)

B = base width [L]

L = base length [L]

H = base block height [L]

δ = effective friction angle alongside the footing [deg]
  = $\phi - 5\ deg$ for rough-sided footings,
  = 0 for smooth-sided footings or where the soil is greatly disturbed

ϕ = soil friction angle, for the drained case $\phi = \bar{\phi}$

For the simple case of a vertical load applied concentrically on a square or circular foundation with no shear keys resting at the seafloor surface, where both the foundation base and seafloor are horizontal, $A' = A$, $B' = B$, $D_f = 0$, $K_\gamma = s_\gamma = 0.6$, and Equation 4-30 reduces to Equation 4-31, where A is the foundation base area:

$$Q_u = 0.3\ A\ \gamma_b\ B\ N_\gamma \qquad (4\text{-}31)$$

4.3.2.10 Rapid Cyclic Loading in COHESIONLESS SOILS

Cyclic loading applied to a foundation on a cohesionless soil may occur without sufficient time for dissipation of generated excess pore pressures. The bearing capacity of a foundation or deadweight anchor under such rapid cyclic loading conditions may be lower or higher than the static bearing capacity. The bearing capacity under rapid cyclic loading can be determined by using Equation 4-30 with bearing capacity factors N_q and N_γ determined using an undrained friction angle ϕ_u obtained from special cyclic undrained laboratory tests (which must be run if this condition is anticipated).

For slower cyclic loading, where the excess pore pressures have time to dissipate, the drained analysis of Section 4.3.2.9 is used with N_q and N_γ obtained using the drained friction angle $\bar{\phi}$ (Table 4.2-1).

4.3.3 Lateral Load Capacity

The lateral load that acts on the foundation or deadweight anchor can result from: downslope gravity force caused by a sloping seafloor, current drag on a foundation and structure, nonvertical mooring line loading, or storm-wave and earthquake loadings.

Shear keys are often incorporated in the foundation base design (Figure 4.1-1), and can be used with a variety of foundation shapes (circular, rectangular, square, etc.). This is done to increase lateral load capacity by forcing the failure surface, the surface on which the foundation will slide, deeper into the seafloor where stronger soils can resist higher lateral loads.

Three possible failure modes that can occur for shallow foundations fitted with shear keys are shown in Figure 4.3-7. Generally, the shear keys should be designed sufficiently close together to force the sliding failure to occur at the base of the keys as shown in Figure 4.3-7a. Procedures for evaluating this base sliding resistance are described in the following two sections. Methods for analyzing the passive wedge failure mode shown in Figure 4.3-7c are addressed in the shear key design section (Section 4.3.5). The deep passive failure mode shown in Figure 4.3-7b does not often occur and is not detailed in this chapter. Maximum lateral load capacity is calculated as explained in this section. This capacity is then compared to the sum of all forces driving the foundation in the downslope direction with an appropriate safety factor applied to these driving forces.

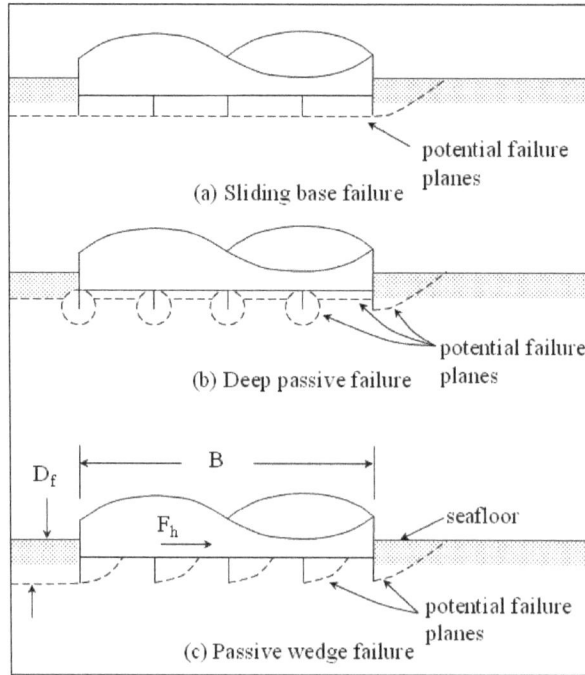

Figure 4.3-7. Possible failure modes when sliding resistance is exceeded (Ref. 4-4).

4.3.3.1 Static Short-Term Loading and All Cyclic Loading in COHESIVE SOILS

Static short-term loading and all cyclic lateral loading of foundations on cohesive soils are treated as undrained failure problems (Table 4.2-1).

The maximum lateral load capacity (parallel to the seafloor and perpendicular to the longitudinal axis), Q_{ul}, for a foundation on cohesive soil under undrained conditions is calculated by:

$$Q_{ul} = s_{uz}A + 2s_{ua}H_sL \tag{4-32}$$

where:

s_{uz} = undrained shear strength at shear key tip (for the case of no keys, see Note below) [F/L^2]

A = foundation base area [L^2]

s_{ua} = undrained shear strength averaged over the side soil contact zone [F/L^2]

H_s = side soil contact height = min (D_f, $H + z_s$) [L]

D_f = depth of foundation base or skirt tip below the seafloor [L]

4-26

H =   height of foundation block [L]

z_s =   height of footing shear key skirt [L]

L =   base length [L]

Note: Shear keys are generally recommended for resisting significant lateral loads. If there are no shear keys, the short-term resistance on the foundation base is limited to $s_{uz} \cdot A$, or $F_n \cdot \mu$, whichever is lower (μ = coefficient of friction ≈ 0.2 for cohesive soil).

4.3.3.2 Static Short- and Long-Term Loading in COHESIONLESS SOILS

For cohesionless soils, lateral load failure is a drained soil failure, and the maximum lateral load capacity in sliding is calculated by:

$$Q_{ul} = \mu\left[\left(W_{bf} + W_{bst} + W_b - F_{ve}\right)\cos\beta - F_h\sin\beta\right] + R_p \qquad (4\text{-}33)$$

where:

μ =   coefficient of friction between foundation base and soil or between soil and soil when shear keys cause this type of sliding failure

W_{bf} =   buoyant weight of foundation [F]

W_{bst} =   buoyant weight of bottom-supported structure [F]

W_b =   buoyant weight of soil contained within the footing skirt [F] = $\gamma_b\, A\, z_s$

γ_b =   buoyant unit weight of soil [F/L^3]

A =   footing base area enclosed by the footing skirt [L^2]

z_s =   height of footing shear key skirt [L]

F_{ve} =   design environmental loading and mooring line loading in the vertical direction (upward is positive) [F]

F_h =   design environmental loading and mooring line loading in the horizontal direction (downslope is assumed positive) [F]

β =   seafloor slope angle [deg]

R_p =   passive soil resistance on leading edge of base and footing shear key skirt (Section 4.3.5.1) [F]

The coefficient of friction depends on soil type and on base material type and material roughness. Table 4.3-2 lists coefficient of friction values for typical construction materials and marine cohesionless soils.

Table 4.3-2. Coefficient of Friction Between Cohesionless Soils and Marine Construction Materials

Soil	Friction Coefficient, μ, for:					
	Soil Internal Friction Coefficient	Smooth Steel	Rough Steel	Smooth Concrete	Rough Concrete	Smooth PVC
Quartz Sand	0.67	0.27	0.60	0.60	0.69	0.33
Coralline Sand	0.67	0.20	0.63	0.63	0.66	0.20
Oolitic Sand	0.79	0.23	0.56	0.58	0.74	0.26
Foram Sand-Silt	0.64	0.40	0.66	0.67	--	0.40

Where there is no other guidance, the value of μ can be estimated as follows:

$\mu \;=\; \tan(\phi - 5 \text{ deg})$ for a rough steel or concrete base without shear keys

$\mu \;=\; \tan(\phi)$ for a base with shear keys

where:

$\phi \;=\;$ soil friction angle [deg]

Where shear keys are present or the foundation is embedded deeply, a wedge of soil in passive failure develops in front of the leading foundation edge and provides resistance to sliding. For some foundations this passive wedge can contribute around 10% of the total lateral resistance. However, because sediment comprising this passive wedge may be removed by current scour or by animal burrowing activity (see Chapter 10), the contribution of the passive wedge to sliding resistance is often omitted.

In the design of a shallow foundation on cohesionless soils, the weight of the foundation is often increased to raise the maximum lateral load capacity.   On a slope, this also increases the downslope force acting to cause sliding, as is represented by the sin β term in the relationship discussed in the next paragraph.

To maintain stability against sliding, the maximum lateral load capacity should exceed the sum of forces acting to cause sliding by a suitable factor of safety (ratio of capacity to sum of driving forces) to account for uncertainties in soil data or failure mechanism. The factor of safety in this case is determined by Equation 4-34:

$$F_s = \frac{Q_{ul}}{\left(W_{bf} + W_{bst} + W_b - F_{ve}\right)\sin\beta + F_h \cos\beta} \tag{4-34}$$

The minimum foundation buoyant weight for this case is derived from Equations 4-33 and 4-34 (assuming $R_p = 0$) as:

$$W_{bf} = \frac{(F_s + \mu \tan \beta) F_h}{\mu - F_s \tan \beta} + F_{ve} - W_{bst} - W_b \qquad (4\text{-}35)$$

Note: Where skirts but no shear keys are used, the sliding will more likely occur along the foundation base and not at the depth of the skirt. Therefore, the buoyant weight of soil within the footing skirts, W_b, should not be used. For the special case of $\beta = 0$ where the seafloor is level, Equation 4-35 becomes:

$$W_{bf} = \frac{F_s F_h}{\mu} + F_{ve} - W_{bst} - W_b \qquad (4\text{-}36)$$

Table 4.2-1 lists the recommended factor of safety value, F_s, to be used in Equations 4-35 and 4-36.

4.3.3.3 Static Long-Term Loading in COHESIVE SOILS

The static long-term loading condition on cohesive soils exists after the excess pore water pressures have dissipated. Static long-term bearing loading in cohesive soils commonly is not the limiting design case compared to static short-term loading because cohesive soil tends to gain strength over time due to consolidation. However, the long-term (frictional) case becomes critical under lateral loading when the uplift component of an anchor line load reduces the normal force that creates the anchor's frictional resistance to sliding.

The long-term lateral load capacity of a foundation or anchor on cohesive soil is based on a drained soil failure analysis similar to that described for cohesionless soil with an additional contribution from $\bar{c}$, the effective soil cohesion. The maximum lateral load capacity (parallel to the seafloor in the downslope direction) for a foundation on a cohesive soil in a drained condition is calculated by:

$$Q_{ul} = \bar{c} A + \mu \left[\left(W_{bf} + W_{bst} + W_b - F_{ve} \right) \cos \beta - F_h \sin \beta \right] \qquad (4\text{-}37)$$

The minimum foundation buoyant weight for this case is determined by:

$$W_{bf} = \frac{(F_s + \mu \tan \beta) F_h - \dfrac{\bar{c} A}{\cos \beta}}{\mu - F_s \tan \beta} + F_{ve} - W_{bst} - W_b \qquad (4\text{-}38)$$

4-29

where:

μ = $\tan(\bar{\phi} - 5\ deg)$ for a rough steel or concrete base without shear keys

μ = $\tan(\bar{\phi})$ for a base with shear keys

$\bar{\phi}$ = drained soil friction angle [deg]

This long-term drained cohesive soil condition may control the design in a very few cases where the soils are heavily overconsolidated and very stiff. Normally, the short-term (undrained) case will yield a lower capacity and, therefore, will control lateral load aspects of foundation and deadweight anchor design.

4.3.4   Resultant Normal Force

The resultant force acting normal to the seafloor slope is obtained from Equation 4-39. The weights of the foundation and enclosed soil block are found according to Section 4.3.3.

$$F_n = \left(W_{bf} + W_{bst} + W_b - F_{ve}\right)\cos \beta - F_h \sin \beta \qquad (4\text{-}39)$$

This normal force must be borne by the bearing capacity with an adequate margin of safety. To make this determination, the normal force is multiplied by a factor of safety (Table 4.2-1) and compared to the bearing capacity, Q_u.

4.3.5   Shear Key Design

The depth of shear keys or perimeter skirts is usually limited by the net downward force available to drive the keys. A penetration resistance calculation should show that full skirt penetration is assured under only the submerged weight of the foundation.

When shear keys or perimeter skirts are used, venting holes are required in the base to allow the water and soft surficial soils trapped by them to escape. Sharpening the leading edge of keys will also aid penetration. The actual foundation placement should be smooth and continuous to minimize disturbance to the seafloor soil and the possible resulting creation of an eccentric foundation orientation.

4.3.5.1 Depth and Spacing of Shear Keys

In cohesive soils, the shear key height, z_s, and spacing of shear keys are dictated by the need to force a failure to occur along the base of the shear keys as shown in Figure 4.3-7a. The recommended maximum depth of shear keys on cohesive soils is $0.1B$ (Ref. 4-5). This depth may have to be reduced if full penetration cannot be achieved (Section 4.3.5.2). Other steps to reach a satisfactory design for adequate lateral load capacity include making the base larger and making it heavier – if necessary to insure shear key penetration.

In cohesionless soils, a value of $z_s = 0.05B$ is appropriate for internal shear keys. However, the shear key around the edge of the foundation, called a perimeter skirt, also provides the benefit of preventing undermining of the foundation by scour and animal burrowing. For this reason, the perimeter skirt will normally be deeper than the internal keys, so a height of $z_s = 0.1B$ is recommended for the perimeter skirt.

The maximum number of shear keys (and the corresponding minimum spacing) is limited to the number needed to force a failure to occur along the base of the shear keys as shown in Figure 4.3-7a. A greater number (and a lesser spacing) do not add to the sliding resistance of the soil block enclosed by the perimeter skirt, but only hinder the full penetration of the shear keys. The minimum recommended shear key spacing is $1.0z_s$ in cohesive soils and $2.0z_s$ in cohesionless soils.

The number of shear keys, n, required in each direction is computed by comparing the design load parallel to the seafloor to the passive resistance developed per key. This number, n, is computed by Equation 4-40:

$$n \geq \frac{\left[F_s F_{hp} + (W_{bf} + W_{bst}) \sin \beta \right]}{R_p} + 1 \tag{4-40}$$

where:

F_{hp} = resultant of applied loads in the downslope direction [F]

R_p = resistance developed by one key against movement along the longitudinal axis [F]; see Equation 4-41 for cohesive soil, or Equation 4-42 for cohesionless soil

For resistance to movement along the longitudinal axis, the shear keys perpendicular to that axis have a width B, and their spacing is equal to $L/(n-1)$. For resistance to movement perpendicular to the longitudinal axis, the shear keys parallel to that axis have a width L, and their calculated spacing is equal to $B/(n-1)$.

For cohesive soils, the resistance developed by one key against movement along the longitudinal axis (perpendicular to a key of width B), R_p, is calculated by Equation 4-41. For motion perpendicular to the longitudinal axis, the width is replaced by the length, L, of the key.

$$R_p = \left[\frac{\gamma_b z_s^2}{2} + 2 s_{ua} z_s \right] B \tag{4-41}$$

where:

s_{ua} = undrained shear strength averaged over the shear key height [F/L^2]

For cohesionless soils, the passive resistance developed by one shear key against longitudinal movement is calculated as shown in Equation 4-42. In Equation 4-42, K_p is the coefficient of passive lateral earth pressure as computed by Equation 4-43 (ϕ is in degrees).

4-31

$$R_p = \frac{K_p \gamma_b z_s^{\ 2} B}{2} \tag{4-42}$$

$$K_p = \tan^2 (45 + \phi/2) \tag{4-43}$$

4.3.5.2 Penetration of Shear Keys

The embedment force required to ensure full penetration of shear keys and perimeter skirts, Q_e, can be calculated as Q_u for the skirts at their full embedment depth using methods presented in Section 4.3.2. The skirt is treated as a long, narrow footing, with its width equal to the skirt thickness and its length equal to the total length of the skirts and shear keys, if any. The effective area is the product of this width and length. In making the analysis, the highest expected values of soil strength properties should be used to be on the conservative side.

For cohesive soils, Equation 4-28 is applicable. Undisturbed soil shear strengths should be used, and the soil sensitivity should be set equal to 1.0. In the last term, the width (thickness) may be neglected. For cohesionless soils, Equations 4-24 through 4-27 and 4-30 are applicable. In the last term in Equation 4-30, the friction angle between skirts/shear keys and sand should be set equal to the soil friction angle, and the width (thickness) may be neglected. A detailed example of these calculations is in Section 4.4.2.2, Step 10.

After Q_e has been calculated, it is compared to the sum of $W_{bst} + W_{bf}$, the forces (without the line load) driving penetration, to check if it is smaller than the driving forces. If it is larger, then the skirts must be made smaller or fewer, or the foundation weight will need to be increased. If the foundation weight is increased, a check must be made to see that this increase does not create a bearing capacity problem.

4.3.6 Foundation Settlement

The bearing capacity of the surficial seafloor soil is assumed to be sufficient to support the foundation. If the bearing capacity is not sufficient, then the foundation will immediately penetrate into the seafloor until its weight and the supported structure weight are balanced by the soil resistance (bearing capacity determined according to Section 4.3.2). A method for predicting rapid penetration is described in Chapter 8.

Foundation settlements due to elastic deformations and soil consolidation may still pose a significant problem, even in the absence of a bearing capacity failure, because such settlements are rarely uniform. The occurrence of differential settlement is greatly aggravated by eccentric loading. The resulting tilting could impair structure function.

Settlement of a deadweight anchor is, in contrast, not normally considered a problem because the holding capacity is unaffected, or is sometimes increased by such embedment. In some cases, however, even excessive embedment of a deadweight anchor is not desirable because it limits the ability to inspect and maintain the mooring line connections to the anchor.

Settlement is the summation of initial and consolidation settlements, which are discussed in the following sections.

4.3.6.1 Initial Settlement

Initial settlement is the instantaneous response of the soil to the foundation loading and results primarily from elastic soil deformations. Its value is obtained for a foundation that is not so heavily loaded as to cause bearing capacity failure. The general expression for the initial vertical settlement of a foundation, δ_i, can be calculated for both cohesive and cohesionless soils by Equation 4-44 (Ref. 4-1):

$$\delta_i = \frac{F_v}{A_b} \cdot \frac{BC_s(1-v^2)}{E} \tag{4-44}$$

where:

F_v = total downward vertical load on the foundation base [F]

A_b = footing area [L^2]

B = footing width [L]

C_s = shape factor (Table 4.3-3 or Equation 4-45)

v = Poisson's ratio of the soil (assume v = 0.35 for cohesionless soils and 0.45 for cohesive soils)

E = Young's modulus of the soil [F/L^2], a property which must be determined by soils testing or can be estimated using Equation 4-47 for cohesive soils, and Equations 4-48 and 4-47 for cohesionless or mixed soils

Table 4.3-3 lists the shape factors for various rigid footings on a soft seafloor soil (Ref. 4-1). In the table, the aspect ratio, L/B, is the ratio of the footing length to footing width. For aspect ratios not listed in Table 4.3-3, Equation 4-45 may be used to compute the shape factor for rigid rectangular footings on a very deep soil of uniform elastic modulus:

$$C_s = 1.46 \log_{10}\left(\frac{L}{B} + \frac{2}{3}\right) + \frac{1}{2} \tag{4-45}$$

For the specific case of a shallow circular foundation, the initial vertical settlement is given by (Ref. 4-6):

$$\delta_i = \frac{(1-v)}{4GR} \cdot F_v \tag{4-46}$$

where:

R = radius of the base [L]

G = $(E/2)/(1+v)$ and is called the elastic shear modulus of the soil [F/L²]

Table 4.3-3. Shape Factors for Rigid Footings on Soft Seafloor

Shape	Aspect Ratio, L/B	Shape Factor, C_s
Circular	1	0.79 (= π/4)
Square	1	0.82
Rectangular	2	1.12
Rectangular	5	1.60
Rectangular	10	2.00

Because it is usually difficult to obtain elasticity data for seafloor soils, an approximate expression for the elastic modulus is given in terms of shear strength of the soil and the load on the footing base relative to its bearing capacity (Equation 4-47). Data from References 4-1 and 4-6 were used to develop a model in terms of shear strength and shear strain associated with elastic settlement, as well as the effects of soil plasticity, and an approximate relationship in terms of footing load was fitted to the model. The result is:

$$E = s_u \frac{\{2/\varepsilon_{po}\}}{\{1+50x^3 y \exp(-s_u/50\,psf)\}} \left\{ \frac{25,000\,psf}{s_u} \right\}^{y^{1/2}}$$

(4-47)

where:

ε_{po} = baseline peak strain, = 0.035 * $(PI/35\%)^{1/2}$

PI = plasticity index = $w_L - w_P$

w_L = liquid limit = water content for specified slump test behavior (% of dry weight)

w_P = plastic limit = minimum water content for specified ductility test behavior (% of dry weight)

x = $1/F_s(bc)$ = inverse factor of safety for bearing capacity = F_v/Q_u

y = $1-x$ = $1-[1/F_s(bc)]$

At footing loads equal to or exceeding the bearing capacity, the inverse factor of safety for bearing capacity is equal to one, and the modulus remains at its minimum value of $(2S_u / \varepsilon_{po})$.

Typical resulting values of modulus for a 5-foot by 10-foot footing on cohesive soil ($PI = 45\%$, $\varepsilon_{po} = 0.04$) are:

Soil Strength	Modulus E (*psf*) for	
s_u (*psf*)	$F_s(bc) = 10$	$F_s(bc) = 2$
30	864,000	64,200
45	888,000	86,500
60	905,000	110,000

Equation 4-47 is based upon the soil's shear strength and is thereby limited to cohesive soils. For cohesionless soils, and for soils mobilizing both cohesive and frictional shearing resistance, Equation 4-47 may be employed by assuming $\varepsilon_{po} = 0.04$ and replacing the shear strength s_u with an effective shear strength given by:

$$S_{uef} = s_u + \frac{\gamma_b D_f \{1 + (N_q K_q - 1) f_z\} + \gamma_b (B/2) N_\gamma K_\gamma + (PH_s / A)\{s_u / S_t + \gamma_b (D/2) \tan \delta\}}{N_c K_c} \tag{4-48}$$

Typical resulting values of modulus for a 5-foot by 10-foot footing on cohesionless soil on-grade ($D_f = H_s = 0$, $N_c = 5.14$, $s_c = 1.10$, $d_c = 1$, $s_\gamma = 0.8$, $d_\gamma = 1$) are:

Soil Density	Friction Angle	BC Factor	Modulus E (*psf*) for	
γ_b (*pcf*)	ϕ (*deg*)	$N\gamma$ (-)	$F_s(bc) = 10$	$F_s(bc) = 2$
60 (loose)	30	27.66	1,031,000	417,000
75 (medium)	35	61.47	1,087,000	562,000
90 (dense)	40	145.19	1,146,000	763,000

4.3.6.2 Consolidation Settlement

In cohesive soils, after installation, a foundation will gradually settle as the excess pore pressure which developed in response to the foundation loading dissipates and the soil consolidates. In cohesionless soils there is no significant amount of consolidation settlement as pore pressure dissipates almost immediately. In evaluating this time-dependent consolidation settlement for cohesive soils, the soil to a depth of $2B$ below the foundation base should be considered. This is the depth where the effective stress change after consolidation is about 10% of the change immediately below the foundation base (for a square foundation) (Figure 4.3-8).

Only consolidation of the cohesive soil layers within the depth $2B$ needs to be calculated. The calculations are made by breaking the distance $2B$ into several incremental layers. If the soil over the depth $2B$ is homogeneous, a fairly small number of layers (8 to 12) will provide reasonably accurate results. If there are multiple cohesive soil layers having different properties, each layer should be represented by at least a few incremental layers. The total consolidation settlement is the sum of the settlement of the individual incremental layers and is obtained from:

$$\delta_c = \sum_{i=1}^{n} \left(\frac{H_i C_{ci}}{1 + e_{ci}} \right) \log_{10} \left(\frac{\overline{p_{ci}} + \Delta \overline{p_{ci}}}{\overline{p_{ci}}} \right) \qquad (4\text{-}49)$$

where:

δ_c = consolidation settlement [L]

n = number of incremental layers within distance $2B$

H_i = thickness of incremental layer i [L]

C_{ci} = compression index of incremental layer i

$\Delta \overline{p_i}$ = added effective vertical stress at midpoint of incremental layer i [F/L^2]

$\overline{p_{oi}}$ = initial effective overburden stress at midpoint of incremental layer i [F/L^2]

e_{oi} = initial void ratio of incremental layer i

An accurate computation of consolidation settlement requires considerable knowledge of soils properties. For instance, high quality soil samples and at least one laboratory consolidation test per cohesive soil layer are required to determine C_{ci} for these computations. Other soil properties must be measured for each cohesive soil layer to perform a settlement analysis. A rough estimate for these settlements can be made by using some estimated properties. Values for C_c may be determined empirically as outlined in Chapter 3. The initial void ratio, e_o, can be determined from soil specific gravity and water content values by the relationship $e = G_s w$. The value of $\Delta \overline{p_i}$ can be estimated from Figure 4.3-8. The value of $\overline{p_{oi}}$ is determined by using the soil buoyant unit weight. It is suggested that a maximum soil layer thickness of 0.25B be used.

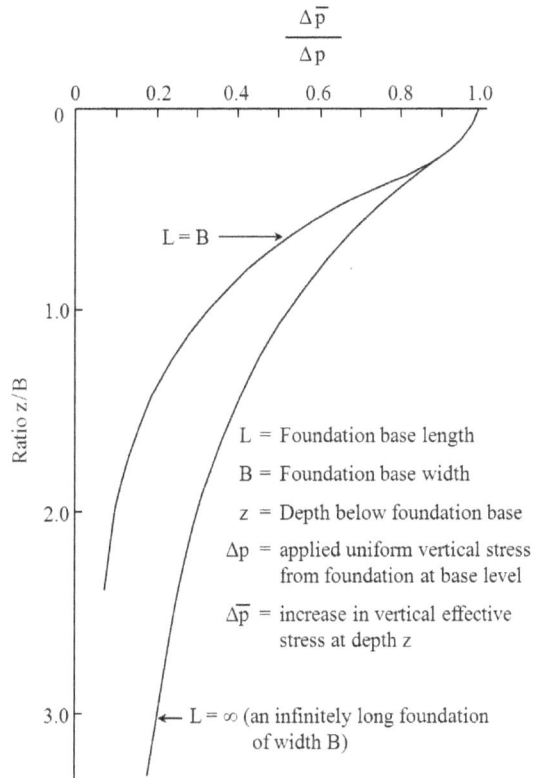

Figure 4.3-8. Soil stress increase beneath a rectangular foundation.

4.3.7 Installation and Removal

The installation of shallow foundations should be planned so that the foundation can be properly set down at the intended site without excessive disturbance to the supporting soil.

The maximum lowering rate of the installation line should not exceed the free-fall velocity of the package to avoid unstable lowering and possible entangling of the lowering line. A rough estimate for maximum lowering rate is:

$$v_{max} = \frac{W_{bv}}{4 A_v} \tag{4-50}$$

where:

v_{max} = maximum lowering velocity [fps]

W_{bv} = submerged weight of installation [lb]

A_v = vertical projected area of the package [ft^2]

4-37

As the installation approaches the seafloor, the maximum lowering velocity must be reduced by at least a factor of four to prevent too hard an impact with the bottom. A hard impact may result in bearing failure, excessive tilting, or an instability failure. It is usually desirable to approach the seafloor as slowly as possible and not to lift the foundation off the seafloor once the initial touchdown has occurred.

The recovery of a shallow foundation also requires careful consideration to insure that adequate lifting force for breakout is available. Prediction of breakout forces and a discussion of techniques used to minimize the breakout force are presented in Chapter 9.

4.4 EXAMPLE PROBLEMS

4.4.1 Problem 1 – Simple Foundation on Cohesive Soil

4.4.1.1 Problem Statement

Determine the dimensions of a concrete and steel square foundation, essentially a deadweight anchor, with shear keys to resist given environmental loadings at a deep ocean site. The seafloor is mildly sloping and is composed of a cohesive soil which has well-established properties. Use a factor of safety of 1.5 and minimize the base horizontal size.

Data: The foundation is to be placed where the seafloor slopes at 5° and must resist loads from a mooring line that may reach 20,000 pounds in uplift and 20,000 pounds in any horizontal direction. Figure 4.4-1 shows a sketch for Problem 1 and for Problem 2. The seafloor is a cohesive silty clay material whose properties have been determined by laboratory tests on cored samples. The data for undrained shear strength (s_u), buoyant unit weight (γ_b), sensitivity (S_t), and the drained parameters of cohesion ($\bar{c}$) and friction angle ($\bar{\phi}$) are shown to the right of Figure 4.4-1.

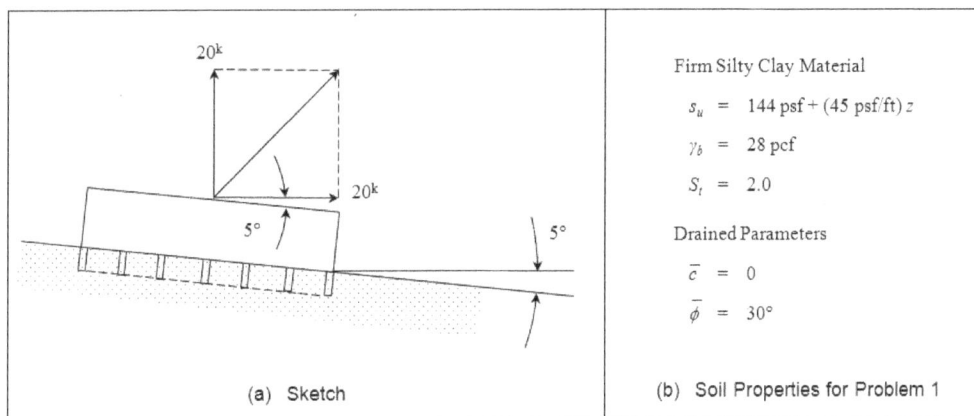

Figure 4.4-1. Foundation sketch for example Problems 1 and 2.

4.4.1.2 Problem Solution

The analytical and computational procedures for the problem's solution are shown below. They follow the method presented for the design of a square foundation to resist the loads under the existing conditions. The forces acting on the foundation are shown in Figure 4.4-2.

4-39

Problem 4.4-1

ANALYTICAL PROCEDURES	COMPUTATIONS
1. Calculate F_{hp}, the load component parallel to the slope (downslope) from existing forces and the safety factor. $F_{hp} = F_h \cos \beta - F_{ve} \sin \beta$	$F_{hp} = (20,000 \ lb) \cos 5° - (20,000 \ lb) \sin 5°$ $F_{hp} = 18,180 \ lb$ $F_s = 1.5$ (given) $F_s \ F_{hp} = (1.5)(18,180 \ lb) = 27,300 \ lb$
2. Select a trial foundation size. The foundation has a side length B and shear keys of height $z_s = 0.1B$ (Section 4.3.5.1). Assume the foundation will be placed parallel to the downslope direction and fully embed (D_f = full key height = z_s).	Say $B = L = 12$ ft. Therefore, $z_s = (0.1)(12 \ ft) = 1.2 \ ft$ $D_f = z_s = 1.2 \ ft$
3. Calculate the foundation's resistance to sliding in the short-term (undrained) condition (Equation 4-32). $Q_{ul} = s_{uz} A + 2 \ s_{ua} H_s L$ $H_s = \min(D_f, H + z_s)$	$s_{uz} = s_u$ @ depth D_f $\quad = 144 \ psf + (45 \ psf/ft)(1.2 \ ft) = 198 \ psf$ $A = (12 \ ft)(12 \ ft) = 144 \ ft^2$ $s_{ua} = s_u$ @ depth $z_s /2$ $\quad = 144 \ psf + (45 \ psf/ft)(0.6 \ ft) = 171 \ psf$ $Q_{ul} = (198 \ psf)(144 \ ft^2)$ $\quad + 2(171 \ psf)(1.2 \ ft)(12 \ ft) = 33,400 \ lb$
4. Calculate weight of soil, W_b, within the footing skirt and overlying the potential failure plane (Equation 4-33), and the minimum foundation weight, W_{bf}, required to resist sliding, long-term (Equation 4-38). $W_b = \gamma_b \ A \ z_s$ $W_{bf} = \dfrac{(F_s + \mu \tan \beta)F_h - \dfrac{\bar{c}A}{\cos \beta}}{\mu - F_s \tan \beta}$ $\qquad\qquad + F_{ve} - W_{bst} - W_b$	$W_b = (28 \ pcf)(144 \ ft^2)(1.2 \ ft) = 4,840 \ lb$ $\mu = \tan\bar{\phi}$ (Section 4.3.3.3) = $\tan(30°) = 0.577$ $W_{bf} = \dfrac{[1.5 + (0.577)\tan(5°)]\cdot 20,000 \ lb - 0}{0.577 - 1.5\tan(5°)}$ $\quad + 20,000 \ lb - 0 - 4,840 \ lb$ $\quad = 84,670 \ lb$

Problem 4.4-1

ANALYTICAL PROCEDURES	COMPUTATIONS
5. Is the 12-ft square foundation's resistance to (undrained) sliding greater than the forces driving it downslope? Is $Q_{ul} \geq F_s\,[F_{hp} + (W_{bf} + W_{bst} + W_b)\sin\beta]$?	$(1.5)[18{,}180\ lb + (84{,}670\ lb + 0$ $\quad + 4{,}840\ lb)\sin(5°)] = 38{,}970\ lb$ NO, $Q_{ul} = 33{,}400\ lb\ < 38{,}970\ lb$ The foundation size is not adequate to resist downslope sliding. The foundation trial size will have to be increased.
6. Try a 13-ft square foundation, which has $0.1B$ deep shear keys. Calculate the new foundation's sliding resistance to (undrained) sliding (Equation 4-32). $Q_{ul} = s_{uz}\,A + 2\,s_{ua}\,H_s\,L$ $H_s = \min(D_f, H + z_s)$ Is the 13-ft square foundation's resistance to sliding greater than the forces driving it downslope?	$B = L = 13\ ft$ $z_s = (0.1)(13\ ft) = 1.3\ ft$ $D_f = z_s = 1.3\ ft$ $A = (13\ ft)(13\ ft) = 169\ ft^2$ $s_{uz} = s_u$ @ depth z_s $\quad = 144\ psf + (45\ psf/ft)(1.3\ ft) = 202\ psf$ $s_{ua} = s_u$ @ depth $z_s/2$ $\quad = 144\ psf + (45\ psf/ft)(0.65\ ft) = 173\ psf$ $Q_{ul} = (202\ psf)(169\ ft^2)$ $\quad + 2(173\ psf)(1.3\ ft)(13\ ft) = 40{,}080\ lb$ (The driving forces are the same, as an increase in z_s causes W_b to increase and W_{bf} to decrease by an equal amount, making the total unchanged in step 5) YES, $Q_{ul} > 38{,}970\ lb$

Problem 4.4-1

ANALYTICAL PROCEDURES	COMPUTATIONS
7. Recalculate W_b (Equation 4-33) and W_{bf} (Equation 4-38) for the 13-ft foundation. $W_b = \gamma_b\, A\, z_s$ $W_{bf} = \dfrac{(F_s + \mu \tan \beta)F_h - \dfrac{\bar{c}A}{\cos \beta}}{\mu - F_s \tan \beta}$ $\qquad\qquad + F_{ve} - W_{bst} - W_b$	$W_b = (28\ pcf)(169\ ft^2)(1.3\ ft) = 6{,}150\ lb$ $W_{bf} = \dfrac{[1.5 + (0.577)\tan(5^\circ)] \cdot 20{,}000\ lb - 0}{0.577 - 1.5 \tan(5^\circ)}$ $\quad + 20{,}000\ lb - 0 - 6{,}150\ lb = 83{,}360\ lb$
8. Calculate the height of a concrete trial foundation necessary to yield an underwater weight of W_{bf}. From the relationship: $W_{bf} = \gamma_b\, A\, H$ Check that H does not exceed the recommended limit. Is $H < 0.25B$?	$\gamma_{b(conc)} \cong 150\ pcf - 64\ pcf = 86\ pcf$ $H = \dfrac{W_{bf}}{\gamma_{b(conc)}A} = \dfrac{83{,}360\ lb}{(86\ pcf)(169\ ft)} = 5.74\ ft$ $0.25B = (0.25)(13\ ft) = 3.25\ ft$ NO, $H > 0.25B$ Therefore, the foundation base is too high, and this may promote overturning instability.
9. Recalculate foundation block density, limiting the height to the recommended maximum, and required amount of scrap steel embedded in concrete. Required buoyant density and embedded steel are: $V_f = AH$ $\gamma_{bf} = W_{bf}\,/\,V_f$ $V_{steel} = V_f\,(\gamma_{bf} - \gamma_{bc})\,/\,(\gamma_{bs} - \gamma_{bc})$ $V_{conc} = V_f - V_{steel}$	Let $H = 0.25B = 0.25(13\ ft) = 3.25\ ft$ $V_f = (169\ ft^2)\ (3.25\ ft) = 549\ ft^3$ $\gamma_{bf} = \dfrac{83{,}360\ lb}{549\ ft^3} = 152\ pcf$ To remain within the recommended foundation height limitation, the foundation must have an average $\gamma_{bf} = 152\ pcf$. This can be accomplished by using a concrete foundation with embedded scrap steel to raise the weight, in the following amounts: $V_{steel} = (549\ ft^3)\ (152\ pcf - 86\ pcf)$ $\qquad\quad /\ (426\ pcf - 86\ pcf) = 106\ ft^3$ $V_{conc} = 549\ ft^3 - 106\ ft^3 = 443\ ft^3$

Problem 4.4-1

ANALYTICAL PROCEDURES	COMPUTATIONS
10. Calculate the trial foundation's resultant normal force F_n (acting perpendicular to the slope) using Equation 4-39. $F_n = (W_{bf} + W_{bst} + W_b - F_{ve}) \cos \beta - F_h \sin \beta$	F_n = (83,360 lb + 0 lb + 6,150 lb – 20,000 lb) $\quad \cdot \cos (5°) - (20,000 \, lb) \sin (5°)$ = 67,500 lb
11. Calculate e_2, the (downslope) eccentricity by summing moments around the center of the shear key base (see Figure 4.4-1). $\Sigma M_o = W_b(z_s/2) \sin \beta + W_{bf}(z_s + H/2) \sin \beta$ $\quad - F_{ve}(z_s + H) \sin \beta + F_h(z_s + H) \cos \beta$ Because ΣM_o is equal to $(e_2) F_n$, $e_2 = \Sigma M_o / F_n$	ΣM_o = (6,150 lb)(1.3 ft/2) sin (5°) $\quad$ +83,360 lb (1.3 ft + 3.25 ft/2) sin(5°) $\quad$ – 20,000 lb (1.3 ft + 3.25 ft) sin(5°) $\quad$ + 20,000 lb (1.3 ft + 3.25 ft) cos(5°) ΣM_o = 104,320 ft-lb e_2 = 104,320 ft-lb / 67,500 lb = 1.55 ft
12. Is eccentricity acceptable? The maximum recommended $e = B/6$ Is $e_2 \leq B/6$?	maximum e = 13.0 ft/6 = 2.17 ft e_2 = 1.55 ft YES, $e_2 < B/6$
13. Calculate the bearing area reduced for the eccentricity (Equations 4-9 and 4-10): $L' = B - 2e_1$ $B' = B - 2e_2$ $A' = B' L'$	e_1 = 0 L' = 13.0 ft – 0 = 13.0 ft B' = 13.0 ft – 2(1.55 ft) = 9.9 ft A' = (9.9 ft)(13.0 ft) = 128.8 ft^2
14. Calculate s_{uz} for the bearing capacity equation (average s_u over the depth 0.7B′ below the shear keys).	s_u at z_s = 144 psf + (45 psf/ft)(1.3 ft) = 202 psf s_u at $(z_s + 0.7B')$ = 144 psf $\quad$ + [45 psf/ft]*[1.3 ft $\quad\quad$ +0.7(9.9 ft)] = 515 psf s_{uz} = (202 psf + 515 psf)/2 = 359 psf

Problem 4.4-1

ANALYTICAL PROCEDURES	COMPUTATIONS
15. Calculate the correction factor K_c for bearing capacity (Equations 4-18 through 4-23). $N_q = 1$ (Equation 4-15) $N_c = 5.14$ (Equation 4-16) $K_c = i_c\, s_c\, d_c\, b_c\, g_c$ $i_c = 1 - \dfrac{mF_h}{B'L's_u N_c}$ $m = \left[\dfrac{2+(L'/B')}{1+(L'/B')}\right]\cos^2\theta$ $\quad + \left[\dfrac{2+(B'/L')}{1+(B'/L')}\right]\sin^2\theta$ $s_c = 1 + (B'/L')(N_q/N_c)$ $d_c = 1 + 2\,(1-\sin\phi)^2\arctan\left(\dfrac{D_f}{B'}\right)\cdot\dfrac{N_q}{N_c}$	$\overline{\phi} = 0°$ (undrained) $\theta = 90°$ (from Section 4.3.2.5) $m = 0 + \left[\dfrac{2+(9.9\ ft\,/13.0\ ft)}{1+(9.9\ ft\,/13.0\ ft)}\right](1)^2 = 1.57$ $i_c = 1 - \dfrac{(1.57)(20,000\ lb)}{(9.9\ ft)(13\ ft)(359\ psf)(5.14)} = 0.868$ $s_c = 1 + \left(\dfrac{9.9\ ft}{13.0\ ft}\right)\left(\dfrac{1}{5.14}\right) = 1.148$ $d_c = 1 + 2(1-\sin 0°)\arctan\left(\dfrac{1.3\ ft}{9.9\ ft}\right)\cdot\left(\dfrac{1}{5.14}\right)$ $d_c = 1.051$ $b_c = g_c = 1.0$ (from Section 4.3.2.5) $K_c = (0.868)(1.148)(1.051)(1)(1) = 1.047$
16. Calculate the short-term bearing capacity (Equation 4-28). $Q_u = A'\big(s_{uz}N_c K_c + \gamma_b D_f\big)$ $\quad + (2B + 2L\big(\min(D_f, H + z_s)\big)\dfrac{s_{ua}}{S_t}$ $F_v = W_{bf} + W_b - F_{ve}$ $F_n = F_v\cdot\cos(\beta) - F_h\cdot\sin(\beta)$ Is there sufficient bearing capacity? Is $Q_u > F_s\, F_n$?	$Q_u = (128.8\ ft^2)[(359\ psf)(5.14)(1.047)$ $\qquad + (28\ pcf)(1.3\ ft)]$ $\qquad + [2(13\ ft) + 2(13\ ft)](1.3\ ft)(173\ psf/3)$ $Q_u = 259,260\ lb$ $F_v = 83,360\ lb + 6,150\ lb - 20,000\ lb$ $\quad = 69,510\ lb$ $F_n = (69,510\ lb)\cos(5°) - (20,000\ lb)\sin(5°)$ $\quad = 67,500\ lb$ $F_s\, F_n = (1.5)(67,500\ lb) = 101,260\ lb$ YES, $Q_u > F_s\, F_n$

Problem 4.4-1

ANALYTICAL PROCEDURES	COMPUTATIONS
17. Calculate the short-term (undrained) bearing capacity for the foundation when the mooring line load is not applied. Compare this to the unloaded normal force, F_n. $F_v = W_{bf} + W_b$ $F_n = F_v \cdot \cos(\beta)$ Is there sufficient bearing capacity? Is $Q_u > F_s F_n$?	$F_v = 83,360\ lb + 6,150\ lb = 89,510\ lb$ $F_n = (89,510\ lb)\cos(5°) = 89,170\ lb$ $\sum M_o = (6,150\ lb)(1.3\ ft/2)\sin(5°)$ $\qquad +83,360\ lb(1.3\ ft + 3.25\ ft/2)\sin(5°)$ $\qquad = 21,600\ ft\text{-}lb$ $e_2 = 21,600\ ft\text{-}lb\ /\ 89,170\ lb = 0.24\ ft$ $B' = 13\ ft - 2(0.24\ ft) = 12.5\ ft$ $A' = (12.5\ ft)(13.0\ ft) = 162.7\ ft^2$ s_u at $z_s = 202\ psf$ s_u at $(z_s + 0.7\ B') = 144\ psf + [45psf/ft]*$ $\qquad\qquad [1.3\ ft+0.7(12.5\ ft)] = 597\ psf$ $s_{uz} = (202\ psf + 597\ psf)/2 = 400\ psf$ $m = 0 + \left[\dfrac{2+(12.5\ ft\,/\,13\ ft)}{1+(12.5\ ft\,/\,13\ ft)}\right]\sin^2 90° = 1.509$ $i_c = 1 - \dfrac{(1.509)(0\ lb)}{(12.5\ ft)(13\ ft)(400\ psf)(5.14)} = 1$ $s_c = 1 + \left(\dfrac{12.5\ ft}{13.0\ ft}\right)\left(\dfrac{1}{5.14}\right) = 1.187$ $d_c = 1 + 2(1-0)^2 \arctan\left(\dfrac{1.3\ ft}{12.5\ ft}\right)\cdot\left(\dfrac{1}{5.14}\right)$ $\qquad = 1.040$ $K_c = (1.0)(1.187)(1.040)(1)(1) = 1.235$ $Q_u = (162.7\ ft^2)[(400\ psf)(5.14)(1.235)$ $\qquad + (28\ pcf)(1.3\ ft)]$ $\qquad + [2(13\ ft)+2(13\ ft)](1.3\ ft)(173\ psf/3)$ $\qquad = 424,650\ lb$ $F_s F_n = (1.5)(89,170lb) = 133,760\ lb$ YES, $Q_u > F_s F_n$

Problem 4.4-1

ANALYTICAL PROCEDURES	COMPUTATIONS
18. Determine the minimum number shear keys (from Equations 4-40 and 4-41). $$n \geq \frac{[F_s F_{hp} + (W_{bf} + W_{bst})\sin\beta]}{R_p} + 1$$ $$R_p = \left[\frac{\gamma_b z_s^2}{2} + 2 s_{ua} z_s\right] B$$ Note: Section 4.3.5.1 recommends a minimum spacing equal to z_s for cohesive soils. Shear key spacing = $B/(n\text{-}1)$	$s_{ua} = s_{ua}$ at depth $z_s/2 = 173$ psf $$R_p = \left[\frac{28\ pcf(1.3\,ft)^2}{2} + 2(173\,psf)(1.3\,ft)\right]$$ $$\cdot (13\,ft)$$ = 6,163 lb (per shear key) $$n \geq \frac{[1.5(18,180\,lb) + (83,360\,lb + 0)\sin(5°)}{6,163\ lb}$$ + 1 = 7 (rounded up from 6.6) And, required spacing = $13\,ft/(7-1) = 2.17\,ft$ The recommended minimum is $z_s = 1.3\,ft$. Therefore, use 7 shear keys spaced 2.17 ft apart.

SUMMARY

The foundation has been checked against bearing capacity failure, sliding, and overturning. A check may also be made (not shown in this example, but shown for Example Problem 2, Section 4.4.2.2) to see if the 83,360-lb buoyant foundation weight is sufficient to cause full shear key embedment (this is a slow penetration calculation that may be done by methods described in this chapter or the simplified versions in Chapter 8). Initial and consolidation settlements cannot be calculated because there is a lack of sufficient data. They are also not considered important for a foundation or deadweight anchor, which is placed on the seafloor to resist only mooring line loads.

The soil-related aspects of the foundation design process have resulted in a foundation designed with the following dimensions:

Side..13.0 ft square
Height...3.25 ft (excluding shear keys)
Buoyant Weight.............................83,360 lb
Shear Key Depth ....1.3 ft
Shear Key Spacing........................2.17 ft

It is noted that the design is most critical with respect to sliding stability.

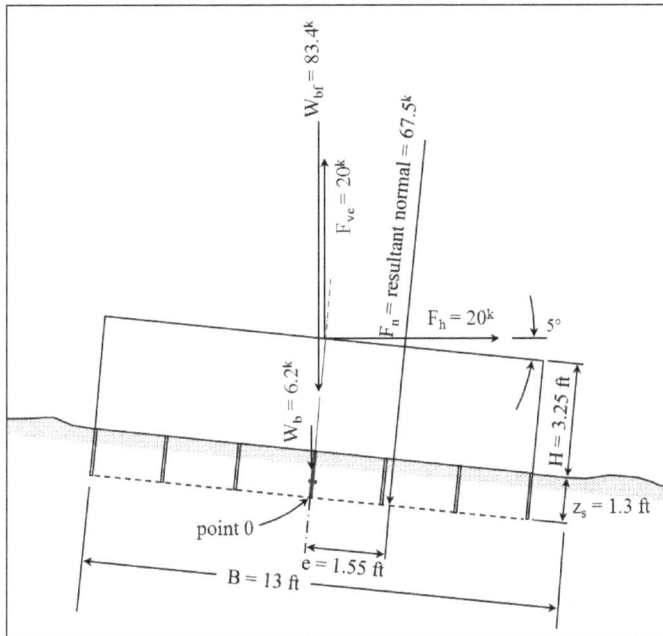

Figure 4.4-2. Forces considered in the overturning analysis for example Problem 1.

4.4.2 Problem 2 – Simple Foundation on Cohesionless Soil

4.4.2.1 Problem Statement

Determine the dimensions of a simple concrete square foundation, essentially a deadweight anchor, with a perimeter skirt to resist given environmental loadings, at a shallow ocean site. The seafloor is mildly sloping and is composed of a cohesionless soil which has well-established properties. Use a factor of safety of 1.5.

Data: The foundation is to be placed where the seafloor slopes at 5° and must resist loads from a mooring line that may reach 20,000 pounds in uplift and in any horizontal direction (see Figure 4.4-1). The area is sheltered and no cyclic effects are expected from surface waves. The seafloor is composed of a medium-dense well-graded sand (cohesionless), with the following properties having been determined from in-situ or laboratory testing: buoyant unit weight of 60 pcf and friction angle of 35°. (This is the same problem as 4.4.1 except that the soil is different and skirts without interior shear keys are to be used.)

4.4.2.2 Problem Solution

The analytical and computational procedures for the problem's solution are shown below. They follow the method presented for the design of a square foundation to resist the loads under the existing conditions. Figure 4.4-3 shows the forces acting on the foundation.

Problem 4.4-2

ANALYTICAL PROCEDURES	COMPUTATIONS
1. Identify known foundation parameters. Because the foundation will be only concrete, its size will be determined by the suggested limiting relationship $H \leq 0.25B$ (Section 4.3.2.3), and determined by the necessary weight of concrete to prevent sliding (Step 3). The foundation has a side length B. Assume a skirt depth z_s of $0.1B$ (Section 4.3.5.1) and that the foundation will be placed parallel to the downslope direction and embedded to a depth of z_s.	F_h = 20,000 lb F_{ve} = 20,000 lb $D_f = z_s$ $\beta = 5°$ γ_b = 60 pcf (given) $\bar{\phi} = \phi = 35°$ F_s = 1.5 (given)
2. Calculate W_{bf}, the minimum buoyant foundation weight necessary to resist sliding (Equation 4-35 without the W_b term). Note: This is both the short-term and long-term condition for a cohesionless soil. $$W_{bf} = \frac{(F_s + \mu \tan \beta)F_h}{\mu - F_s \tan \beta} + F_{ve} - W_{bst}$$ $\mu = \tan \phi$ (Section 4.3.3.2)	$\mu = \tan(35°) = 0.70$ F_h = 20,000 lb F_{ve} = 20,000 lb $W_{bst} = 0$ $$W_{bf} = \frac{[1.5 + (0.70)\tan(5°)] \cdot 20{,}000 \ lb}{0.70 - 1.5\tan(5°)}$$ $+$ 20,000 lb $-$ 0 = 74,900 lb

Problem 4.4-2

ANALYTICAL PROCEDURES	COMPUTATIONS
3. Calculate the dimensions of the trial foundation necessary to yield an underwater weight of W_{bf}. $W_{bf} = \gamma_{b(conc)} AH$ $H = 0.25B$ $A = B^2$ Therefore, $B = \left(\dfrac{4W_{bf}}{\gamma_b}\right)^{1/3}$	γ (concrete) $\approx 150\ pcf$ $\gamma_{b(conc)} = 150\ pcf - 64\ pcf = 86\ pcf$ $B = \left[\dfrac{(4)(74,900\ lb)}{86\ pcf}\right]^{1/3} = 15.16\ ft$ Say $B = 15\ ft$ $z_s = D_f = 0.1(15\ ft) = 1.5\ ft$ $A = (15\ ft)^2 = 225\ ft^2$ $H = \dfrac{W_{bf}}{\gamma_b A} = \dfrac{74,900\ lb}{(86\ pcf)(15\ ft)^2} = 3.87\ ft$
4. Calculate F_n, the resultant force acting normal to the slope (Equation 4-39, with $W_{bst} = 0$ for no structure, and W_b retained because D_f is taken at the skirt bottom for bearing capacity comparison). $F_n = (W_{bf} + W_b - F_{ve})\cos\beta - F_h\sin\beta$	$W_b = \gamma\, z_s\, A = (60\ pcf)(1.5\ ft)(225\ ft^2)$ $= 20,200\ lb$ $F_n = (74,900\ lb + 20,200\ lb - 20,000\ lb)$ $\cdot \cos(5°) - (20,000\ lb)\sin(5°)$ $F_n = 73,100\ lb$
5. Calculate e_2, the (downslope) eccentricity by summing moments around the center of the skirt-line. $\sum M_o = [W_{bf}(H/2 + z_s) + W_b(z_s/2) - F_{ve}(H + z_s)]$ $\cdot \sin\beta + F_h(H + z_s)\cos\beta$ $e_2 = \sum M_o / F_n$	$\sum M_o = (74,900\ lb)(3.87\ ft/2 + 1.5\ ft)$ $+ (20,200\ lb)(1.5\ ft)/2$ $- (20,000\ lb)(3.87\ ft + 1.5\ ft)]\sin(5°)$ $+ 20,000\ lb(3.87\ ft + 1.5\ ft)\cos(5°)$ $\sum M_o = 121,400\ ft\text{-}lb$ $e_2 = 121,400\ ft\text{-}lb/73,100\ lb = 1.66\ ft$
6. Calculate the effective bearing area (Equations 4-9 and 4-10 with $L = B$): $B' = B - 2e_2$ and $L' = B - 2e_1$ $A' = B'L'$	$e_1 = 0$ $B' = 15.0\ ft - 2(1.66\ ft) = 11.7\ ft$ $L' = 15.0\ ft - 0 = 15.0\ ft$ $A' = (11.7\ ft)(15.0\ ft) = 175.2\ ft^2$

Problem 4.4-2

ANALYTICAL PROCEDURES	COMPUTATIONS
7. Calculate the correction factors K_γ and K_q for bearing capacity (Equations 4-18 through 4-22), and the bearing capacity factors N_q and N_γ (Equations 4-15 and 4-17).	$\theta = 90°$ (from Section 4.3.2.5)
	$\bar{c} = 0$ (cohesionless soil)
$K_\gamma = i_\gamma\, s_\gamma\, d_\gamma\, b_\gamma\, g_\gamma$	$F_v = W_{bf} + W_b - F_{ve} = 74{,}900\ lb + 20{,}200\ lb$ $- 20{,}000\ lb = 75{,}100\ lb$
$K_q = i_q\, s_q\, d_q\, b_q\, g_q$	$m = 0 + \left[\dfrac{2+(11.7\ ft/15\ ft)}{1+(11.7\ ft/15\ ft)}\right](1)^2 = 1.56$
$i_\gamma = \left(1 - \dfrac{F_h}{F_v + B'\,L'\,\bar{c}\cot\bar{\phi}}\right)^{m+1}$	$i_\gamma = \left(1 - \dfrac{20{,}000\ lb}{75{,}100\ lb + 0}\right)^{2\,56} = 0.452$
$m = \left[\dfrac{2+(L'/B')}{1+(L'/B')}\right]\cos^2\theta$ $+ \left[\dfrac{2+(B'/L')}{1+(B'/L')}\right]\sin^2\theta$	$s_\gamma = 1 - 0.4(11.7\ ft/15\ ft) = 0.689$ $d_\gamma = 1.0$
$s_\gamma = 1 - 0.4(B'/L')$	$b_\gamma = g_\gamma = 1.0$
$d_\gamma = 1.0$ (from Section 4.3.2.5)	$K_\gamma = (0.452)(0.689)(1)(1)(1) = 0.312$
$b_\gamma = g_\gamma = 1.0$ (from Section 4.3.2.5)	$i_q = \left(1 - \dfrac{20{,}000\ lb}{75{,}100\ lb + 0}\right)^{1\,56} = 0.617$
$i_q = \left(1 - \dfrac{F_h}{F_v + B'\,L'\,\bar{c}\cot\bar{\phi}}\right)^{m}$	$s_q = 1 + (11.7\ ft/15\ ft)\tan(35°) = 1.545$
$s_q = 1 + (B'/L')\tan\bar{\phi}$	$d_q = \left[1 + 2(1-\sin 35°)^2\arctan\left(\dfrac{1.5\,ft}{11.7\,ft}\right)\tan 35°\right] = 1.033$
$d_q = \left[1 + 2(1-\sin\phi)^2\arctan\left(\dfrac{D_f}{B'}\right)\tan\phi\right]$	$b_q = g_q = 1.0$
$b_q = g_q = 1.0$ (from Section 4.3.2.5)	$K_q = (0.617)(1.545)(1.033)(1)(1) = 0.984$
$N_\gamma = 2(1+N_q)\tan\phi\,\tan(\pi/4 + \phi/5)$	$N_\phi = [\tan(180°/4 + 35°/2)]^2 = 3.69$
$N_q = \exp[\pi\tan(\phi)]\, N_\phi$	$N_q = \exp[\pi\tan(35°)]\ 3.69 = 33.30$
$N_\phi = [\tan(\pi/4 + \phi/2)]^2$	$N_\gamma = 2(1 + 33.29)\tan(35°)\tan(180°/4 + 35°/5)$ $= 61.47$

Problem 4.4-2

ANALYTICAL PROCEDURES	COMPUTATIONS
8. Calculate the short-term and long term bearing capacity (Equation 4-30). $$Q_u = A'\gamma_b[D_f\{1+(N_qK_q-1)f_z\}+(B'/2)N_\gamma K_\gamma f_z]$$ $$+(2B+2L)\big(\min(D_f,H+z_s)\big)\gamma_b\tan\delta$$ $$\cdot[(D_f+\max(0,D_f-H-z_s))/2]$$ where: $$f_z = \frac{\arctan\left(D_f/D_t\right)}{\left(D_f/D_t\right)}$$ $$D_t = \frac{q_{f\max}}{(\pi/2)\gamma_b\left[(N_qK_q-1)+\{(B'/2)/D_f\}N_\gamma K_\gamma\right]}$$ $q_{f\max} = s_{ucr}\,N_{cclay}\,K_{cclay}$ $N_{cclay} = 2 + \pi$ $K_{cclay} = [1 + (B'/L')/(2 + \pi)]$ $\quad\quad \cdot [1 + \{2/(2+\pi)\}\arctan(D_f/B')]$ $s_{ucr} = \sigma_{cr}\sin(\phi)\,/\,\{1-\sin(\phi)\}$ $\sigma_{cr} \approx D_r^{1.7}\cdot 20,000\ psf$ $D_r \approx (\gamma_b - 56.5\ pcf)\,/\,11.5\ pcf$ Is there sufficient bearing capacity? Is $Q_u > F_s\,F_n$?	$D_r \approx (60\ pcf - 56.5\ pcf)\,/\,11.5\ pcf = 0.304$ $\sigma_{cr} \approx (0.304)^{1.7}\cdot 20,000\ psf = 2,650\ psf$ $s_{ucr} = (2,650\ psf)\sin(35°)\,/\,\{1-\sin(35°)\}$ $\quad\quad = 3,560\ psf$ $N_{cclay} = 5.14$ $K_{cclay} = [1 + (11.7\ ft\,/15\ ft)\,/\,5.14]$ $\quad\quad\quad \cdot[1 + \{2\,/\,5.14\}\arctan(1.5\ ft\,/\,11.7\ ft)]$ $\quad\quad\quad = 1.209$ $q_{f\max} = (3,560\ psf)\,(5.14)\,(1.209) = 22,130\ psf$ $D_t = [(22,130\ psf)\,/\,\{(1.57)(\,60\ pcf)\}]$ $\quad\quad /[\{(33.30)\,(0.984) - 1\}$ $\quad\quad +\{(11.7\ ft\,/\,2)\,/\,(1.5\ ft)\}(61.47)(0.312)]$ $\quad\quad = 2.20\ ft$ $f_z = \arctan(1.5\ ft\,/\,2.20\ ft)\,/\,(1.5\ ft\,/\,2.20\ ft)$ $\quad\quad = 0.878$ $Q_u = (175.2\ ft^2)(60\ pcf)$ $\quad\quad \cdot [(1.5\ ft)\{1+((33.30)(0.984) - 1)(0.878)\}$ $\quad\quad\quad + (11.7ft/2)(61.47)(0.312)(0.878)]$ $\quad\quad + [(2)(15\ ft)+(2)(15\ ft)]\,(1.5\ ft)$ $\quad\quad\quad \cdot [(60\ pcf)\,(\tan 30°)\,\{(1.5\ ft)+0\}\,/\,2)]$ $Q_u = 1,493,465\ lb$ $F_s\,F_n = (1.5)(73,100\ lb) = 109,650\ lb$ YES, $Q_u > F_s\,F_n$
9. Is eccentricity acceptable? The maximum recommended $e = B/6$. Is $e_2 \le B/6$?	maximum $e = 15.0\ ft/6 = 2.50\ ft$, $e_2 = 1.66\ ft$ YES, $e_2 < B/6$

Note: Removal of the line load would cause an increase in bearing load of about 20,000 lb and an increase in capacity due to a reduction of eccentricity. Since the bearing capacity is so much greater than the applied load, no detailed calculations are needed to verify that the bearing capacity is adequate with no line load applied.

Problem 4.4-2

ANALYTICAL PROCEDURES	COMPUTATIONS
10. Is the 74,900-lb buoyant foundation weight, without the line load, sufficient to cause full skirt embedment? (Equations 4-30 and 4-24 through 4-27, using the full soil friction angle per Section 4.3.5.2, and B_s and L_s in place of B' and L'). B_s = skirt thickness (assume 0.1 ft) L_s = skirt length = 2 $(B + L - 2 B_s)$ $$Q_u = B_s L_s \gamma_b \left[D_f \{1+(N_q K_q - 1) f_z\} + \left(\frac{B_s}{2}\right) N_\gamma K_\gamma f_z \right]$$ $$+ (2L_s) D_f \gamma_b \tan\phi (D_f / 2)$$ $i_q = i_\gamma = 1$ (no lateral load), so: $K_q = s_q\, d_q$ $K_\gamma = s_\gamma\, d_\gamma$ $s_q = 1 + (B_s / L_s) \tan\phi$ $s_\gamma = 1 - 0.4\,(B_s / L_s)$ $$d_q = \left[1 + 2\left(1 - \sin\phi\right)^2 \arctan\left(\frac{D_f}{B_s}\right) \tan\phi \right]$$ $$f_z = \frac{\arctan\left(D_f / D_t\right)}{\left(D_f / D_t\right)}$$ $$D_t = \frac{q_{f\max} / \{(\pi/2)\gamma_b\}}{(N_q K_q - 1) + \{(B_s / 2) / D_f\} N_\gamma K_\gamma}$$ $q_{f\max} = s_{ucr}\, N_{cclay}\, K_{cclay}$ $K_{cclay} = [1 + (B'/L') / (2 + \pi)]$ $\quad\quad \cdot [1 + \{2 / (2+\pi)\} \arctan(D_f/B')]$	B_s = 0.1 ft L_s = 2 {(15 ft) + (15 ft) − 2 (0.1 ft)} = 59.6 ft K_{cclay} = [1 + (0.1 ft / 59.6 ft) / 5.14] $\quad\quad \cdot$ [1 + {2 / 5.14} arctan(1.5 ft / 0.1 ft)] = 1.586 $q_{f\max}$ = (3,560 psf) (5.14) (1.586) = 29,030 psf s_q = 1 + (0.1 ft / 59.6 ft) tan (35°) = 1.001 $d_q = \left[1 + 2\left(1 - \sin 35°\right)^2 \arctan\left(\dfrac{1.5\,ft}{0.1\,ft}\right) \tan 35° \right] = 1.383$ K_q = (1)(1.001)(1.383)(1)(1) = 1.385 s_γ = 1 − 0.4(0.1 ft / 59.6 ft) = 0.999 K_γ = (1)(0.999)(1)(1)(1) = 0.999 D_t = [(29,030 psf) / {(1.57)(60 pcf)}] $\quad\quad$ /[{(33.30) (1.385) − 1} $\quad\quad\quad$ +{(0.1 ft / 2) / (1.5 ft)}(61.47)(0.999)] $\quad\quad$ = 6.53 ft f_z = arctan(1.5 ft / 6.53 ft) / (1.5 ft / 6.53 ft) $\quad\quad$ = 0.983 Q_u = (0.1 ft) (59.6 ft) (60 pcf) $\quad\quad \cdot$ [(1.5 ft) {1 + ((33.30) (1.385) − 1) (0.983)} $\quad\quad\quad$ + (0.1 ft / 2) (61.47) (0.999) (0.983)] $\quad\quad$ + 2(59.6ft)(1.5ft)(60pcf)(tan35º)(1.5ft)/2 Q_u = 31,030 lb YES, $Q_u < W_{bf}$

Problem 4.4-2

ANALYTICAL PROCEDURES	COMPUTATIONS

SUMMARY

The trial foundation has been checked against sliding, bearing capacity failure, and overturning. A check also has been made to ensure that the 74,900-lb buoyant foundation weight, without the line load, is sufficient to cause full skirt embedment. Consolidation settlements were not calculated, as they would be negligible for a cohesionless soil.

The soil-related aspects of the foundation design process have resulted in a concrete foundation designed with the following dimensions:

Side .... 15.0 ft
Height .... 3.87 ft
Skirt Depth .... 1.5 ft
Buoyant Weight .... 74,900 lb

It is noted that the design is most critical with respect to downslope sliding, which was used to figure the required minimum buoyant weight.

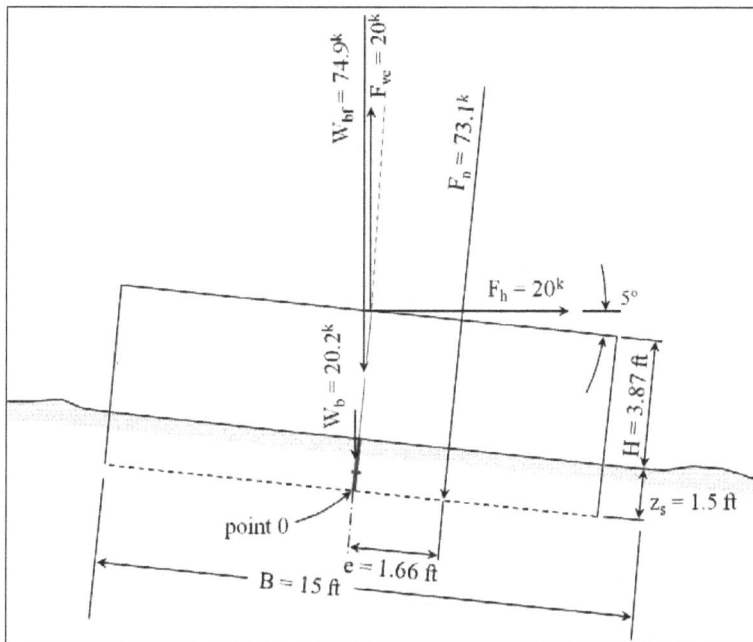

Figure 4.4-3. Forces considered in the overturning analysis for example Problem 2.

4.5 REFERENCES

4-1.   H.Y. Fang (ed). Foundation Engineering Handbook, 2nd edition. New York, NY, Van Nostrand Reinhold, 1991.

4-2.   G. G. Meyerhof. "The Bearing Capacity of Foundations Under Eccentric and Inclined Loads," in Proceedings of the 3rd International Conference on Soil Mechanics and Foundation Engineering, Zurich, Vol. I, pp. 440-445, 1953.

4-3.   Recommended Practice for Planning, Designing, and Constructing Fixed Offshore Platforms – Working Stress Design, American Petroleum Institute, API RP 2A-WSD. Washington, DC, Oct 2007.

4-4.   A.G. Young et al. "Geotechnical Considerations in Foundation Design of Offshore Gravity Structures," in Proceedings of the 1975 Offshore Technology Conference (OTC 2371), Houston, TX, 1975.

4-5.   P.J. Valent et al. OTEC Single Anchor Holding Capacities in Typical Deep Sea Sediments, Civil Engineering Laboratory, Technical Note N-1463. Port Hueneme, CA, Dec 1976.

4-6.   T.W. Lambe and R.W. Whitman. Soil Mechanics. New York, NY, John Wiley, 1969.

4.6 SYMBOLS

A	Foundation or anchor base area [L^2]
A'	Effective foundation or anchor base area depending on load eccentricity [L^2]
B	Minimum foundation/anchor base dimension (usually called foundation width) [L]
B'	Effective base width depending on eccentricity [L]
b_α, b_q, b_γ	Correction factor for inclination of foundation or anchor base
C_c, C_{ci}	Compression index
$\bar{c}$	Effective cohesion intercept as determined by Mohr-Coulomb envelope [F/L^2]
D_f	Depth of embedment of foundation or anchor (depth of foundation base or skirt tip below the seafloor) [L]
D_r	fractional relative density
D_t	transition characteristic depth related to onset of grain crushing behavior [L]
d_α, d_q, d_γ	Correction factor for depth of base embedment
E	Young's modulus of soil [F/L^2]
e, e_1, e_2	Eccentricity [L]
e_o, e_{oi}	Initial void ratio
F_h	Applied horizontal load [F]
F_{hp}	Resultant of applied loads in the downslope direction [F]
F_n	Normal bearing load [F]
F_s	Factor of safety
F_v	Vertical load component (downward is positive) [F]
F_{ve}	Design environmental loading in vertical direction (upward is positive) [F]
f_z	attenuation factor for bearing capacity stress due to friction at depth
G	Elastic shear modulus of soil [F/L^2]
G_s	Specific gravity of the soil grains
g_α, g_q, g_γ	Correction factor for inclination of ground surface
H	Height or thickness of the foundation block or anchor [L]
H'	Vertical distance from point of application of F, to the point of rotation (assumed to be at the shear key tip, or foundation base if no shear keys) [L]
H_i	Thickness of individual cohesive soil layer [L]
H_s	side soil contact height = (base block plus shear key) [L]
i_α, i_q, i_γ	Correction factors for inclination of resultant load
K_α, K_q, K_γ	Bearing capacity correction factors

K_p	Coefficient of passive earth pressure
K_{qd}	the product of K_q with depth, at great depth
L	Length of foundation or anchor base [L]
L'	Effective length of foundation or anchor base [L]
M, M_o, M_1, M_2	Overturning moment [FL]
M_s	(Stabilizing) resisting moment [FL]
m	Exponential term used in Equations 4-19 and 4-20
$N_c, N_q, N_\gamma, N_\phi$	Bearing capacity factors
n	Number of shear keys oriented in one direction, or number of incremental layers in a consolidation settlement calculation
o	Assumed point of rotation
P	base perimeter = $2B + 2L$ [L]
PI	plasticity index = Liquid Limit minus Plastic Limit [% of dry weight]
$\bar{p}_{oi}$	Initial effective overburden stress at midpoint of ith soil layer [F/L^2]
$\bar{p}_v$	Effective vertical stress [F/L^2]
Δp	Applied uniform vertical stress from foundation at base level
$\Delta\bar{p}$	Increase in vertical effective stress at depth z (Figure 4.3-6)
$\Delta\bar{p}_i$	Added effective vertical stress at midpoint of ith soil layer [F/L^2]
Q_e	Embedment force necessary to fully penetrate shear keys [F]
Q_u	Ultimate bearing load resistance (bearing capacity) [F]
Q_{ul}	Ultimate lateral load resistance of the foundation or deadweight [F]
q_c	bearing capacity stress for cohesion [F/L^2]
q_{fmax}	frictional portion of the bearing capacity stress evaluated at great depth [F/L^2]
q_q	bearing capacity stress for overburden [F/L^2]
q_{qf}	bearing capacity stress for overburden, net after subtracting soil buoyancy [F/L^2]
q_γ	bearing capacity stress for friction [F/L^2]
R	Radius of circular foundation base [L]
R_p	Passive soil resistance on leading edge of base & shear key skirt [F]
R_s	Normal soil reaction
S_t	Sensitivity of cohesive soil
s_c, s_q, s_γ	Correction factors for shape of base
s_u	Undrained shear strength of cohesive soil [F/L^2]
s_{ua}	Undrained shear strength averaged over side soil contact zone (to tip of key), or as specified in specific analyses [F/L^2]

s_{ucr}	effective shear strength at critical confining pressure [F/L^2]
s_{uz}	Undrained shear strength at shear key tip (or at base, if no shear keys), or as specified in specific analyses [F/L^2]
V_{conc}	Volume of concrete in foundation block [L^3]
V_f	Volume of foundation block [L^3]
V_{steel}	Volume of steel in foundation block [L^3]
v_{max}	Maximum allowable lowering velocity max during installation [L/T]
w	Water content
w_L	liquid limit: water content for specified slump test behavior [% of dry weight]
w_P	plastic limit: min water content for specified ductility test behavior [% of dry weight]
W_b	Effective or weight of the soil trapped within the footing shear key skirt [F]
W_{bf}	Effective or buoyant weight of the foundation or deadweight anchor [F]
W_{bst}	Effective or buoyant weight of structure supported by a foundation [F]
W_{bv}	Submerged weight of installation being lowered
z_{avg}	Depth averaged over side contact zone to tip of shear key [L]
z_s	Depth of shear key tip below the foundation base (height of footing shear key skirt) [L]
β	Slope of seafloor [deg]
γ_b	Buoyant unit weight of soil [F/L^3]
γ_{bc}	Buoyant unit weight of concrete [F/L^3]
γ_{bf}	Buoyant unit weight of foundation [F/L^3]
δ	Effective friction angle between the soil and the side of the foundation
δ_c	Total consolidation settlement [L]
δ_i	Initial settlement [L]
θ	Angle between the line of action of F. and the long axis of the foundation
μ	Coefficient of friction between soil and foundation or between soil and soil when shear keys cause this type of sliding failure
ν	Poisson's ratio of soil
σ_{cr}	critical confining pressure [F/L^2]
ϕ	Soil friction angle [deg]
$\bar{\phi}$	Effective friction angle for drained analysis [deg]
ϕ_u	Undrained friction angle of cohesionless soil [deg]

[This page intentionally left blank]

5   PILE FOUNDATIONS AND ANCHORS

5.1   INTRODUCTION

Piles are deep foundation elements installed by driving or by drilling-and-grouting. High installation costs dictate that piles be used only where surface foundation or anchor elements, such as shallow foundations, deadweight anchors, or drag or plate anchors, cannot supply the support required. Use of piles as foundations or anchors becomes less frequent as water depth increases due to associated rapidly increasing installation costs. This chapter describes pile types, designs, installation considerations, and presents a simplified design procedure.

5.2   PILE DESCRIPTIONS

5.2.1   Pile Types

Steel is the most common pile material used offshore. Pre-stressed concrete and wood are used in nearshore and harbor applications but are rarely used in deeper water because of construction handling, and splicing difficulties with long piles. For use as foundations and anchors in deep water, pipe sections and H-piles are the most commonly used pile types. A number of specially designed piles are also in use as anchors (Ref. 5-1). These special piles are designed to increase lateral or uplift capacity of the anchorage. Table 5.2-1 lists features of the more common pipe and H-piles and of several types of specialty piles.

5.2.2   Mooring Line Connections

Table 5.2-2 shows four common types of mooring line connections. The selection of a type of mooring line connection should be based on the relative importance of the factors listed and the cost. Connection points, especially, must be sturdy enough to withstand installation stresses as well as service loads.

5.2.3   Modifications for Increasing Lateral Load Capacity

The lateral load capacity of a pile anchor can be increased in several ways (as shown in Table 5.2-3): by lowering the attachment point along the pile length, by lowering the pile head beneath the soil surface into stronger soils, or by attaching fins or shear collars near the pile head to increase the lateral bearing area. The expected increase in pile capacity must be weighed against increases in cost for fabrication and for the higher complexity of the installation procedure.

Table 5.2-1. Pile Types

Characteristics	Pipe and H-Piles	Umbrella Piles[a] or Plate Anchors	Chain-in-Hole[a]	Rock Bolts[a]
Applications	Foundations and Anchors	Anchors	Anchors	Anchors
Approximate Maximum Capacity	Axial: 20,000 kips Lateral: 1,500 kips	300 kips in sand 100 kips in mud	550 kips	260 kips
Installation Methods	Driven or drilled and grouted	Driven	Drilled and grouted	Drilled and grouted, or mechanically wedged
Applicable Soil Type	Soil and rock	Soils without boulders and other obstructions	Rock, with overlaying soil strata	Rock
Advantages	Easy to splice; high capacity; can penetrate through light obstructions	High capacity in uplift	High capacity	Very low cost, no heavy mechanical equipment necessary
Disadvantages	High cost; vulnerable to corrosion	Maximum depth limited by hammer; soils must be homogeneous; inspection of connection not possible	Inspection of connection not possible	Rock must be competent, non-fractured (shallow water only), low capacity
Remarks	Pipe piles resist bending in any direction	Resistance developed similar to plate-embedment anchor (Chapter 6)		Diver hand-installed, much smaller size than normal pipes
Illustration	(anchor pile shown) 	(in-service position) 		(wedged bolt shown)

[a] Special anchor pile

Table 5.2-2. Mooring Line Connections

Type of Connection	End Padeye	Side Padeye	Bridle	Swivel
Advantages	Omni-directional loading. Easily inspected and repaired. Simple construction.	Applicable to H-piles. Simple construction.	Distributes load around pile.	Omni-directional loading. Eliminates torsional stresses in pile.
Disadvantages	Can introduce torsional stresses in pile.	Applicable for uni-directional loadings.	Uni-directional loading. Cannot be inspected and repaired.	Design must protect against fretting corrosion. Complex construction.
Illustration				

Table 5.2-3. Techniques to Improve Pile Lateral Load Capacity

Technique	Lowered Attachment Point	Buried Pile Head	Attaching Fins	Shear Collars With Anchor Plates
Advantages	Lateral load is reduced. Lateral resistance is higher.	Lateral load is reduced. Provides for scour in sand.	Increases lateral resistance. Limits pile head deflection and bending moment.	Omni-directional loading. Eliminates torsional stresses in pile.
Disadvantages	Uni-directional loading. Inspection and repair of connection impratical. Soil in front of pile may be weakened.	Inspection and repair of connection impractical.	More costly fabrication.	Complex install. More costly fabrication. Limited experience with system.
Illustration				

5.3   DESIGN PROCEDURES FOR SIMPLE PILES IN SOIL SEAFLOORS

5.3.1   General

Pile foundations or anchors may be subjected to one or more of the following loads:

- Axial downward loads (compression)
- Axial uplift loads (tension)
- Lateral loads
- Bending moments

For simple pipe and H-piles, axial forces are resisted by soil friction developed along the pile shaft and by bearing on the pile tip (for downward loads). Lateral forces and moments are resisted primarily by the pile shaft bearing on the near-surface soils. For a foundation pile, design is normally controlled by downward axial and lateral loads. However, significant moment and uplift loads may also be present. For an anchor pile, design is normally controlled by uplift and lateral loads. A significant moment may be present, depending on the point of application of the lateral load. A simplified procedure for the design of uniform cross-section piles embedded in a non-layered seafloor consisting of sand, clay, or calcareous soils is presented in Section 5.3.2. The objective of the procedure is to determine if the pile length, width, and stiffness are capable of resisting applied moments, lateral and axial forces at the seafloor without excessive movement and without exceeding the allowable stresses for the pile material. Design is a trial-and-error procedure, where a pile is selected and then is evaluated for its ability to resist those forces and movements under the existing soil conditions. A check is then made to determine if the allowable pile material stresses are exceeded. In the procedure, the pile is assumed to be a beam on an elastic foundation with an elastic modulus that increases linearly with depth. Further description of this type of analysis is found in Reference 5-2.

The procedure applies to piles loaded at the pile head. (Modifications to this procedure to account for other conditions are discussed in Section 5.3.7.)

In the following text sections, steps for pile design are denoted to the left of where the procedure step is presented. Figure 5.3-1 summarizes the design procedure. Note that in this design a factor of safety is applied by increasing underlined expected loads to the underlined design loads before the pile is designed.

STEP 1 — DETERMINE SOIL PROPERTIES
SAND: ϕ, γ_b, and D_r
CLAY: s_u and γ_b

STEP 2 — DETERMINE LOADS AT SEAFLOOR
ANCHORS: T_a and T_h
FOUNDATIONS: P_h, P_t, and P_s

STEP 3 — IS THIS AN ANCHOR PILE?
YES → CALCULATE DESIGN LOADS AT PILEHEAD T_h' and T_t' — EQUATIONS 5-2 and 5-3
NO ↓

STEP 4 — SELECT PILE: D, t_w and EI

ARE THERE LATERAL LOADS?
NO →
YES ↓

STEP 5 — SELECT MAXIMUM DEFLECTION: y_{max}

STEPS 6 & 7 — DETERMINE: n_h and T — EQUATION 5-4

STEP 8 — SELECT PILE LENGTH L_p

STEP 9 — DETERMINE A_y and B_y

STEP 10 — CALCULATE LATERAL LOAD: P_h (calc) — EQUATION 5-5

STEP 11 — IS P_h (calc) ≥ P_h?
NO → INCREASE D or T_w / INCREASE L_p / INCREASE z_0
YES ↓
IS P_h (calc) >> P_h?
YES → DECREASE D or t_w / DECREASE L_p
NO ↓

STEPS 12 & 13 — CALCULATE UNIT SKIN FRICTION RESISTANCE, f_s — EQUATIONS 5-7 and 5-8

STEP 14 — CALCULATE TOTAL SKIN FRICTION RESISTANCE, Q_s — EQUATION 5-9

IS Q_s ≥ P_t?
NO → INCREASE L_p
YES ↓

IS THERE A COMPRESSIVE LOAD?
NO →
YES ↓

STEP 15 — CALCULATE PILE TIP BEARING CAPACITY, Q_p — EQUATION 5-10

STEP 16 — CALCULATE PILE CAPACITY IN COMPRESSION, Q_n — EQUATION 5-13

IS Q_p ≥ P_s?
NO → INCREASE L_p
YES ↓

STEP 17 — DETERMINE A_m and B_m — EQUATION 5-14

CALCULATE M_{max}

STEP 18 — CALCULATE f_{maxo} and f_{maxi} — EQUATIONS 5-15 and 5-16

STEP 19 — IS f_{maxo} ≤ ALLOWABLE f_{maxi} ≤ ALLOWABLE?
NO → INCREASE S
YES ↓

CHECK P_h, P_t and P_s FOR FINAL PILE DESIGN

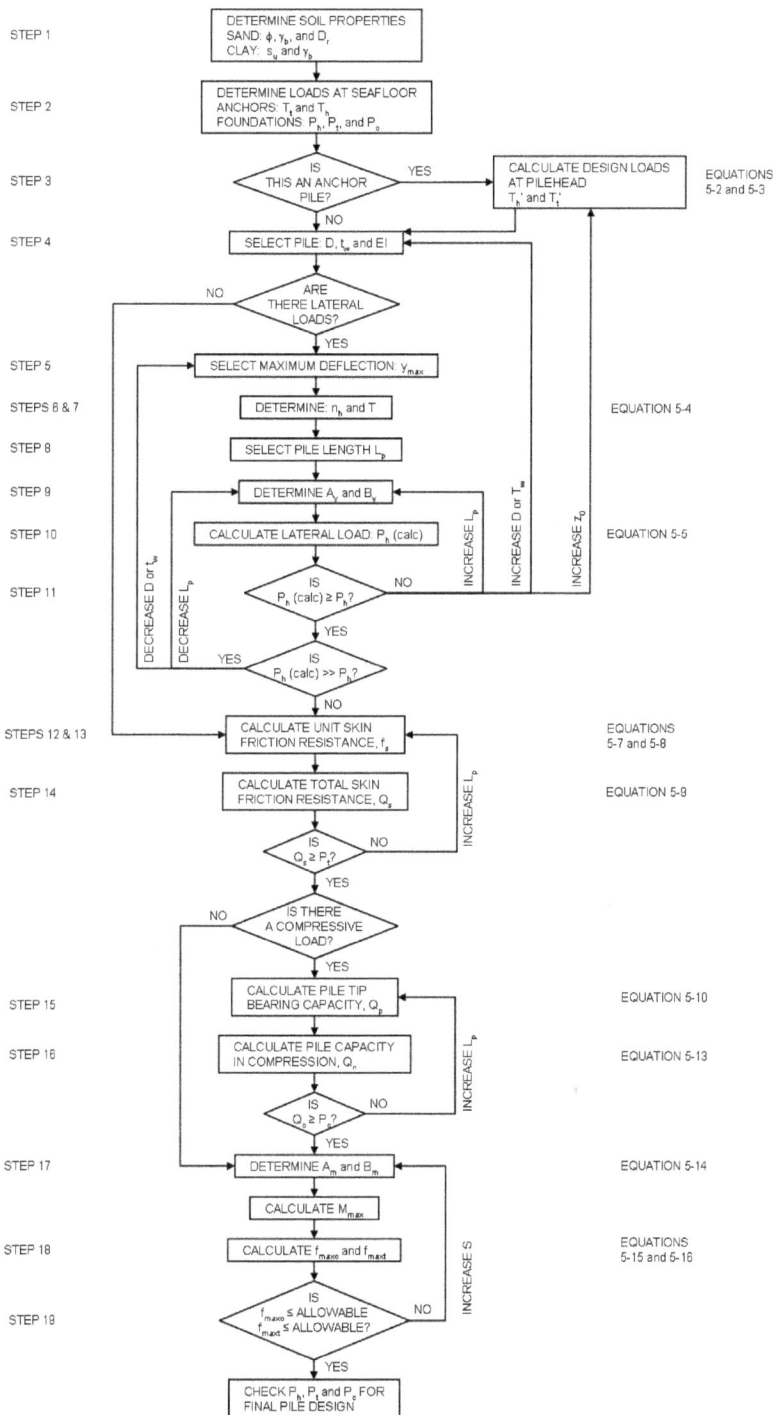

Figure 5.3-1. Flow chart for the pile design procedure.

5.3.2 Soil Properties

STEP 1: Determine required soil engineering properties.

These properties should be site-specific, based on in-situ or laboratory tests, or both. The soil properties required for design are:

For cohesionless soil (sands and silts),

ϕ = drained (effective) friction angle [deg]

γ_b = buoyant unit weight [F/L^3]

D_r = relative density [%]

Cohesive soils (clays),

s_u = undrained shear strength [F/L^2]

γ_b = buoyant unit weight [F/L^3]

Calcareous soils,

Carbonate content, degree of cementation, and degree of crushing (in addition to the properties required for cohesionless soils).

If site-specific soil data are not available, it may be possible to extrapolate from other property data using geologic and geophysical information from similar nearby areas. In the absence of any test-determined data, estimates of soil properties may be made based on the geologic and depositional environment and geophysical data. Chapter 2 provides guidelines for estimating engineering properties and identifies some engineering properties of major sediment types. Additional soil properties important to pile design are given in Table 5.3-1 through Table 5.3-3 for cohesionless and cohesive soils based on soil density or degree of consolidation.

Where soil properties vary significantly with depth, average properties in the uppermost four pile diameters are used for the lateral load analysis (STEPS 4 through 11) and average properties over the pile length are used for the axial load analysis (STEPS 12 through 18).

Table 5.3-1. Properties of Cohesionless Soil Useful in Pile Design

Type	Standard Penetration Blow Count, N	ϕ (deg)	D_r (%)	γ_b (pcf)
Very loose to loose	<10	28-30	0-50	45-55
Medium dense	10-30	30-36	50-70	55-65
Dense	30-50	35-42	70-85	60-70
Very dense	50+	40-45	85-100	60-70

Table 5.3-2. Properties of Cohesive Soils Useful in Pile Design

Type	S_u (psf)	ε_{50} (%)	γ_b (pcf)
Underconsolidated clays (very soft to soft)	50-150	2	20-25
Normally consolidated soils at depth z, ft. (firm)	150 + 10z	2-1	25-50
Overconsolidated soils based on consistency (very firm) (hard) (very hard)	300-600 600-1,500 1,500	1.0 0.7 0.5	50-65

Table 5.3-3. Properties of Calcareous Soil Useful in Pile Design

Characteristic	Property
Carbonate content	low, 0 to 30% medium, 30 to 45% high, 45+
Degree of cementation	uncemented or lightly cemented or well cemented
Degree of crushing	crushes easily resistant to crushing

5.3.3 Pile Design Loads

STEP 2: Determine the maximum load or load combinations at the seafloor surface.

Actual loads are determined from an analysis of what is attached to the pile. A safety factor, F_s, is applied to the actual loads to determine those loads used in the design. If the necessary soils data are accurately known (from in-situ testing or laboratory testing of core samples), a safety factor of 1.5 to 2.0 is recommended. If soil properties are not accurately known, a higher safety factor of 2.0 to 3.0 should be used. Within these recommended ranges, higher values should be used where the installation is more critical and where the soil properties are more questionable.

For anchor piles the design loads are:

T_h = F_s * horizontal component of mooring line tension at the seafloor

T_t = F_s * vertical component of mooring line tension at the seafloor

where:

T_h = design horizontal load at the anchor pile [F]

T_t = design vertical load at the anchor pile [F]

For foundation piles the design loads are:

P_h = F_s * horizontal component of load at the pile

P_t = F_s * vertical uplift load at the pile

P_c = F_s * vertical compression load at the pile

where:

P_h = design horizontal load at the foundation pile [F]

P_t = design vertical uplift load at the foundation pile [F]

P_c = design vertical compression load at the foundation pile [F]

In unusual cases, a moment may be applied to the foundation pile. This actual moment should similarly be multiplied by the factor of safety to yield a design moment.

M_a = F_s * moment applied to the pile head

where:

M_a = design applied moment [L·F]

STEP 3: Calculate load at a submerged anchor pile head.

For anchor piles to be driven below the seafloor surface, the mooring line angle at the pile is not the same as that angle at the seafloor, due to soil bearing resistance against the mooring line. The actual angle at the pile becomes higher and the force exerted on the pile becomes more of an axial load. To account for this, horizontal and vertical design loads determined in STEP 2 must be corrected. The general effect of this correction is to decrease the horizontal load a pile must resist and increase the vertical or axial load. The force corrections include several simplifying conservative assumptions and are made as follows.

The correction to the horizontal force is the soil force that is exerted on the mooring line in a horizontal direction:

For cohesionless soils,

$$F_{cb} = z_c^2 d_b \gamma_b \overline{N}_q \qquad (5\text{-}1a)$$

For cohesive soil,

$$F_{cb} = (11.0)\, s_u d_b z_c \qquad (5\text{-}1b)$$

where:

F_{cb} = horizontal force exerted on the mooring line by the soil [F]

z_c = depth of pile connection below seafloor [L]

d_b = characteristic mooring line size [L] (for chain, use 3 times the chain size; for wire, use wire diameter)

$\overline{N}_q$ = bearing capacity factor, from Table 5.3-4

The corrected anchor pile design loads are:

$$T_h' = T_h - F_{cb} F_s \qquad (5\text{-}2)$$

$$T_t' = (T_t^2 + 2 T_h F_{cb} F_s - F_{cb}^2 F_s^2)^{1/2} \qquad (5\text{-}3)$$

where T_h' and T_t' are the values for T_h and T_t, corrected for the effects of a pile head being driven to a depth z_c below the seafloor.

Table 5.3-4. Bearing Capacity Factors for Chain Lateral Force in Sand (Ref. 5-3)

Soil Friction Angle, ϕ (deg)	$\overline{N}_q$
20	3
25	5
30	8
35	12
40	22
45	36

5.3.4   Lateral Load Analysis

STEP 4: Select a trial pile size.

The trial-and-error pile selection process begins with a pile selection--usually made on the basis of availability. This determines pile diameter, D; wall thickness, t_w; and stiffness, EI.

STEP 5: Select deflection criteria.

For anchor piles, a suggested maximum lateral deflection criterion is y_{max}/D = 10%, where y_{max} is the maximum allowable lateral deflection. For foundation piles, a more rigid deflection criterion (a lower value of y_{max}/D) is used if justified by structural requirements.

STEP 6: Determine coefficient of subgrade reaction.

The coefficient of subgrade soil reaction, n_h, is determined for the selected value of y_{max}/D. For cohesionless soils, this is obtained from Figure 5.3-2 and D_r, the soil relative density. For cohesive soils, first Figure 5.3-3 and s_u are used to determine a value for the coefficient K_1. Then n_h is determined from the equation on that figure. For calcareous soils, Figure 5.3-2 and a value of D_r = 35% are used.

STEP 7: Determine pile-soil stiffness.   The pile-soil relative stiffness, T, is computed by:

$$T = \left(\frac{EI}{n_h} \right)^{0.2} \qquad (5\text{-}4a)$$

$$I_{pipe} = \pi r^3 t \qquad (5\text{-}4b)$$

where:

r =   pipe radius [L]

t =   pipe wall thickness [L]

STEP 8: Select pile length.

A pile length, L_p, is assumed. (A length of L_p= 3T is suggested as a minimum.)

STEP 9: Determine deflection coefficients.

The maximum value of the depth coefficient z_{max} = L_p /T is computed. Then deflection coefficients A_y and B_y at the ground surface are obtained from Figure 5.3-4.

STEP 10: Calculate lateral load capacity.

The lateral load capacity of the trial pile, $P_h(calc)$, for the value of y_{max} selected in STEP 5 is computed.

$$P_h(calc) = \frac{y_{max}(EI)}{A_y T^3 + aB_y T^2} \tag{5-5}$$

where a is the distance of the pile load attachment point above the seafloor surface for foundation piles [L] ($a \geq 0$, as foundation piles will not be driven below the surface).

STEP 11: Compare pile capacity to required design lateral load capacity.

If $P_h(calc)$ = P_h (or P_h' or T_h) or is slightly higher, then the trial pile is adequate for resisting lateral loads.

If $P_h(calc)$ >> P_h (or P_h' or T_h) then the pile is overdesigned. One or more of the following changes are made and computations are repeated on the new trial pile size.

   (a)   Reduce pile stiffness by reducing diameter or thickness (repeat calculations from STEP 4).

   (b)   Reduce pile length, unless already very short (repeat calculations from STEP 9).

If $P_h(calc)$ << P_h (or P_h' or T_h) then the pile is underdesigned. One or more of the following changes are made and computations are repeated on the new trial pile size. Note: this is done until $P_h(calc)$ is only slightly greater than P_h.

   (a)   Increase pile stiffness by increasing diameter and/or thickness (repeat calculations from STEP 4).

   (b)   Increase pile length, unless already very long (repeat calculations from STEP 9).

   (c)   Increase design depth of pile head (repeat calculations from STEP 3).

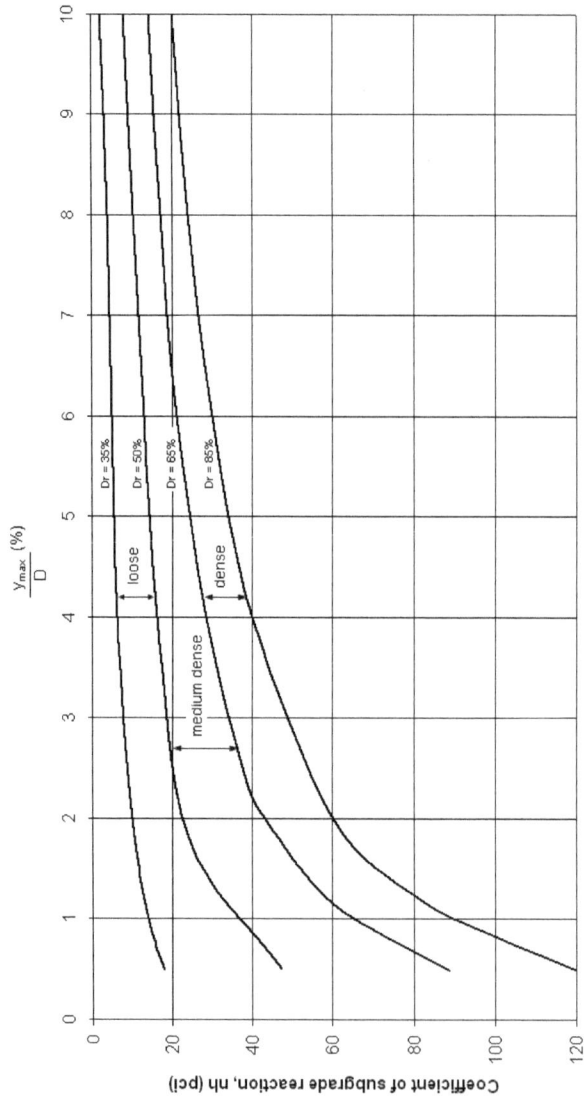

Figure 5.3-2. Design values for n_h for cohesionless soils (Ref. 5-4).

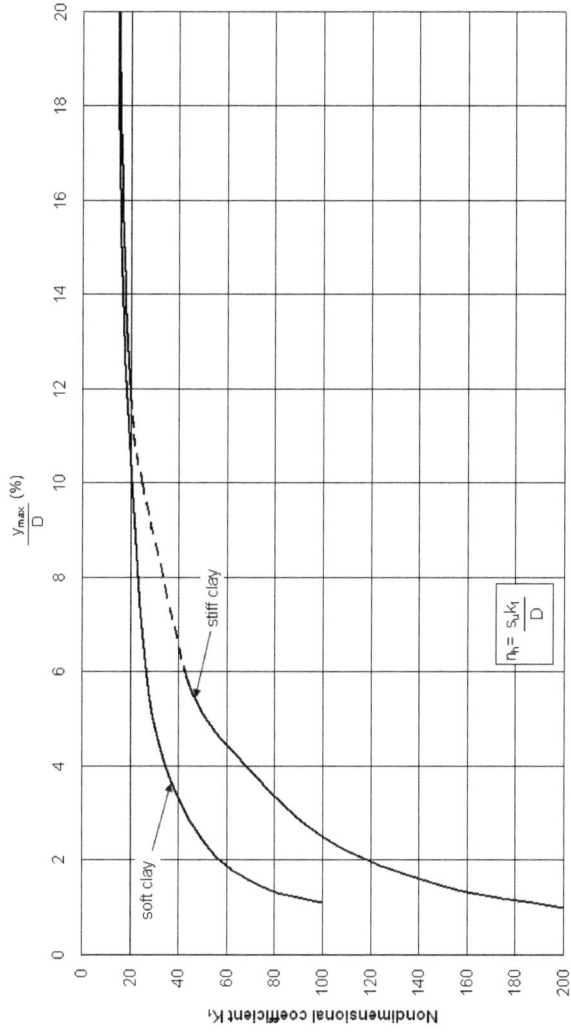

Figure 5.3-3. Design values for n_h for cohesive soils.

5-13

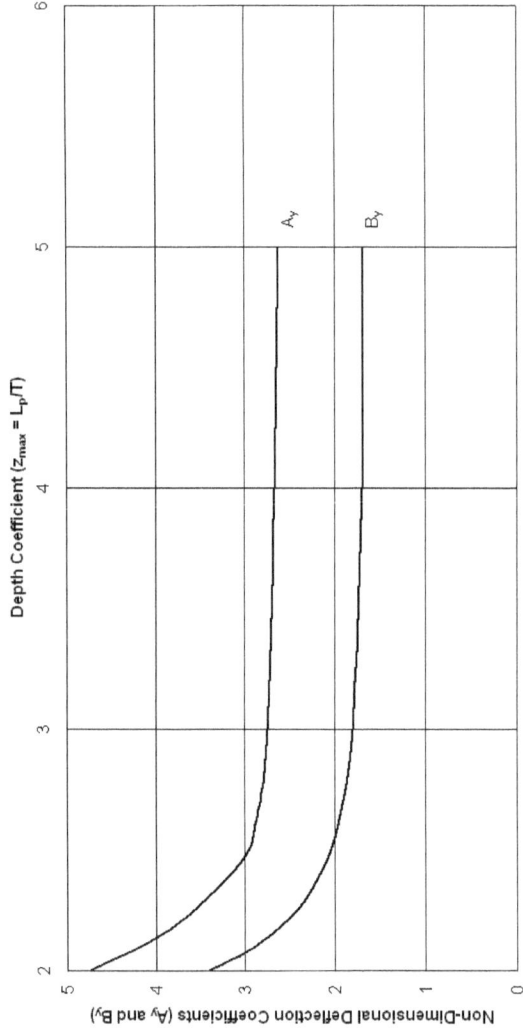

Figure 5.3-4. Deflection coefficients A_y and B_y at the ground surface.

5-14

5.3.5   Axial Load Analysis

STEP 12: Calculate average effective overburden pressure.

The average effective overburden soil pressure $\left(\overline{p}_{vo}\right)$ at the pile midpoint is computed:

$$\overline{p}_{vo} = \frac{\gamma_b L_p}{2} \qquad (5\text{-}6)$$

If the pile is not fully buried, find $\left(\overline{p}_{vo}\right)$ at the midpoint of the buried length. If the density changes, average the densities along the pile.

STEP 13: Calculate skin frictional resistance per unit length of pile.

For cohesionless soils, the average unit skin frictional resistance, f_s (in uplift and compression), is calculated from:

$$f_s = k\,\overline{p}_{vo}\tan(\phi - 5\deg) \qquad (5\text{-}7)$$

where k is 0.7 for compression and 0.5 for uplift.

Table 5.3-5 presents limiting or maximum values for f_s. It must be checked to ensure that calculated values of f_s do not exceed those limiting values. For piles driven into calcareous soils, the table's limiting values shown should be used unless higher values are justified by on-site testing.

For cohesive soils, first determine if the soil is normally consolidated or overconsolidated. To do this, an average value for $s_u/\overline{p}_{vo}$ over the pile length is computed. If $s_u/\overline{p}_{vo} \le 0.4$, the soil is considered normally consolidated (NC); if $s_u/\overline{p}_{vo} > 0.4$, the soil is considered overconsolidated (OC).

For NC soils, the average unit skin friction resistance, f_s, is equal to:

$$f_s = \overline{p}_{vo}[0.468 - 0.052\ln(L_p/2.0)] \qquad (5\text{-}8a)$$

where L_p is in feet. If Equation 5-8a exceeds s_u then:

$$f_s = s_u$$

For OC soils:

$$f_s = [0.458 - 0.155 \ln(s_u / \overline{p}_{vo})] s_u \qquad\qquad (5\text{-}8b)$$

If $s_u / \overline{p}_{vo}$ exceeds 2.0 then:

$$f_s = 0.351 s_u$$

Table 5.3-5. Recommended Limiting Values for Unit Skin Friction
and End Bearing for Cohesionless Soils

Soil Type	ϕ (deg)	δ (deg)	N_q	f_s (max) (ksf)	q_p (max) (ksf)
Noncalcareous Soils					
Sand	35	30	40	2.0	200
Silty Sand	30	25	20	1.7	100
Sandy Silt	25	20	12	1.4	60
Silt	20	15	8	1.0	40
Calcareous Soils					
Uncemented calcareous sand (easily crushed)	30	20	20	0.3[a]	60

Partially cemented calcareous sands with carbonate content:
  0 to 30%
  30 to 45%
  Above 45%
	--
	--
	--
		--
		--
		--
			--
			--
			--
				2.0
				0.64[a]
				0.56[a]
					100
					160
					140

Highly cemented calcareous soils such as chalk	--	--	--	1.1	140

[a] For drilled and grouted piles, the value may approach 2.0 ksf – the value for quartz sand; actual value depends upon installation technique.

STEP 14: Compute uplift capacity and compare to design load.

For both anchor and foundation piles, the frictional resistance of the pile, Q_s, must exceed the design load for uplift, P_t. Pile frictional resistance is:

$$Q_s = A_s f_s \qquad (5\text{-}9)$$

where A_s is the surface area of the pile below the seafloor [L²].

If $Q_s \geq P_t$, then the design is adequate for resisting uplift forces.

If $Q_s < P_t$, then the trial pile length is increased. Although increasing the pile diameter is another approach will increase Q_s, increasing pile length is usually preferred.

STEP 15: Compute soil bearing capacity for foundation pile tip.

For foundation piles, resistance to compressive loading comes from frictional resistance along the pile and from resistance to tip or end penetration. For closed-ended piles, the soil bearing capacity for the pile tip, Q_p, is computed as follows:

$$Q_p = A_p q_p \qquad (5\text{-}10)$$

where:

A_p = gross end area of the closed pile [L²]

q_p = unit soil bearing capacity of that pile tip [F/L²]

For cohesionless soils,

$$q_p = \overline{p}_{vo}(tip) N_q \qquad (5\text{-}11)$$

For cohesive soils,

$$q_p = 9 s_u (tip) \qquad (5\text{-}12)$$

where:

$\overline{p}_{vo}(tip)$ = effective vertical stress at pile tip [F/L²]

N_q = bearing capacity factor from Table 5.3-5

$s_u(tip)$ = soil undrained shear strength at pile tip [F/L²]

5-17

Piles that are not closed-ended will develop a soil plug when installed. (When this soil plug is removed and replaced with concrete, the pile is considered closed-ended.) The soil plug will limit the value of Q_p that can develop to the force required to push a soil plug up into a thin-walled pipe. This limiting value is approximately equal to the frictional capacity of the pile, previously computed as Q_p. Therefore, for open-ended piles, Q_p is equal to Q_s.

STEP 16: Compare pile capacity in compression with required capacity.

The pile capacity in compression, Q_c, is computed, and this is compared with the design load under vertical compression, P_c.

$$Q_c = Q_s + Q_p \qquad (5\text{-}13)$$

If $Q_c \geq P_c$, the pile is adequate in compression.

If $Q_c < P_c$, then the pile is not adequate in compression. The pile length must be increased (repeat the calculations from STEP 14).

If using an open-ended pile and Q_p was significantly limited by the value of Q_s, it may be of significant benefit to plug the pile end. Recompute Q_c for a closed-ended trial pile and again check for adequacy in compression.

5.3.6   Steel Stress Analysis

STEP 17: Calculate maximum compression.

The maximum moment (M_{max}) in the pile is determined by max combining any applied (design) bending moments, M_a, and moments created by the design horizontal load. The latter is calculated by recomputing T (STEP 7) and L_p/T (STEP 9) for the current trial pile; selecting the influence coefficients A_m and B_m from the appropriate curves on Figure 5.3-5; then computing the maximum moment, M_{max}, at a point along the pile by:

$$M_{max} = A_m (P_h T) + B_m (M_a) \qquad (5\text{-}14)$$

It may be necessary to determine M_{max} at several locations (z) along the pile in order to find the maximum value for M_{max}. This is done by entering Figure 5.3-5 with the several values of z/T.

STEP 18: Calculate maximum steel stress.

Maximum stress in the pile under tension (f_{maxt}) and (f_{maxc}) is calculated by:

$$f_{\max t} = -P_t / A_{ps} - M_{\max} / S \qquad\qquad (5\text{-}15)$$

$$f_{\max c} = P_c / A_{ps} + M_{\max} / S \qquad\qquad (5\text{-}16)$$

where:

A_{ps} = cross-sectional area of the pile [L^2]

S = section modulus of the pile [L^3]

<u>Note</u>: The terms A_{ps} and S and allowable maximum stress in tension and compression used in STEP 19 are available from steel design manuals or manufacturers literature containing data on these pile shapes.

STEP 19: Compare calculated and allowable steel stress.

The values for f_{maxt} and f_{maxc} are compared with the allowable steel stress in tension and compression for the pile being used. For most common structural shapes f_a, the allowable maximum stress in tension and in bending (tension and compression), is at least $0.6F_y$, where F_y is the minimum yield point of the steel being used.

If f_{maxt} and f_{maxc} ≤ allowable, then the pile is adequate.

If f_{maxt} or f_{maxc} > allowable, then the pile is inadequate. A new trial pile size with a larger section modulus (larger t_w or D) is selected (repeat the calculations from STEP 18).

It is also possible to reinforce the pile over that length where high moments exist. This will result in a larger section modulus and may be done rather than using a larger pile. While this is often done because it is cost-effective, it is beyond the scope of this handbook.

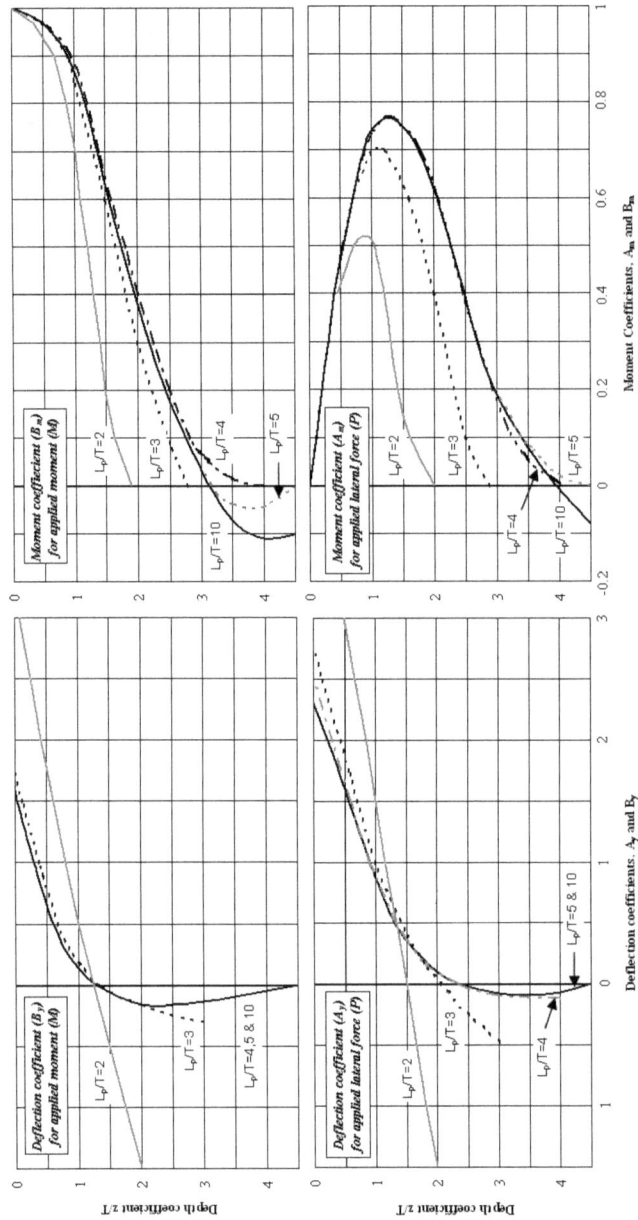

Figure 5.3-5. Influence values for a pile with applied lateral load or moment (Ref. 5-2).

5.3.7 Special Cases

5.3.7.1 Load Applied Below the Anchor Pile Head

When the mooring line is connected to the side of an anchor pile at a distance of more than five pile diameters from the head, the simplified approach presented in Section 5.3.4 for lateral load is not adequate. For the anchor pile with such a buried side connection, a lateral load analysis must be made by other means, such as with the assistance of a computer program (Ref. 5-5). For the same lateral load, pile deflection and bending moments are reduced significantly as the mooring line connection point is lowered from the pile top to a point midway down the pile. The mooring line will also approach the pile at a decreasing angle as the connection point is lowered, thus greatly lowering the lateral force and raising the uplift force.

Where the mooring line chain is almost parallel to the pile, the presence of chain alongside the pile and the influence of repeated mooring loads may also reduce the soil strength and, therefore, reduce soil resistance above the connection point.

5.3.7.2 Piles with Enlarged Cross Section

Pile size may be increased near the seafloor surface to increase resistance to lateral loads and bending moments. Analysis of the response of piles with variable cross section to lateral loads is much more complex and is usually done with the aid of computer programs (Refs. 5-6 and 5-7). In general, a large increase in pile diameter over a small depth is more efficient in reducing deflections at the seafloor than is a small increase in diameter over a larger depth. When enlarged pile sections at the pile head are used, the length of the enlarged section should be limited to three times the larger diameter.

5.3.7.3 Special Seafloor Conditions

A number of seafloor conditions can be considered hazardous to the placement or functioning of piles in the offshore environment. Some of these are listed below.

Steeply sloping seafloors and accompanying down-slope soil movements can cause additional lateral pile forces in the downslope direction. Also, steep sloping makes the area more subject to instability problems such as might be initiated by seismic activity of wave forces.

The presence of rocks, cobbles, or cemented zones can make installation of driven piles difficult. Drilled and grouted piles may be the best method of installation in these areas.

Scour of near-surface soils can occur and can be accentuated in the vicinity of piles. Scour extent depends on the velocity of seafloor currents, the type of soil, and the size and configuration of pile groups. Generally, granular soils are more susceptible to scour than cohesive soils. For granular soils, the upper 5 to 10 feet may be subject to scour. Removal of this surface material can significantly affect pile behavior, and the possibility of this occurring should be taken into account during pile design. Scour is discussed in more detail in Chapter 10.

Earthquake-related hazards should be assessed in seismically active areas. Earthquake motions may cause partial loss of strength or complete liquefaction in loose granular soil zones, essentially removing soil support developed in these zones. Additionally, a weakened zone may slide downslope and overload the pile due to high lateral force applied by the moving soils. For important structures in areas where such problems can be anticipated, the design should be based on a thorough evaluation of such hazards and their influence on the structure.

5.4 DESIGN OF PILE ANCHORS IN ROCK SEAFLOORS

An approach to pile anchor design in rock is presented in this section. A detailed procedure for pile anchor design in rock is not presented because this cannot be done simply, primarily because of the difficulty in characterizing the material failure mode.

Rock failure modes are illustrated in Figure 5.4-1. Failure of a pile anchor in rock may occur in lateral bearing due to rock crushing under lateral loads. It may also fail in uplift because of a failure of grout-to-rock bonding or because of a rock-mass failure in fractured material. Uplift failure can also occur due to the pile's loosening and the loss of resistance as a result of repeated lateral loads.

The mode of failure is difficult to establish or predict for a specific location. The strength of a cored sample may be misleading when applied to the prediction of pile anchor holding capacity in jointed, bedded, faulted, or weathered rock masses. Gross rock characteristics will likely govern pile behavior, and these gross characteristics must be thoroughly evaluated at each pile location.

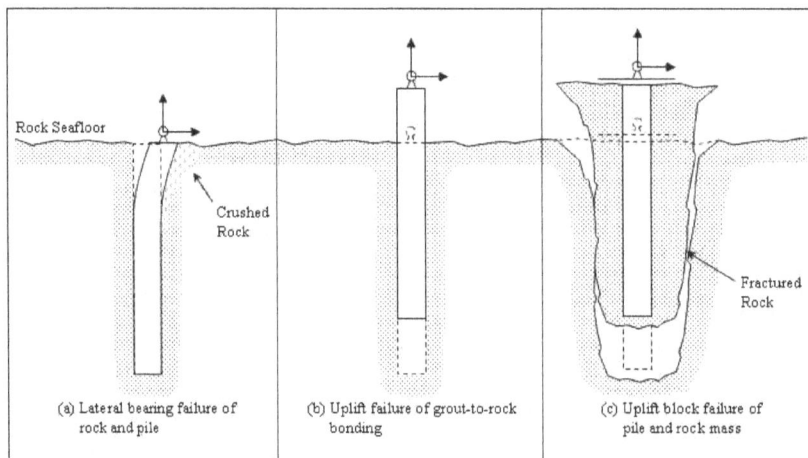

Figure 5.4-1. Failure modes for pile anchors in a rock seafloor.

5.4.1 Lateral Capacity

In a rock or hard cemented soil seafloor, a soil cover may be present above the rock, or the rock/cemented zone may be underlain by soil. For layered soil-rock sites, available computer programs should be used (Refs. 5-6 and 5-7) to account for the complexities introduced by these nonuniform conditions.

5.4.1.1 Soil Overlying Rock

For soil overlying rock, pile capacity can be roughly estimated by the following procedure. The relative depth to the rock, z_s/T, is determined, where z_s is the thickness of the soil over the rock and T is the pile relative stiffness. If $z_s/T \geq 3.0$, the pile can likely be designed to develop all support from the soil layer. The pile design should be made with the soil analysis procedures presented in Section 5.3.

If $z_s/T < 3.0$, then the pile can be conservatively designed as if the soil were not present. That is, the pile is considered cantilevered out of the rock surface and resists all forces without assistance from the soil layer. If this is done, a check must be made to determine if the compressive strength of the rock, s_c, is exceeded by stresses from the applied lateral load. That is, a check should be made to ensure that:

$$\frac{P_h}{DL_e} < s_c \tag{5-17}$$

where:

P_h = lateral force applied to the pile head [F]

D = width or diameter of the pile [L]

L_e = effective length of pile bearing on the rock strata equal to the smaller of: (1) the rock layer thickness, z_r, or (2) a depth interval equal to the pile diameter, D

s_c = rock compressive strength (see Table 5.4-1 for typical values of s_c) [F/L^2]

5.4.1.2 Rock Layer Overlying Soil

For the case of rock or other hard layer overlying soil, if the rock layer thickness, z_r, is less than $0.2T$, where $T= (EI/n_h)^{0.2}$ and n_h is the coefficient of soil reaction of the underlying soil, the influence of the rock layer may be ignored and the pile designed by using the procedure of Section 5.3.2. For rock layer thickness, z_r, greater than $0.5T$, the influence of underlying soil may be ignored and the pile designed for rock compressive strength as under Section 5.4.1.1. For intermediate values of z_r, the designer must judge whether to use the soil or rock procedure; the decision depends upon pile diameter, knowledge of rock layer strength and fracturing, and the layer thickness.

Table 5.4-1. Rock Properties (Ref. 5-8)

Rock Type	Compressive Strength, s_c (ksi)	Buoyant Saturated Unit Weight, γ_b (pcf)
Dolerite	28.4 – 49.8	123
Gabbro	25.6 – 42.7	123
Gneiss	7.1 – 28.4	117
Basalt	21.3 – 42.7	111
Quartzite	21.3 – 42.7	101
Granite	14.2 – 35.6	98
Marble	14.2 – 35.6	98
Slate	14.2 – 28.4	98
Dolomite	11.4 – 35.6	92
Limestone	4.3 – 35.6	73
Sandstone	2.8 – 24.1	61
Shale	1.4 – 14.2	61
Coal	0.7 – 7.1	5

5.4.2 Uplift Capacity

Failure due to uplift load may occur: (1) at the grout-pile interface, (2) at the grout-rock interface, or (3) along a rock fracture zone outside the grouted area. In (3), a block of rock containing the pile is assumed to be lifted free of the surrounding rock as shown in Figure 5.4-1(c). The following guidelines for design are suggested.

5.4.2.1 Massive, Competent Rock

For massive, competent rock, uplift capacity will be governed by frictional forces developed by the grout bonding strength along the pile, by the grout shear strength itself, or, less likely, by the grout bonding strength along the drilled shaft wall. For failures of these types, the uplift capacity, R_a, of the rock anchor is given by:

$$R_a = s_b L_r C_p \tag{5-18}$$

where:

s_b = the lesser of (1) the grout-to-pile bonding strength, or (2) the shear strength of the grout [F/L^2]

L_r = length of pile embedded in rock [L]

C_p = minimum perimeter transmitting the uplift load [L]

Unless higher bond strengths are verified by testing, the grout-to-steel bonding strength should be limited to 27 psi. The grout-to-rock bond strength may vary from 0.3 to 1.0 times the rock shear strength, depending on cleanliness of the drilled hole, type of rock, and grouting procedure.

5.4.2.2 Fractured Rock

For fractured rock, anchor uplift capacity is determined by the mass of the blocks of rock which move with the anchor and by the frictional force developed between the attached blocks and adjacent blocks. Because of the difficulty in estimating the normal forces acting on vertical joints and cracks, this frictional force is normally ignored.

5.5 PILE INSTALLATION

Piles are installed in the offshore environment by one or a combination of the following methods:

- Driving
- Drilling and grouting
- Jacking
- Jetting

5.5.1 Driven Piles

Piles may be driven by impact hammers operated above the water surface, by underwater impact hammers, or by vibratory hammers.

5.5.1.1 Conventional Hammers

Piles for piers, harbor structures, bridges, and many offshore structures in shallow water are driven from above the water surface with conventional hammers used in pile driving on land. The pile is made long enough to extend above the water surface when driven to its design penetration depth or, alternatively, a pile follower is used (Figure 5.5-1). The piles are commonly guided by a template that rests on the seafloor, although floating templates may be used for small, shallow-water installations. The pile-driving operation is conducted from a carefully moored work barge that supports the necessary cranes and auxiliary equipment.

The pile hammers used for this construction are scaled-up versions of land hammers. The types include single-acting steam, compressed air, diesel, and hydraulic hammers. The rated energy of these hammers varies from less than 100,000 ft-lb per blow to over 1,500,000 ft-lb per blow. Surface-operated pile drivers have been used in water depths in excess of 1,000 feet.

The success of the surface-driven method of pile emplacement in deep water is dependent upon the presence of the template to act as a guide for the piles. Without the restraint offered by the template, most of the driving energy would be dissipated by lateral deflection of the pile. For anchor piles driven from the surface without lateral restraint, a reasonable maximum water depth appears to be about 250 feet.

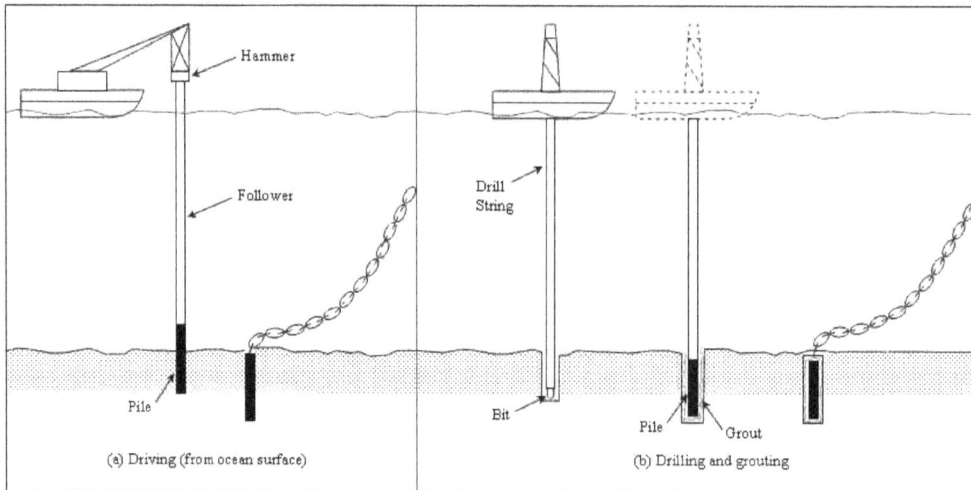

Figure 5.5-1. Pile installation techniques.

5.5.1.2 Underwater Hammers

Several standard terrestrial pile hammers may be modified for operation underwater. One manufacturer makes a total of 12 steam/compressed air hammers, with rated energies in air of 8,750 ft-lb to 60,000 ft-lb. These may be operated while submerged with little loss of efficiency. The modifications consist primarily of providing exhaust hoses that extend to the water surface. Because steam cools too much when the hoses go underwater, compressed air is usually used to operate the hammers.

A number of large hammers have been developed over the last 10 years that are capable of both above and underwater operation. Some of these hammers have rated striking energies of up to 1,700,000 ft-lb.

5.5.1.3 Underwater Vibrators

Vibratory pile drivers are becoming more common in American practice as experience is gained with their use and as more powerful machines are developed. The machines usually use counter-rotating eccentric weights powered by electric or hydraulic motors to produce the vibratory forces. The major depth-limiting factors on present systems are the difficulty in handling long lengths of large-diameter, high-pressure hydraulic lines and the large friction losses in the line. These factors limit the maximum practical depth of a surface-powered, hydraulic vibratory drive to about 1,000 feet.

5.5.1.4 Selection of Hammers

When selecting a hammer for "lighter" offshore tasks, where required hammer energies are less than 150,000 ft-lb, a general rule of thumb for hammer selection is:

- For steam/air hammers, the weight of the pile should be no more than two times the weight of the hammer ram.

- For diesel hammers, the weight of the pile should be no more than four times the weight of the ram.

- For vibratory drivers, the driving amplitude of the driver/pile system should be between 0.25 and 0.5 inch.

For "heavier" tasks, where hammer energies in excess of 150,000 ft-lb are required, the maximum rated energy of a steam/air hammer required to drive a steel pile should be estimated by:

$$E_h = 2000 A_{ps} \tag{5-19}$$

where:

E_h = maximum rated energy of the hammer [ft-lb]

A_{ps} = area of steel cross section [in²]

This recommendation is based on the allowable stress in steel of 12,000 psi under working loads.

For preliminary selection of air/steam hammers, it is recommended that Equation 5-19 be used to estimate maximum hammer energy. (For diesel hammers, the maximum energy of the hammer may be 20 to 35% higher than the value given by Equation 5-19.) These recommendations assume that the pile will be driven to the maximum axial capacity. In cases where lateral load governs pile design and full axial capacity of the pile cross-section is not mobilized, hammers with significantly less rated energy than given by Equation 5-19 may be adequate to drive the pile to its design penetration depth. For all important installations, pile driveability must be investigated beyond the level of Equation 5-19 by the use of wave equation analyses using available computer programs (Refs. 5-9 and 5-10) to evaluate the proper combination of pile-hammer-cushion system.

5.5.2   Drilled and Grouted Piles

By use of drilling and grouting procedures, piles up to 8 feet in diameter have been placed in water depths in excess of 600 feet. The method is essentially identical to that used to set a casing for an oil well. A hole of somewhat larger diameter than the pile is drilled to proper depth using rotary drilling tools. The hole is cleaned out by pumping seawater through the drill string, and the pile is placed over the drill string and lowered into the hole. Portland cement grout is pumped down the drill string and forced up outside of the pile to fill the annular void and bond the pile to the soil. Then, the interior of the pile is filled with grout as the drill string is withdrawn.

5.5.3   Jack-in Piles

Piles may be pushed or jacked into the seafloor if an adequate reaction force can be supplied. To develop a satisfactory degree of safety against bearing capacity failure under design loads, jacking loads equal to two to three times the design load must be applied. With mobile offshore jack-up platforms, water ballast is used to develop this surcharge. However, when the entire installation is submerged, water ballasting or the addition and removal of deadweight ballast is generally not practical.

The actual jacking of the piles can be accomplished by a number of systems. A rack-and-pinion system may be used, with the rack being an integral part of the pile and running its entire length. A chain acted on by a chain jack or a cable acted on by a hydraulic cable puller may be used, with the chain or cable applying load to the top of the pile. A short-stroke hydraulic jack equipped with a means of gripping the wall of a pile may also be used.

5.5.4   Jetted Piles

Jetting is used to place piles in cohesionless soils. The piles are pushed or lowered into the soil area, which has been greatly weakened by jetting. The jetting action is generally confined to the inside of a pile or to portions of the outside of the pile several diameters above its tip. Jetting can also be used in a form of reverse circulation in which both air and water are forced down a pipe inside or outside the pile. The air/water mixture is used to lift the displaced soil materials to the surface or seafloor.

5.6 EXAMPLE PROBLEMS

5.6.1 Problem 1 – Pile Design in a Cohesive Soil

5.6.1.1 Problem Statement

Design an anchor pile for specified lateral and uplift loads. Also, determine its axial capacity in compression and the maximum steel stresses. Can a pile which is on-hand be used for this mooring?

Data: A closed-ended anchor pile is to be designed for a floating drydock which will be placed in an area with a soft clay seafloor. The drydock will be used for repair of nuclear submarines. The pile must resist lateral loads up to 50,000 pounds and axial uplift loads of 25,000 pounds applied by a chain mooring system at the seafloor. At some later date, the pile may be used as a foundation pile and see pure compression loads. A sketch of this pile is shown in Figure 5.6-1. The soils data at the site are fairly well known and are shown in Figure 5.6-1. Available pile sizes are very limited. A supply of 24-inch diameter and 48-inch diameter steel piles (F_y = 36,000 psi) of several wall thicknesses is on-hand.

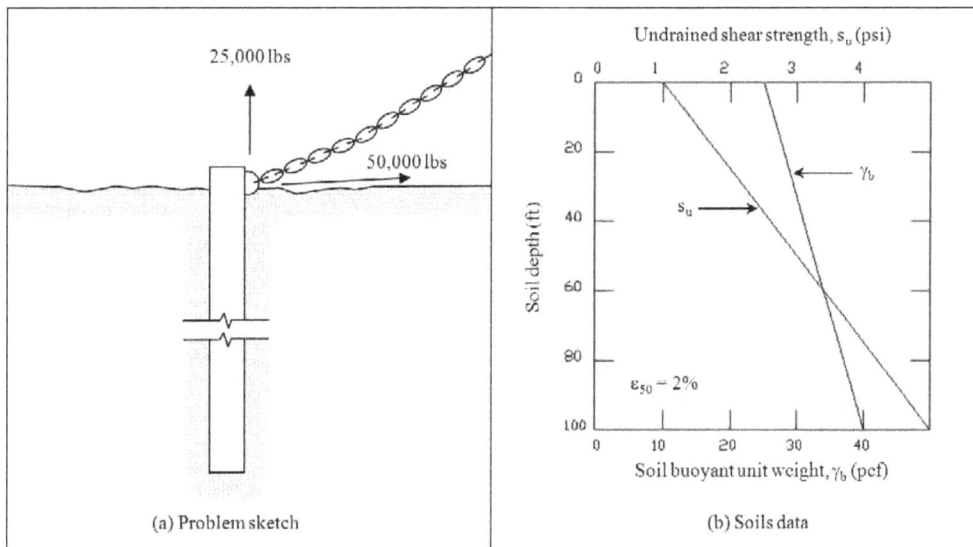

Figure 5.6-1. Problem sketch for example Problems 1 and 2, and soils data for example Problem 1.

5.6.1.2 Problem Solution

The trial-and-error computational procedures for the problem's solution are presented below. These follow the method presented in this chapter and outlined by the flow chart in Figure 5.3-1.

Problem 5.6-1

ANALYTICAL PROCEDURES	COMPUTATIONS
1. Determine soil properties (STEP 1) s_u and γ_b (cohesive soil).	s_u and γ_b are given (Figure 5.6-1).
2. Determine design loads at the seafloor (STEP 2) (F_s from Section 5.3.3). $T_h = F_s$ * horizontal load at seafloor $T_t = F_s$ * vertical load at seafloor	$F_s = 2.0$ (for a critical case where soil properties are well known) $T_h = (2.0)(50,000 \ lb) = 100,000 \ lb$ $T_t = (2.0)(25,000 \ lb) = 50,000 \ lb$
First, find a pile that will have sufficient lateral load capacity.	
3. Select a trial pile size (STEP 4) (24- and 48-in. diam steel piles are available). Data on the piles are taken from a pile design manual.	Try a 24-in pile with a 0.5-in wall thickness. $D = 24 \ in$ $t_w = 0.5 \ in$ $EI = 7.647 * 10^{10} \ in^2\text{-}lb$ $A_{ps} = 36.9 \ in^2$
4. Select deflection criteria (STEP 5). From Section 5.3.4: $\dfrac{y_{max}}{D} = 10\%$	$y_{max} = (0.10)(24 \ in) = 2.4 \ in$

Problem 5.6-1

ANALYTICAL PROCEDURES	COMPUTATIONS
5. Determine the coefficient subgrade reaction, n_h (STEP 6). K_1 determined from Figure 5.3-3 and s_u. s_u is averaged over 0 ft to a depth of four pile diameters. $$n_h = \frac{s_u K_1}{D}$$	4D = (4)(24 in) = 96 in = 8 ft From Figure 5.6-1, the equation of the line for s_u vs. soil depth (z) is given by: s_u = (0.04z + 1) psi , where z is given in ft s_u at 0 ft = (0 + 1) psi = 1.0 psi s_u at 8 ft = [0.04(8) +1] psi = 1.32 psi avg. s_u = (1.0 psi + 1.32 psi)/2 = 1.16 psi $K_1 \cong 20$ (soft clay) $$n_h = \frac{(1.16\ psi)(20)}{24\ in} = 0.967\ pci$$
6. Determine stiffness (STEP 7) (from Equation 5-4). $$T = \left(\frac{EI}{n_h}\right)^{0.2}$$	$$T = \left(\frac{7.647 \times 10^{10}\ in^2 - lb}{0.967\ lb/in^3}\right)^{0.2} = 151.2\ in$$
7. Select pile length (STEP 8). Guidance given in Section 5.3.4.	Say $L_p = 3T$ L_p = 3(151.2 in) = 453.6 in $\cong$ 38 ft
8. Determine deflection coefficients (STEP 9). From Figure 5.3-4, and the calculated deflection coefficient $z_{max} = L_p/T$.	z_{max} = 453.6 in / 151.2 in = 3 A_y = 2.7 and B_y = 1.8
9. Calculate lateral load capacity (STEP 10). From Equation 5-5 $$P_h(calc) = \frac{y_{max}(EI)}{A_y T^3 + aB_y T^2}$$	a = 0 (because load applied at seafloor) $$P_h(calc) = \frac{(2.4\ in)(7.647 \times 10^{10}\ in^2 - lb)}{(2.7)(151.2\ in)^3 + 0}$$ = 19,700 lb

Problem 5.6-1

ANALYTICAL PROCEDURES	COMPUTATIONS
10. Check lateral load against pile capacity (STEP 11). Is $P_h(calc) \geq T_h$?	T_h = 100,000 lb NO, $P_h(calc) \ll T_h$ THEREFORE, BECAUSE $P_h(calc)$ IS TOO SMALL, THE 24-IN PILE IS TOO SMALL.
11. Try another trial pile size (STEP 4). (Because $P_h(calc)$ is very much smaller, increase both the pile size and thickness.)	Try a 48-in diam pile with a 1.0-in wall thickness. D = 48 in = 4 ft t_w = 1.0 in EI = 1.224 * 10^{12} $lb\text{-}in^2$ A_{ps} = 147.6 in^2 S = 1,699.6 in^3
12. Select deflection criteria (STEP 5). $\dfrac{y_{max}}{D} = 10\%$	y_{max} = (0.10)(48 in) = 4.8 in
13. Determine n_h (STEP 6). From s_u, Figure 5.3-3, K_l, and $n_h = \dfrac{s_u K_1}{D}$	s_u at 0 ft = 1.0 psi s_u at 4D = s_u at 16 ft = [0.04(16) +1] = 1.64 psi avg. s_u = (1.0 psi + 1.64 psi)/2 = 1.32 psi $K_l \cong 20$ $n_h = \dfrac{(1.32\,psi)(20)}{48\,in} = 0.55\,pci$
14. Determine stiffness (STEP 7). From Equation 5-4. $T = \left(\dfrac{EI}{n_h}\right)^{0.2}$	$T = \left(\dfrac{1.224 \times 10^{12}\,in^2 - lb}{0.55\,lb/in^3}\right)^{0.2} = 295\,in$

Problem 5.6-1

ANALYTICAL PROCEDURES	COMPUTATIONS
15. Select pile length (STEP 8).	Say $L_p = 3T = 3(295\ in) = 885\ in \cong 74\ ft$
16. Determine deflection coefficients (STEP 9) from Figure 5.3-4 and $z_{max} = L_p/T$.	$z_{max} = 3$ $A_y = 2.7$ (B_y not needed)
17. Calculate lateral load capacity (STEP 10). From Equation 5-5	$a = 0$ $P_h(calc) = \dfrac{(4.8\,in)(1.224 \times 10^{12}\ in^2 - lb)}{(2.7)(295\,in)^3 + 0}$ $= 84{,}800\ lb$
18. Check lateral load against pile capacity (STEP 11). Is $P_h(calc) \geq T_h$?	$T_h = 100{,}000\ lb$ NO, $P_h(calc) < T_h$ THEREFORE, BECAUSE $P_h(calc)$ IS TOO SMALL, THE 48-IN PILE IS TOO SMALL.
19. Try another trial pile size (STEP 4). (Because $P_h(calc)$ is only a little smaller, don't increase the pile diameter but increase the wall thickness.)	Try a 48-in diameter pile with a 1.25-in wall thickness. $D = 48\ in$ $t_w = 1.25\ in$ $EI = 1.506 * 10^{12}\ lb\text{-}in^2$ $A_{ps} = 183.6\ in^2$ $S = 2{,}091\ in^3$
20. Determine stiffness (STEP 7). <u>Note</u>: The deflection criteria and n_h remain the same. $T = \left(\dfrac{EI}{n_h}\right)^{0.2}$	$n_h = 0.55\ pci$ $T = \left(\dfrac{1.506 \times 10^{12}\ in^2 - lb}{0.55\,lb/in^3}\right)^{0.2} = 307\,in$
21. Select pile length (STEP 8).	Say $L_p = 3T = 3(307\ in) = 921\ in \cong 77\ ft$

Problem 5.6-1

ANALYTICAL PROCEDURES	COMPUTATIONS
22. Determine deflection coefficients (STEP 9) from Figure 5.3-4 and $z_{max} = L_p /T$.	$z_{max} = 3$ $A_y = 2.7$ (B_y not needed)
23. Calculate lateral load capacity (STEP 10) from Equation 5-5.	$a = 0$ $$P_h(calc) = \frac{(4.8\,in)(1.506 \times 10^{12}\ in^2 - lb)}{(2.7)(307\,in)^3 + 0}$$ $= 92{,}500\ lb$
24. Check lateral load against pile capacity (STEP 11). Is $P_h(calc) \geq T_h$?	$T_h = 100{,}000\ lb$ NO, $P_h(calc) < T_h$ THEREFORE, THIS PILE IS ALSO TOO SMALL.
25. Try another trial pile size (STEP 4). (Because $P_h(calc)$ was close to P_h, try a slightly longer pile. A pile with a larger t_w could also be tried.)	Because the same pile size is used, D, t_w, EI, A_{ps}, S, n_h, and T remain the same. Say $L_p = 4T = 4(307\ in) = 1228\ in \cong 102\ ft$
26. Determine deflection coefficients (STEP 9) from Figure 5.3-4 and $z_{max} = L_p /T$.	$z_{max} = 4$ $A_y = 2.4$ (B_y not needed)
27. Calculate lateral load capacity (STEP 10) from Equation 5-5.	$$P_h(calc) = \frac{(4.8\,in)(1.506 \times 10^{12}\ in^2 - lb)}{(2.4)(307\,in)^3 + 0}$$ $= 104{,}100\ lb$
28. Check lateral load against pile capacity (STEP 11). Is $P_h(calc) \geq T_h$?	$T_h = 100{,}000\ lb$ YES, $P_h(calc) > T_h$ THEREFORE, THE PILE IS ADEQUATE FOR LATERAL LOAD.
Now, check the pile for its uplift capacity (axial upward load analysis).	

Problem 5.6-1

ANALYTICAL PROCEDURES	COMPUTATIONS
29. Calculate the average overburden pressure at the pile midpoint (STEP 12). From Figure 5.6-1 and Equation 5-6 $$\overline{p}_{vo} = \frac{\gamma_b L_p}{2}$$	Pile midpoint = $(102\,ft)/2 = 51\,ft$ From Figure 5.6-1, the equation of the line for γ_b vs. soil depth (z) is given by: $\gamma_b = (0.15z + 25)\,pcf$, where z is given in ft γ_b at $51\,ft = [(0.15)(51) + 25]\,pcf = 32.7\,pcf$ $$\overline{p}_{vo} = \frac{(32.7\,pcf)(102\,ft)}{2} = 1,670\,psf$$
30. Calculate the skin frictional resistance per unit length of pile (STEP 13). First: Is this an NC clay? If $s_u / \overline{p}_{vo} \le 0.4$, it is. From Equation 5-8a (NC clay), $$f_s = \overline{p}_{vo}[0.468 - 0.052\ln(L_p/2)]$$ (not to exceed $f_s = s_u$)	s_u at $51\,ft = [0.04(51) +1]\,psi = 3.0\,psi$ $= 432\,psf$ $$\frac{s_u}{\overline{p}_{vo}} = \frac{432\,psf}{1,670\,psf} = 0.26$$ YES, it is a NC clay. $f_s = (1,670\,psf)[0.468 - 0.052 \ln (102\,ft/2]$ $f_s = 440\,psf$ but $s_u = 3\,psi = 432\,psf$ Therefore, $f_s = 432\,psf$
31. Calculate the uplift capacity (STEP 14). From Equation 5-9 $$Q_s = A_s f_s$$	$A_s = \pi D L_p = \pi(4\,ft)(102\,ft) = 1282\,ft^2$ $Q_s = (1282\,ft^2)(432\,psf) = 554,000\,lb$
32. Is the pile capacity in uplift greater than the uplift load? (STEP 14) Is $Q_s > T_t$?	$T_t = 50,000\,lb$ YES, $Q_s > 50,000\,lb$ THEREFORE, THE PILE IS ADEQUATE FOR UPLIFT CAPACITY.
Now, calculate the pile load capacity in compression (axial downward load).	

Problem 5.6-1

ANALYTICAL PROCEDURES	COMPUTATIONS
33. Calculate the pile tip bearing capacity (STEP 15) from Equations 5-12 and 5-10 $q_p = 9\,s_u\,(tip)$ $Q_p = A_p\,q_p$	Extrapolate from Figure 5.6-2 for s_u $s_u\,(tip) = s_u$ at $102\,ft \cong [0.04(102)+1]\,psi$ $\qquad = 5.08\,psi$ $q_p = 9(5.08\,psi) = 45.7\,psi = 6{,}580\,psf$ $$A_p = \frac{\pi D^2}{4} = \frac{\pi (4\,ft)^2}{4} = 12.57\,ft^2$$ $Q_p = (12.57\,ft^2)(6580\,psf) = 83{,}000\,lb$
34. Calculate the pile capacity in compression (STEP 16) from Equation 5-13. $Q_c = Q_s + Q_p$	$Q_c = 554{,}000\,lb + 83{,}000\,lb = 637{,}000\,lb$
35. Calculate the maximum moment in the pile (STEP 17). Compute the actual T, L_p/T, and find A_m and B_m from Figure 5.3-5. Then from Equation 5-14 $M_{max} = A_m(P_h T) + B_m(M_a)$	$T = 307\,in = 25.6\,ft$ $L_p/T = 102\,ft\,/\,25.6\,ft = 3.98 \cong 4$ $A_m = 0.77$ (maximum value of A_m for $L_p/T = 4$) $B_m = 1.00$ (maximum value of B_m for $L_p/T = 4$) $(P_h = T_h = 100{,}000\,lb)$ $M_a = 0$ $M_{max} = (0.77)(100{,}000\,lb)(307\,in) + (1.00)(0)$ $\qquad = 2.36 * 10^7\,in\text{-}lb$

Problem 5.6-1

ANALYTICAL PROCEDURES	COMPUTATIONS
36. Calculate the maximum steel stresses in tension and compression (STEP 18) and check if they are below f_a, the maximum allowable stress. From Equations 5-15 and 5-16 $f_{maxt} = -P_t / A_{ps} - M_{max} / S$ $f_{maxc} = P_c / A_{ps} + M_{max} / S$ $f_a = 0.6 F_y$	$P_t = T_t = 50,000\ lb$ $f_{max\,t} = -\dfrac{50,000\,lb}{183.6\,in^2} - \dfrac{2.36\times10^7\,in-lb}{2,091\,in^3}$ $= -11,600\ psi$ Compute f_{maxc} for $P_c = 0$ (pile in tension), and for $P_c = Q_c = 637,000\ lb$ (at max capacity) $f_{max\,c} = 0 + \dfrac{2.36\times10^7\,in-lb}{2,091\,in^3} = 11,300\,psi$ $f_{max\,c} = \dfrac{637,000\,lb}{183.6\,in^2} + \dfrac{2.36\times10^7\,in-lb}{2,091\,in^3}$ $= 14,800\ psi$ $F_y = 36,000\ psi$ (given) $f_a = (0.6)(36,000\ psi) = 21,600\ psi$ YES, STEEL STRESSES IN TENSION AND COMPRESSION ARE BELOW THE MAXIMUM ALLOWABLE.

<u>SUMMARY</u>

1. The anchor piles on-hand can be used for this mooring.

2. The pile to be used has the following dimensions:

    $D = 48$ in

    $t_w = 1.25$ in

    $L_p = 102$ ft

3. The pile's axial capacity in compression is 637,000 lb.

4. The maximum steel stress under the actual mooring load is 11,600 psi in tension and 11,300 psi in compression. The steel compressive stress at the pile's axial capacity is 14,800 psi.

5.6.2 Problem 2 – Pile Design in a Cohesionless Soil

5.6.2.1 Problem Statement

Design an anchor pile for specified lateral and uplift loads. Also, determine its axial capacity in compression, and the maximum steel stresses. Can piles which are on-hand be used for this mooring?

Data: The anchor pile is to be used for the floating drydock mentioned in Problem 1, but the use of the drydock is different and the seafloor conditions are different. The drydock is to be moored while not in use, but it again must resist lateral loads up to 50,000 pounds and axial uplift loads up to 25,000 pounds at the seafloor. (The pile sketch in Figure 5.6-1 is the same for this problem.) The soils data at this site are not well known. The soils have not been tested but are believed to be a cohesionless silty sand of medium to dense relative density. The 24- and 48-inch diameter steel piles (F_y = 36,000 psi) of several wall thicknesses are also available for this mooring.

5.6.2.2 Problem Solution

The trial-and-error computational procedures for the problem's solution are summarized below. These follow the method presented in this chapter and outlined by the flow chart in Figure 5.3-1.

Problem 5.6-2

ANALYTICAL PROCEDURES	COMPUTATIONS
1. Determine soil properties (STEP 1) ϕ, γ_b, and D_r (cohesionless soil). Because the properties have not been measured, they must be estimated. Table 5.3-1.	For a cohesionless sand of medium to dense relative density $\phi \approx 35$ deg $\gamma_b \approx 60$ pcf $D_r \approx 65\%$
2. Determine design loads at the seafloor (STEP 2) (F_s from Section 5.3.3). $T_h = F_s *$ horizontal load at seafloor $T_t = F_s *$ vertical load at seafloor	$F_s = 2.0$ (for a noncritical case where the soils information is not well known) $T_h = (2.0)(50,000\ lb) = 100,000\ lb$ $T_t = (2.0)(25,000\ lb) = 50,000\ lb$
First, find a pile that will have sufficient lateral load capacity.	

Problem 5.6-2

ANALYTICAL PROCEDURES	COMPUTATIONS
3. Select a trial pile size (STEP 4) (24- and 48-in diameter steel piles are available). Data on the piles are taken from pile manufacturers' data.	Try a 24-in pile with a 0.5-in wall thickness. D = 24 in t_w = 0.5 in $EI = 7.647 * 10^{10}$ in²-lb A_{ps} = 36.9 in² S = 212.4 in³
4. Select deflection criteria (STEP 5). From Section 5.3.4: $$\frac{y_{\max}}{D} = 10\%$$	y_{max} = (0.10)(24 in) = 2.4 in
5. Determine the coefficient subgrade reaction, n_h (STEP 6) from Figure 5.3-2.	n_h = 13 pci
6. Determine stiffness (STEP 7) from Equation 5-4. $$T = \left(\frac{EI}{n_h}\right)^{0.2}$$	$$T = \left(\frac{7.647 \times 10^{10}\, in^2 - lb}{13\, lb/in^3}\right)^{0.2} = 89.9\, in$$
7. Select pile length (STEP 8).	Say L_p = 3T = 3(89.9 in) = 270 in = 22.5 ft
8. Determine deflection coefficients (STEP 9) from Figure 5.3-4, and z_{max} = L_p/T.	z_{max} = 270 in / 89.9 in = 3 A_y = 2.7 and B_y = 1.8
9. Calculate lateral load capacity (STEP 10). From Equation 5-5 $$P_h(calc) = \frac{y_{\max}(EI)}{A_y T^3 + a B_y T^2}$$	a = 0 (because load applied at seafloor) $$P_h(calc) = \frac{(2.4\,in)(7.647 \times 10^{10}\, in^2 - lb)}{(2.7)(89.9\,in)^3 + 0}$$ = 94,000 lb

Problem 5.6-2

ANALYTICAL PROCEDURES	COMPUTATIONS
10. Check lateral load against pile capacity (STEP 11). Is $P_h(calc) \geq T_h$?	T_h = 100,000 *lb* NO, $P_h(calc) < T_h$ THEREFORE, THE PILE IS TOO SMALL.
11. Try another trial pile size (STEP 4). (Try a longer pile. A pile with a larger t_w could also be tried.)	Say $L_p = 4T = 4(89.9\ in) = 360\ in = 30\ ft$
12. Calculate lateral load capacity (STEP 10) from Equation 5-5.	n_h and T remain the same $z_{max} = L_p/T = 4$ $A_y = 2.4$ (from Figure 5.3-4) $$P_h(calc) = \frac{(2.4\,in)(7.647 \times 10^{10}\ in^2 - lb)}{(2.4)(89.9\,in)^3 + 0}$$ $= 105,000\ lb$
13. Check lateral load against pile capacity (STEP 11). Is $P_h(calc) \geq T_h$?	T_h = 100,000 *lb* YES, $P_h(calc) > T_h$ THEREFORE, THE PILE IS ADEQUATE FOR LATERAL LOAD.
Now, check the pile for its axial load capacity.	
14. Calculate the average overburden pressure at the pile midpoint (STEP 12). From Equation 5-6, $$\bar{P}_{vo} = \frac{\gamma_b\,L_p}{2}$$	$$\bar{P}_{vo} = \frac{(60\,pcf)(30\,ft)}{2} = 900\,psf$$

Problem 5.6-2

ANALYTICAL PROCEDURES	COMPUTATIONS
15. Calculate the skin frictional resistance per unit length of pile in tension (STEP 13). From Equation 5-7 $f_s = k\,\overline{p}_{vo}\,\tan(\phi - 5\,\text{deg})]$	$k = 0.5$ (for uplift resistance) $k = 0.7$ (for compression) In uplift, $f_s = (0.5)(900\ psf)\ \tan(35°\text{-}5°) = 260\ psf$ In compression, $f_s = (0.7)(900\ psf)\ \tan(35°\text{-}5°) = 364\ psf$ A check with Table 5.3-5 shows these f_s values are below the allowable limit.
16. Calculate the uplift capacity (STEP 14). From Equation 5-9 $Q_s = A_s f_s$	$A_s = \pi D L_p = \pi\,(2\ ft)(30\ ft) = 188\ ft^2$ $Q_s = (188\ ft^2)(260\ psf) = 48{,}900\ lb$
17. Is the pile capacity in uplift greater than the uplift load? (STEP 14) Is $Q_s > T_t$?	$T_t = 50{,}000\ lb$ NO, $Q_s < 50{,}000\ lb$ THEREFORE, THE PILE IS NOT ADEQUATE FOR UPLIFT CAPACITY. THE PILE IS TOO SHORT OR TOO SMALL.
18. Try another trial pile size (STEP 4). (Try a longer pile. A pile with a larger t_w could also be tried.)	The uplift capacity is close to the uplift load, so try slightly increasing the length. Say $L_p = 32\ ft$
19. Calculate the average overburden pressure at the pile midpoint (STEP 12), from Equation 5-6.	$\overline{p}_{vo} = \dfrac{(60\,pcf)(32\,ft)}{2} = 960\,psf$

Problem 5.6-2

ANALYTICAL PROCEDURES	COMPUTATIONS
20. Calculate the skin frictional resistance per unit length of pile in tension (STEP 13) using Equation 5-7.	In uplift, f_s = (0.5)(960 *psf*) tan(35°-5°) = 277 *psf* In compression, f_s = (0.7)(960 *psf*) tan(35°-5°) = 388 *psf* A check with Table 5.3-5 shows these f_s values are below the allowable limit.
21. Calculate the uplift capacity (STEP 14) from Equation 5-9.	$A_s = \pi D L_p = \pi (2 \, ft)(32 \, ft) = 201 \, ft^2$ Q_s = (201 ft^2)(277 *psf*) = 55,700 *lb*
22. Is the pile capacity in uplift greater than the uplift load? (STEP 14) Is $Q_s > T_t$?	T_t = 50,000 *lb* YES, Q_s > 50,000 *lb* THEREFORE, THE PILE IS ADEQUATE FOR UPLIFT CAPACITY.
23. Calculate the pile tip bearing capacity (STEP 15) from Equations 5-10 and 5-11. $q_p = \bar{P}_{vo}(tip) \, N_q$ $Q_p = A_p \, q_p$	$\bar{P}_{vo}(tip) = (60 \, pcf)(32 \, ft) = 1{,}920 \, psf$ N_q = 12 (from Table 5.3-4 for ϕ = 35°) q_p = (1,920 *psf*)(12) = 23,040 *psf* $A_p = \dfrac{\pi D^2}{4} = \dfrac{\pi (2 \, ft)^2}{4} = 3.14 \, ft^2$ Q_p = (3.14 ft^2)(23,040 *psf*) = 72,350 *lb*
24. Calculate the pile capacity in compression (STEP 16) from Equation 5-13. $Q_c = Q_{s(compression)} + Q_p$	Q_c = (388 *psf*) (201 ft^2) +72,350 *lb* Q_c = 77,990 *lb* + 72,350 *lb* = 150,340 *lb*

Problem 5.6-2

ANALYTICAL PROCEDURES	COMPUTATIONS
25. Calculate the maximum moment in the pile (STEP 17). Obtain A_m from Figure 5.3-5. Then from Equation 5-14, $$M_{max} = A_m(P_h T) + B_m(M_a)$$	$T = 89.9$ *in* $= 7.5$ ft $L_p/T = 32$ *ft* / 7.5 *ft* $= 4.3 \cong 4$ $A_m = 0.77$ (maximum value of A_m for $L_p/T = 4$) $(P_h = T_h = 100,000$ *lb*) $M_a = 0$ $M_{max} = (0.77)(100,000$ *lb*$)(89.9$ *in*$) + 0$ $= 6.92\ 310^6$ *in-lb*
26. Calculate the maximum steel stresses in tension and compression (STEP 18) and check if they are below f_a, the maximum allowable stress. From Equations 5-15 and 5-16 $f_{maxt} = -P_t / A_{ps} - M_{max} / S$ $f_{maxc} = P_c / A_{ps} + M_{max} / S$ $f_a = 0.6\ F_y$	$P_t = T_t = 50,000$ *lb* $$f_{max\,t} = -\frac{50,000\,lb}{36.9\,in^2} - \frac{6.92 \times 10^6\,in-lb}{212.4\,in^3}$$ $= -34,000$ *psi* Compute f_{maxc} for $P_c = 0$ (pile in tension only, no compression) $$f_{max\,c} = 0 + \frac{6.92 \times 10^6\,in-lb}{212.4\,in^3} = 32,600\,psi$$ $F_y = 36,000$ *psi* (given) $f_a = (0.6)(36,000$ *psi*$) = 21,600$ *psi* NO, THE STEEL STRESSES IN TENSION AND COMPRESSION EXCEED THE MAXIMUM ALLOWABLE.
27. Try a larger size (thicker) pile.	Try a 24-in diameter pile with a 1.0 in wall thickness.

Problem 5.6-2

ANALYTICAL PROCEDURES	COMPUTATIONS
28. Recalculate the maximum steel stresses in tension and compression (STEP 18) and check if they are below f_a, the maximum allowable stress. From Equations 5-15 and 5-16 $f_{maxt} = -P_t / A_{ps} - M_{max} / S$ $f_{maxc} = P_c / A_{ps} + M_{max} / S$ $f_a = 0.6\ F_y$	$P_t = T_t = 50{,}000\ lb$ $A_{ps} = 72.3\ in^2$ $S = 390\ in^2$ $$f_{max\,t} = -\frac{50{,}000\,lb}{72.3\,in^2} - \frac{6.92 \times 10^6\,in-lb}{390\,in^3}$$ $= -\,18{,}400\ psi$ Compute f_{maxc} for $P_c = 0$ (pile in tension), and for $P_c = Q_c = 150{,}000\ lb$ (at maximum capacity) $$f_{max\,c} = 0 + \frac{6.92 \times 10^6\,in-lb}{390\,in^3} = 17{,}700\,psi$$ $$f_{max\,c} = \frac{150{,}300\,lb}{72.3\,in^2} + \frac{6.92 \times 10^6\,in-lb}{390\,in^3}$$ $= 19{,}780\ psi$ $F_y = 36{,}000\ psi$ (given) $f_a = (0.6)(36{,}000\ psi) = 21{,}600\ psi$ YES, THE STEEL STRESSES IN TENSION AND COMPRESSION ARE BELOW THE MAXIMUM ALLOWABLE.

Problem 5.6-2

ANALYTICAL PROCEDURES	COMPUTATIONS

SUMMARY

1. The anchor piles on-hand can be used for this mooring.

2. The pile to be used has the following dimensions:

   D = 24 in

   t_w = 1.00 in

   L_p = 32 ft

3. The pile's axial capacity in compression is 150,300 lb.

4. The maximum steel stress under the actual mooring load is 18,400 psi in tension and 17,700 psi in compression. The steel compressive stress at the pile's axial capacity is 19,780 psi.

5.7 REFERENCES

5-1.   R.J. Taylor, D. Jones, and R.M. Beard. Handbook for Uplift-Resisting Anchors, Civil Engineering Laboratory. Port Hueneme, CA, Sep 1975.

5-2.   H. Matlock, L. C. Reese, "Generalized Solutions for Laterally Loaded Piles." Journal of the Soil Mechanics and Foundations Division, Vol. 86, No. 5, September/October 1960, pp. 63-94.

5-3.   G.G. Meyerhoff and J.I. Adams. "The Ultimate Uplift Capacity of Foundations," Canadian Geotechnical Journal, Vol. 5, No. 4, Nov 1968, pp. 225-244.

5-4.   K. Terzaghi. "Evaluation of Coefficients of Subgrade Reaction," Geotechnique, Vol. 5, No. 4, 1955, pp. 297-326.

5-5.   LPILE Plus. Computer software. Vers. 5.0. Austin, TX: Ensoft, Inc.

5-6.   AllPile. Computer software. Vers. 6. Seattle, WA: CivilTech.

5-7.   H.L. Gill and K.R. Oemars. Displacement of Laterally Loaded Structures in Nonlinearly Responsive Soil, Naval Civil Engineering Laboratory, Technical Report R-670. Port Hueneme, CA, Apr 1970.

5-8.   I.W. Farmer. Engineering Properties of Rocks. London, England, E. and F.N. Spon Ltd., 1968, p. 57.

5-9.   G.G. Goble and F. Rausch. Wave Equation Analysis of Pile Foundations – WEAP86 Program, Vols. 1-4, Federal Highway Administration, U.S. Department of Transportation, FHWA-IP-86-19. Washington, DC, 1986.

5-10.   L. L. Lowery, Jr. Pile Driving Analysis by the Wave Equation – MICROWAVE Users Manual, Texas A&M University, Department of Civil Engineering. College Station, TX, 1993.

5.8 SYMBOLS

a	Distance of pile load attachment point above seafloor for foundation piles [L]
A_m	Nondimensional moment coefficient
A_p	Gross end area of pile [L^2]
A_{ps}	Cross-sectional area of pile [L^2]
A_s	Surface area of pile below seafloor [L^2]
A_y	Nondimensional deflection coefficient
B_m	Nondimensional moment coefficient
B_y	Nondimensional deflection coefficient
C_p	Minimum perimeter transmitting uplift load [L]
D	Pile width or diameter [L]
d_b	Characteristic mooring line size [L]
D_r	Relative density [%]
E	Modulus of elasticity of pile [F/L^2]
E_h	Maximum rated energy of hammer [LF]
EI	Stiffness
F_{cb}	Horizontal component of mooring line bearing resistance (lateral force exerted by mooring line on the soil) [F]
F_s	Factor of safety
F_y	Minimum yield point of the pile material [F/L^2]
f_a	Allowable stress in pile material
f_{maxc}	Maximum applied compressive stress in pile [F/L^2]
f_{maxt}	Maximum applied tensile stress in pile [F/L^2]
f_s	Unit skin resistance along pile shaft
$f_{s(max)}$	Maximum allowable unit skin friction [F/L^2]
I	Moment of inertia of pile [L^4]
K_1	$= n_h D / s_u$ = nondimensional coefficient for clay
k	Coefficient of lateral earth pressure
L_e	Effective length of pile bearing on rock [L]
L_p	Length of pile [L]
L_r	Length of pile embedded in rock [L]
M_a	Design applied moment [LF]
M_{max}	Maximum bending moment in pile [LF]

NC	Normally consolidate
$N_q, \overline{N_q}$	Sand and a special case sand/chain dimensionless bearing capacity factors
n_h	Coefficient of subgrade soil reaction [F/L^3]
OC	Overconsolidated
P_c	Design vertical compressive load at the foundation pile [F]
P_h	Design horizontal load at the foundlation pile [F]
P_h'	P_h corrected for effects on loading of a pile being driven below the seafloor [F]
$P_h(calc)$	Lateral load capacity of the trial pile [F]
P_t	Design vertical uplift load at the foundation pile [F]
P_t'	P_t corrected for effects on loading of a pile being driven below the seafloor [F]
$\overline{p}_{vo}$	Effective vertical stress in soil [F/L^2]
Q_c	Pile capacity in compression [F]
Q_p	Soil bearing capacity for the pile tip [F]
Q_s	Frictional resistance of pile [F]
q_p	Unit soil bearing capacity of the pile tip [F/L^2]
$q_p(max)$	Maximum allowable q from Table 5.3-5 [F/L^2]
R_a	Uplift capacity of rock anchor [F]
S	Section modulus of pile [L^3]
s_b	Bond strength of grout to pile, or grout shear strength, whichever is less [F/L^2]
s_c	Compressive strength of rock [F/L^2]
s_u	Soil undrained shear strength [F/L^2]
$s_u(z)$	Soil undrained shear strength at depth z [F/L^2]
T	Relative stiffness of pile-soil system [L]
T_h	Design horizontal load at anchor pile [F]
T_h'	T_h, corrected for the effects on loading of a pile being driven below the seafloor
T_t	Design vertical load at anchor Pile [F]
T_t'	T_t corrected for the effects on loading of a pile being driven below the seafloor
t_w	Wall thickness of pile [L]
y	Lateral pile deflection [L]
y_{max}	Maximum allowable lateral pile deflection [L]
z	Depth below seafloor [L]
z_c	Depth of pile connection below seafloor [L]
z_{max}	Maximum value of the depth coefficient
z_o	Depth of pile head below seafloor [L]

z_r	Thickness of rock layer [L]
z_s	Thickness of soil over rock [L]
γ_b	Buoyant unit weight of soil [F/L^3]
ϕ	Drained (effective) friction angle
δ	Friction angle for sand against pile

[This page intentionally left blank]

6 DIRECT-EMBEDMENT ANCHORS

6.1 INTRODUCTION

6.1.1 Purpose

This chapter details the procedures and considerations for use of direct-embedment anchors. It deals primarily with plate-type anchors, specifically the pile-driven types used by the Navy. The design of these anchors is very flexible and the anchors can be suited to a wide variety of soil conditions and holding capacities. The anchors can be driven using surface pile-driving equipment in water depths up to approximately 100 feet (33m) and sub-surface driving equipment is needed in deeper water depths.

6.1.2 Function

Most direct-embedment anchors are installed by inserting the anchor member vertically into the seafloor and then expanding or re-orienting the anchor member to increase its pull out resistance. Direct-embedment anchor types are described in Section 6.2.

6.1.3 Features

The four major types of direct-embedment anchors are: impact/vibratory-driven, jetted-in, and augered-in anchors.

Features of direct-embedment anchors are summarized in Chapter 1. The more significant advantages of the direct-embedment anchors are: (1) their very high holding-capacity-to-weight ratio (100:1 for driven-embedded types) and (2) their resistance to non-horizontal loading, which permits short mooring line scopes and tighter moorings.

6.2 DIRECT-EMBEDMENT ANCHOR TYPES AND SIZES

6.2.1 Impact/Vibratory-Driven Anchor

Various types of impact-, hammer- or vibratory-driven anchors, which expand or rotate to achieve high capacity, have been developed (Ref. 6-1). These anchors typically consist of a rigid plate, a keying flap (if required) and a driving follower. If the water depth is approximately 100 feet (33m) or less then surface driving equipment can be used. For deeper water submerged equipment is used.

The type of hammer depends on the soils and anchoring requirement. Impact or hammer equipment is used in stiff soils. Vibratory hammers may be considered for cohensionless soil or soft mud. The U.S. Navy, for example, installed approximately 400 of these anchors during the period of 1990-2010. Sizes ranged from 2-foot x 3-foot (0.6m x 0.9m) to 6-foot x 12-foot (1.8m x 3.7m) and the anchors had holding capacities of up to 1,000 kips (4,400 kN). Larger anchors may also be practical. Two examples of driven anchors are illustrated in Figure 6.2-1. The major anchor features are summarized in Figure 6.2-2, and a sample anchor design is given in Figure 6.2-3. Details of pile driven plate anchors are discussed in Section 6.9.

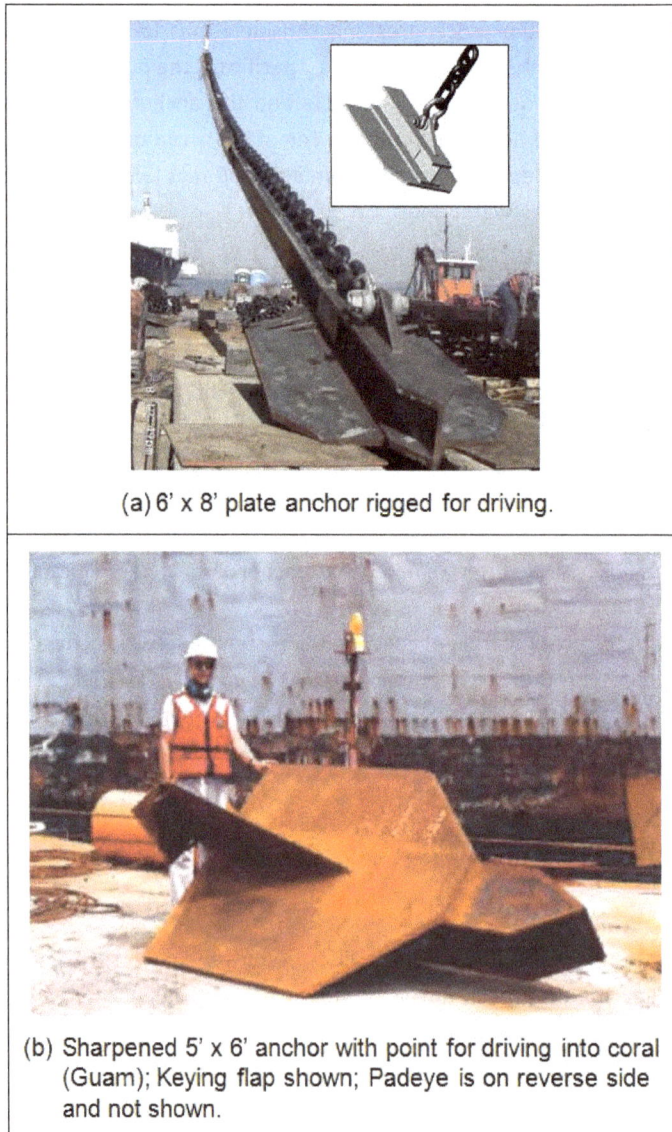

(a) 6' x 8' plate anchor rigged for driving.

(b) Sharpened 5' x 6' anchor with point for driving into coral (Guam); Keying flap shown; Padeye is on reverse side and not shown.

Figure 6.2-1. Sample impact/vibratory driven anchors.

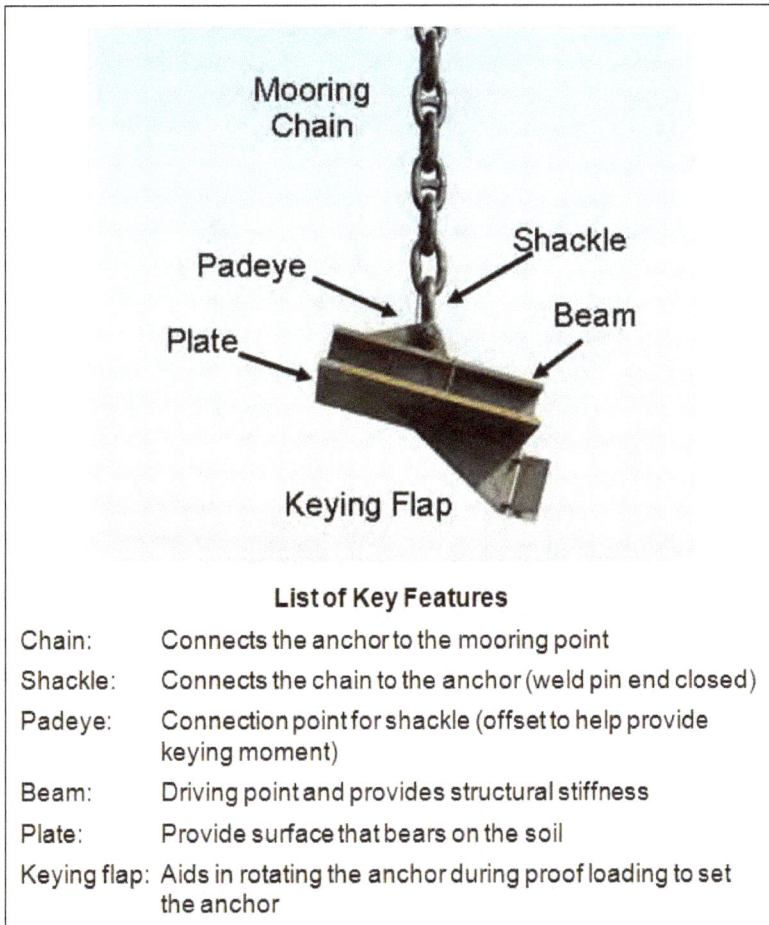

List of Key Features

Chain: Connects the anchor to the mooring point

Shackle: Connects the chain to the anchor (weld pin end closed)

Padeye: Connection point for shackle (offset to help provide keying moment)

Beam: Driving point and provides structural stiffness

Plate: Provide surface that bears on the soil

Keying flap: Aids in rotating the anchor during proof loading to set the anchor

Figure 6.2-2. Key features of impact/vibratory driven anchors.

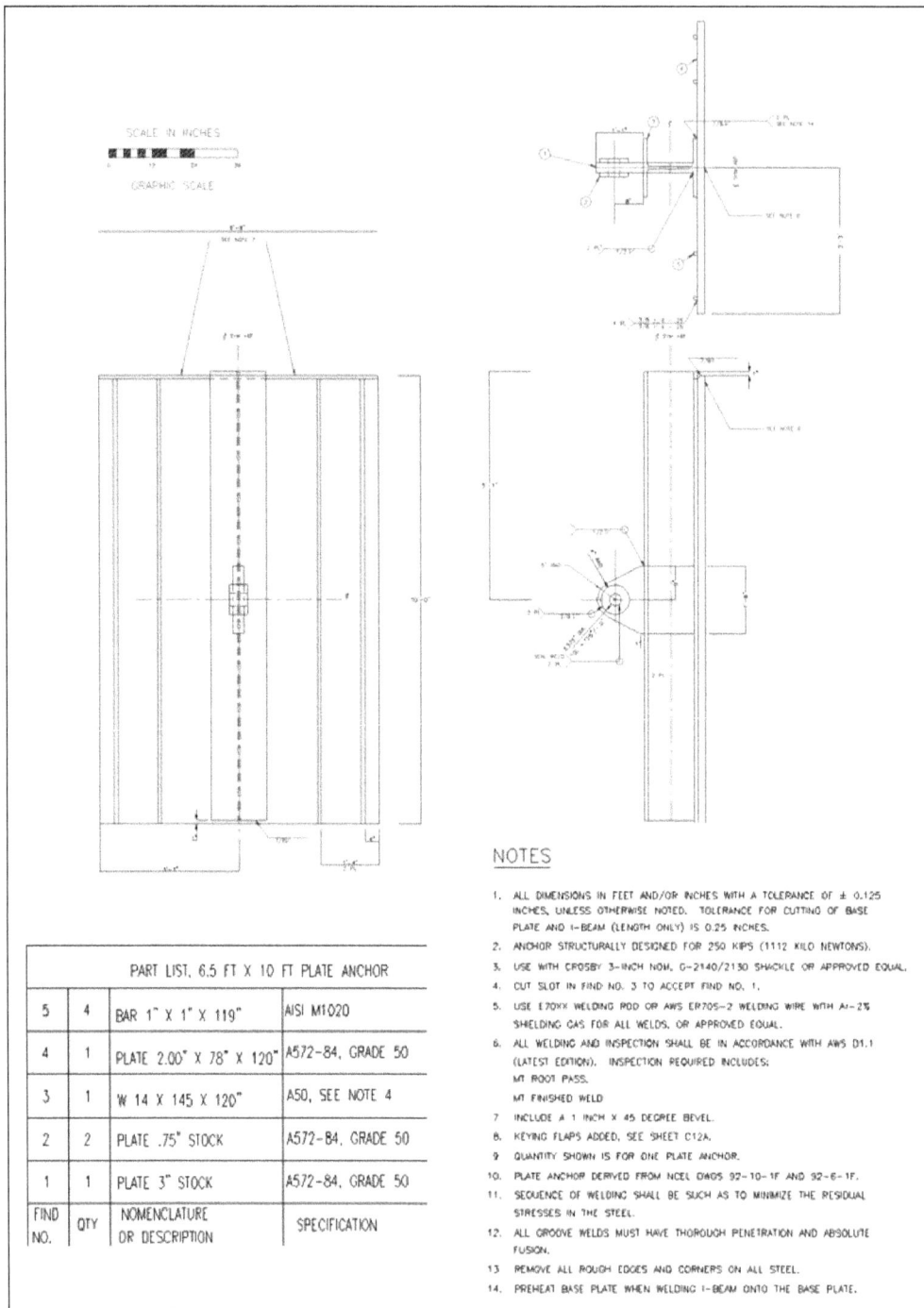

PART LIST, 6.5 FT X 10 FT PLATE ANCHOR

FIND NO.	QTY	NOMENCLATURE OR DESCRIPTION	SPECIFICATION
5	4	BAR 1" X 1" X 119"	AISI M1020
4	1	PLATE 2.00" X 78" X 120"	A572-84, GRADE 50
3	1	W 14 X 145 X 120"	A50, SEE NOTE 4
2	2	PLATE .75" STOCK	A572-84, GRADE 50
1	1	PLATE 3" STOCK	A572-84, GRADE 50

NOTES

1. ALL DIMENSIONS IN FEET AND/OR INCHES WITH A TOLERANCE OF ± 0.125 INCHES, UNLESS OTHERWISE NOTED. TOLERANCE FOR CUTTING OF BASE PLATE AND I-BEAM (LENGTH ONLY) IS 0.25 INCHES.
2. ANCHOR STRUCTURALLY DESIGNED FOR 250 KIPS (1112 KILO NEWTONS).
3. USE WITH CROSBY 3-INCH NOM, G-2140/2130 SHACKLE OR APPROVED EQUAL.
4. CUT SLOT IN FIND NO. 3 TO ACCEPT FIND NO. 1.
5. USE E70XX WELDING ROD OR AWS ER70S-2 WELDING WIRE WITH Ar-2% SHIELDING GAS FOR ALL WELDS, OR APPROVED EQUAL.
6. ALL WELDING AND INSPECTION SHALL BE IN ACCORDANCE WITH AWS D1.1 (LATEST EDITION). INSPECTION REQUIRED INCLUDES:
   MT ROOT PASS.
   MT FINISHED WELD
7. INCLUDE A 1 INCH X 45 DEGREE BEVEL.
8. KEYING FLAPS ADDED, SEE SHEET C12A.
9. QUANTITY SHOWN IS FOR ONE PLATE ANCHOR.
10. PLATE ANCHOR DERIVED FROM NCEL DWGS 92-10-1F AND 92-6-1F.
11. SEQUENCE OF WELDING SHALL BE SUCH AS TO MINIMIZE THE RESIDUAL STRESSES IN THE STEEL.
12. ALL GROOVE WELDS MUST HAVE THOROUGH PENETRATION AND ABSOLUTE FUSION.
13. REMOVE ALL ROUGH EDGES AND CORNERS ON ALL STEEL.
14. PREHEAT BASE PLATE WHEN WELDING I-BEAM ONTO THE BASE PLATE.

Figure 6.2-3. Sample plate anchor design (keying flaps not shown).

6.2.2 Jetted-In Anchors

Jetted-in anchors are buried through water jet disturbance of the soil (Figure 6.2-4). This allows the anchor to be placed, or to be more easily-pushed, into the seafloor. These systems range in holding capacity from small, diver-installed anchors of 1 to 10 kips (Ref. 6-2) to larger systems of 150 kips and greater (Refs. 6-3 and 6-4). Jetted anchors function primarily in sands that are easily liquefied by the water jets. After the jetting action is stopped, the liquefied sands return to a more dense condition over the anchor plate. Penetration in clays is not as easily accomplished, and the resulting backfill is much weaker than the undisturbed material. In hard clays and shell and cobble soils, penetration by jetting will likely be very slow and not economical.

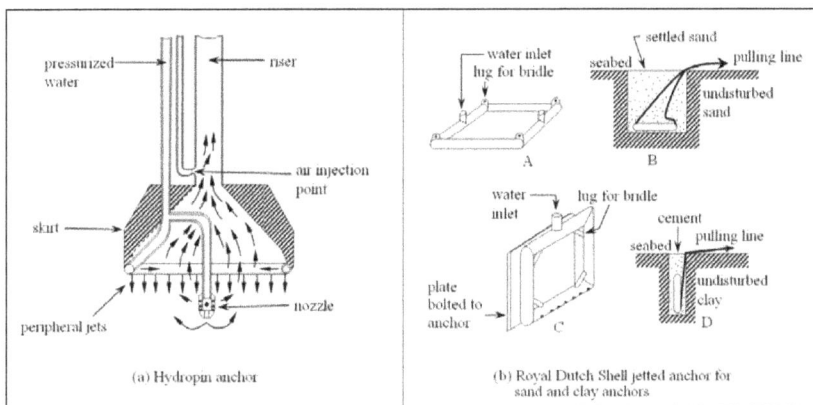

Figure 6.2-4. Jetted-in anchors (Ref. 6-3).

6.2.3 Auger Anchors

Auger anchors are screw-shaped shafts installed under high torque and some vertical load. They have been used for anchoring pipelines on the seafloor (Ref. 6-5). Normally, anchors are installed in pairs, one on each side of the pipeline at the same time to provide a torque reaction for each other. Operational water depth is limited primarily by difficulties in supplying power, usually by hydraulic hoses, at the seafloor. The experience limit is approximately 500 feet.

6.3 SITE DATA NEEDED

6.3.1 General

Figure 6.3-1 presents an outline of steps in the selection of an adequate direct-embedment anchor size for a given set of loading and site conditions. The first steps are the collection of adequate site data.

Use of a direct-embedment anchor requires knowledge of sediment properties over the possible depth of penetration of the anchor fluke. Maximum penetration depth, keying distance, and holding capacity under the specific conditions of applied loadings must be calculated from these properties. With adequate site data, a direct-embedment anchor sufficient for existing soil and loading conditions can be selected.

6.3.2   Preliminary Penetration Estimate

In many cases some preliminary data are available on the general seafloor environment to permit identification of the probable sediment type (see Chapter 2 for sediment data sources). Knowledge of a probable sediment type allows determination of the required anchor penetration depth and keying distances (see Ref. 6-1). These estimates of soil type and penetration depth guide selection of site survey and sampling equipment.

6.3.3   Topography, Strata Thickness, Type

In non-homogeneous soils, strata material type and thickness must be determined. The seafloor material type and approximate consistency must be known in order to select the appropriate anchor design. The thicknesses of the sediment strata must be known to ensure that the anchor fluke has sufficient sediment thickness to develop the design capacity. These data are best obtained over a wide area through acoustic subbottom profiling and coring (see Chapter 2). Seafloor topography, surficial sediment layer thicknesses, and depth to rock are best measured over the wide area by a 3.5-kHz acoustic profiling system. In areas of large relief, such as areas of outcropping rock, erosion, or slumping features, a deep tow profiling system may be necessary to obtain an accurate picture of seafloor topography and distribution of sediment infill between the relief features.

The results of the acoustic surveys should be used during the exploratory program to position coring locations so that: (1) the core samples will provide data representative of the probable anchor locations and (2) the potential for lost time and damage to coring equipment is minimized. Core samples and in-situ tests (see Chapters 2 and 3) are used primarily to identify soil type accurately and to determine soil engineering property data. They also provide a necessary control and calibration for interpretation of the acoustic data.

Some of the specific site survey information required for determining the anchor design include: seafloor material type, layer thickness, and depth to rock. Where consequences of a single anchor failure are not severe (noncritical applications), a lower level of data on sediment type and thicknesses may be sufficient for anchor selection. Geotechnical properties may then be estimated from soil property profiles to make a rough estimate of capacity in lieu of accurate site-specific data.

6.3.4 Engineering Properties

To make accurate predictions of holding capacity, several soil engineering properties must be known. These are sensitivity, natural water content, bulk density, grain size, carbonate content, origin and history, permeability, and shear strength (drained and undrained). Where dynamic loads are significant, other, more specialized tests on core samples may be necessary. At least one good quality sediment core is required to determine these properties. In many cases, only one is required. For example, for a deep water mooring on a large abyssal plain, sediment variability over a 4- to 5-mile diameter of the mooring area may not be great. One core should suffice. However, where the sediment consistency or type may vary across the mooring site, cores should be obtained at each anchor location. Soil cores should be obtained over the full estimated penetration depth of the fluke.

Long, heavy corers and specialized handling systems are required to penetrate and recover cores from the deeper soil depths. However, smaller corers, which achieve penetrations of 10 feet in sands and 30 feet in clays, are often used to obtain the sediment and define the upper portion of the geotechnical property profile. With the guidance of geophysical data, sediments below this sampled depth may be assumed to be similar, and the soil property profile would be extended to the necessary depth. Expendable penetrometers can provide additional data where longer coring is not possible and can thus extend the depth of a survey below sampling depth.

6.3.5 Complicating or Hazardous Conditions

Direct-embedment anchor systems function well in a wide range of seafloor conditions. They can be adapted to function well where drag anchors and pile anchors are inefficient and even nonfunctional. Extreme soil conditions such as very hard or very soft seafloors complicate the use of direct-embedment anchors, making special efforts necessary during site survey, positioning, design, installation, and proof-loading. Table 6.3-1 lists some complicating conditions and describes their impact on direct-embedment anchor performance. The approach to most of these is to avoid the area by relocating the plate anchor so the problem is not acute or to select a different anchor system--one less sensitive to the problem (e.g., using gravity anchors when on rock).

START

DETERMINE LEVEL OF SITE SURVEY NEEDED   (SECTION 6.3)

NONCRITICAL — CRITICAL

(SECTION 6.3.4)   SELECT TYPICAL ENGINEERING PARAMETERS FROM LITERATURE

CONDUCT SURVEY TO OBTAIN ENGINEERING PROPERTIES   (SECTION 6.3.4)

DETERMINE IF HAZARDOUS / UNUSUAL CONDITIONS EXIST   (SECTION 6.3.5)

ESTIMATE AN ANCHOR SIZE FROM AVAILABLE HARDWARE

COHESIVE — COHESIONLESS

(CHAPTER 8)   CALCULATE PENETRATION

CALCULATE PENETRATION   (TABLE 6.2-2)

(EQUATION 6-1)   CALCULATE KEYED DEPTH

CALCULATE KEYED DEPTH   (EQUATION 6-2)

(EQUATION 6-3)   CALCULATE SHORT-TERM STATIC CAPACITY

CALCULATE SHORT-TERM STATIC CAPACITY   (EQUATION 6-7)

DETERMINE ALL ANTICIPATED TYPES OF LOADING   (SECTIONS 6.5-1 AND 6.6-1)

SHORT-TERM EFFECTS

CALCULATE LONG-TERM EFFECTS

CALCULATE CYCLIC EFFECTS

CALCULATE IMPULSE EFFECTS

COHESIONLESS — COHESIVE

COHESIVE — COHESIONLESS

CALCULATE CAPACITY (EQUATION 6-4)

CALCULATE CAPACITY (SECTION 6.6.2.2)

CALCULATE CAPACITY (SECTION 6.6.4.2)

CALCULATE CAPACITY (SECTION 6.6.4.3)

APPLY SAFETY FACTOR (SECTION 6.5.7)

APPLY SAFETY FACTOR (SECTION 6.5.7)

APPLY SAFETY FACTOR (SECTION 6.6.2.4)

APPLY SAFETY FACTOR (SECTION 6.6.4.4)

APPLY SAFETY FACTOR (SECTION 6.6.4.4)

SELECT LARGER SIZED ANCHOR   ← INADEQUATE — IS SELECTED ANCHOR SIZE ADEQUATE FOR DESIGN LOADS? — MUCH MORE THAN ADEQUATE →   SELECT SMALLER SIZED ANCHOR

ADEQUATE

USE THIS SIZE ANCHOR

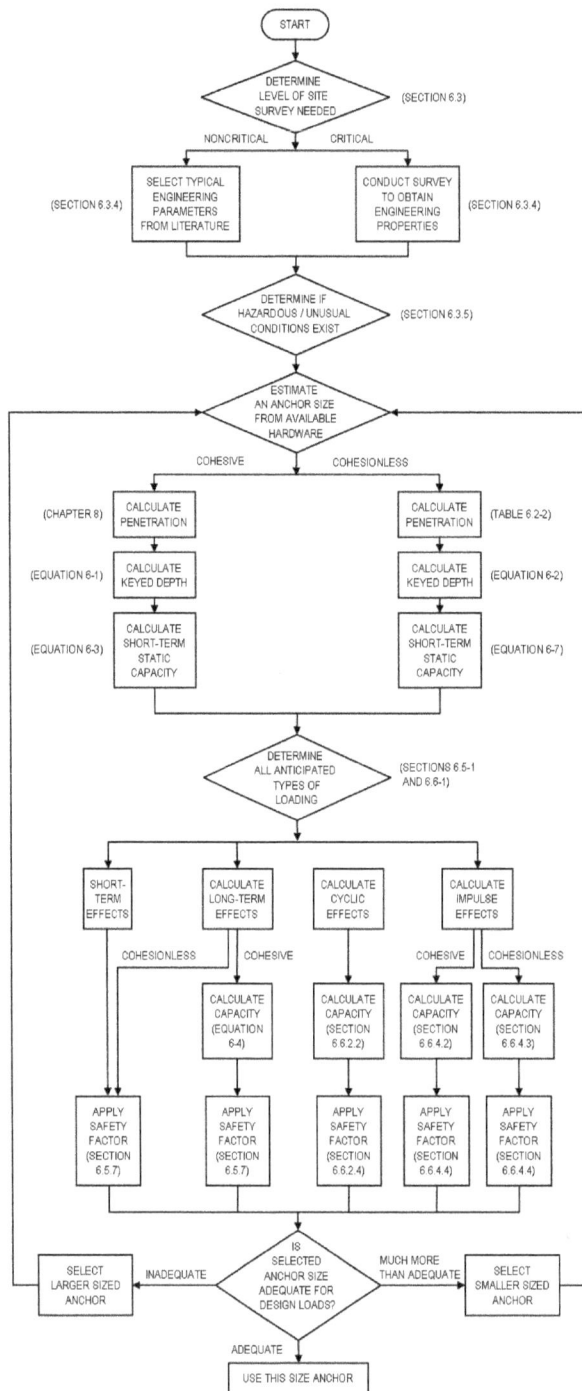

Figure 6.3-1. Flow chart for predicting the holding capacity of a direct-embedment anchor.

Table 6.3-1. Conditions Complicating or Hazardous to Direct-Embedment Anchor Use

Seafloor Condition	Potential Complication
Hard Strata • Thin soil layer over rock, or thin soft clay over sand. • Glacial erratics or residual surficial gravel and cobbles. • Nodule or pavement formations (usually manganese) over soil. • Submarine lava flows.	 • Soil thickness not sufficient to develop fluke capacity, but sufficient to consume most of fluke kinetic energy before it reaches stronger layer • Damages flukes and limits penetration into underlying sands and hard clays. • Same as above. • Extremely irregular and complex. Brittle rock, often fractured, sometimes as "pillows" (i.e., bulbous discrete or welded forms 3 to 6 ft in diameter). To ensure reliability, anchors must be proof-tested to full load.
Sloping Seafloor • Soil slopes over 10 deg. • Hard soil or rock scarps or cliffs.	 • Anchor may affect slope stability (see Table 6.7-1), but direct-embedment anchor will suffer less than most other anchor types on sloping soils. • Improper embedment from deflection or ricochet of fluke.
Scour • Sand waves.	 • Sand waves can be large and move rapidly, removing significant overburden from a shallowly embedded plate anchor.
Sensitive, Soft Soils • Clean calcareous ooze in deep ocean. • Siliceous ooze (deep ocean). • Other cohesive soil with sensitivity of 6 or greater. • Weak, high-void-ratio clays with $s_u/\bar{p}_o$ of 0.10-0.15.	 • Fluke installation can remold and weaken soil, limiting developed holding capacity. • Same as above. • Same as above. • Long-term capacity may be lower than short-term.

6.3.6 Specialized Survey Tools

Specialized site survey tools have been developed that can support the siting and design of embedment anchors. One, the expendable Doppler penetrometer (Ref. 6-6), indirectly measures the undrained shear strength of the soil (see Chapter 2, Section 2.5.5). Although strength is not determined to the level of accuracy as that from in-situ measurement or coring, the method requires very little on-site ship time and may be cost-effective.

6.4 FLUKE PENETRATION AND KEYING

6.4.1 Penetration Requirement

The distance that a plate anchor needs to be driven or vibrated into the seafloor depends on the required holding capacity, anchor size and design, keying distance and soil characteristics.   Prediction of the penetration requirement is not presented in this handbook.

6.4.2 Keying Prediction

The anchor fluke moves upward a certain distance as it rotates into a horizontal orientation – the position of maximum holding capacity. This keying distance, z_k, is a function of fluke geometry, soil type, soil sensitivity, and duration of time between penetration and keying. However, experience has shown that the Navy's anchor flukes key in about 2.0 fluke lengths when embedded in cohesive soil and in about 1.5 fluke lengths in cohesionless soils. Thus, given the initial fluke penetration, the keyed depth for a Navy anchor fluke is estimated as follows:

In cohesive soil,
$$z = z_p - 2L \tag{6-1}$$

In cohesionless soil,
$$z = z_p - 1.5L \tag{6-2}$$

where:

z = depth of the fluke after keying [L]

z_p = maximum penetration depth of the fluke [L]

L = length of the fluke or plate anchor [L]

Although no recommendation is made for altering the above estimate for z, it is believed that keying distance may be longer in highly sensitive soils but may be shorter as more time elapses between penetration and keying.

6.5 STATIC HOLDING CAPACITY

6.5.1 Loading Conditions

Static loading is a relatively constant load maintained for a long period of time. This is in contrast to dynamic loading, where the short length of time the load is applied significantly affects anchor holding capacity. In reality, loadings on seafloor anchors are rarely completely static but often have impulse or repetitive components. However, at dynamic load levels below certain limits, the anchor-soil response will be as if the system were statically loaded. For cyclic repetitive loading such as that caused by wave effects, the loading can be approximated as static if the change in load is less than 5% of the static load component. For impulse loadings such as that caused by a short tug on a mooring line, the loading can be treated as static when the load development occurs over more than 10 minutes duration in clays and more than 10 seconds in sands and coarse silts (Refs. 6-3 and 6-7).

6.5.2 Deep and Shallow Anchor Failure

Holding capacity depends on the soil failure mode, which is dependent on the anchor plate embedment depth and on the soil type and strength. Anchor failure is characterized as being either shallow or deep (illustrated in Figure 6.5-1). A shallow failure occurs when the seabed surface is displaced by the upward motion of the anchor plate and the soil failure surface continues up to the seabed surface. A deep failure is present when the anchor plate is sufficiently deep within the seabed so that the soil failure surface accompanying movement of the anchor does not reach the seabed surface. The transition from shallow to deep behavior has been found to be a function of relative embedment depth (the ratio of embedment depth to anchor minimum dimension, z/B) and soil strength. It occurs over a range of z/B values in cohesive soil from 2 to 5 and in cohesionless soil from 2 to 10 (Ref. 6-8).

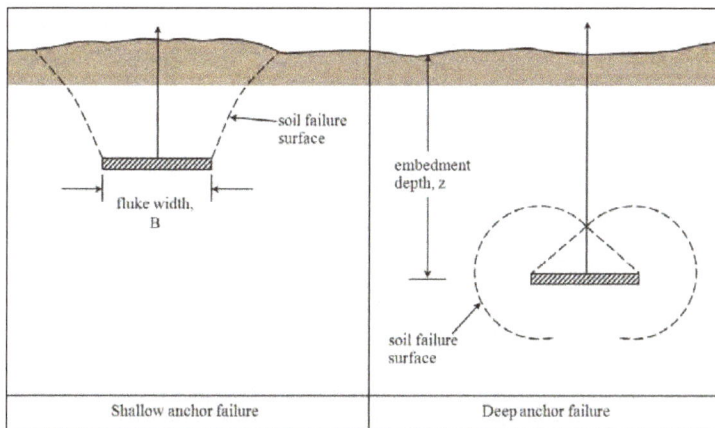

Figure 6.5-1. Failure modes for shallow and deep embedded plate anchors.

6.5.3   Short-Term Capacity in Cohesive Soils

The short-term condition exists when the anchor-caused soil failure is governed by a soil's undrained shear strength. Failure in the short-term condition occurs before significant drainage of pore water can take place. It occurs immediately, or within a few minutes, of full load application. Plate anchor short term holding capacity under static loading conditions in cohesive soils, F_{st}, is:

$$F_{st} = As_u h\overline{N}_{cs}[0.84 + 0.16(B/L)] \tag{6-3}$$

where:

A = projected maximum fluke area perpendicular to direction of pullout [L^2]

s_u = soil undrained shear strength [F/L^2]

h = correction factor for soil disturbance due to penetration and keying

$\overline{N}_{cs}$ = short-term holding capacity factor in cohesive soil

B = plate minimum dimension, usually width [L]

L = plate maximum dimension, usually length [L]

The value of the holding capacity factor, $\overline{N}_{cs}$, is obtained from Figure 6.5-2. It is a function of the soil's undrained shear strength and the relative embedment depth. For the deep failure mode, $\overline{N}_{cs}$ = 15. The disturbance correction is explained in Section 6.5.6.

In some instances, $\overline{N}_{cs}$ from Figure 6.5-2 may be too high for the existing conditions. If drainage vents or tubes allow water to flow rapidly to the underside of the plate anchor, then the suction formed on the underside of the plate will be relieved. If this happens, the value of $\overline{N}_{cs}$ should be reduced to that for the long-term holding capacity factor, $\overline{N}_c$, from Figure 6.5-3 (Ref. 6-8).

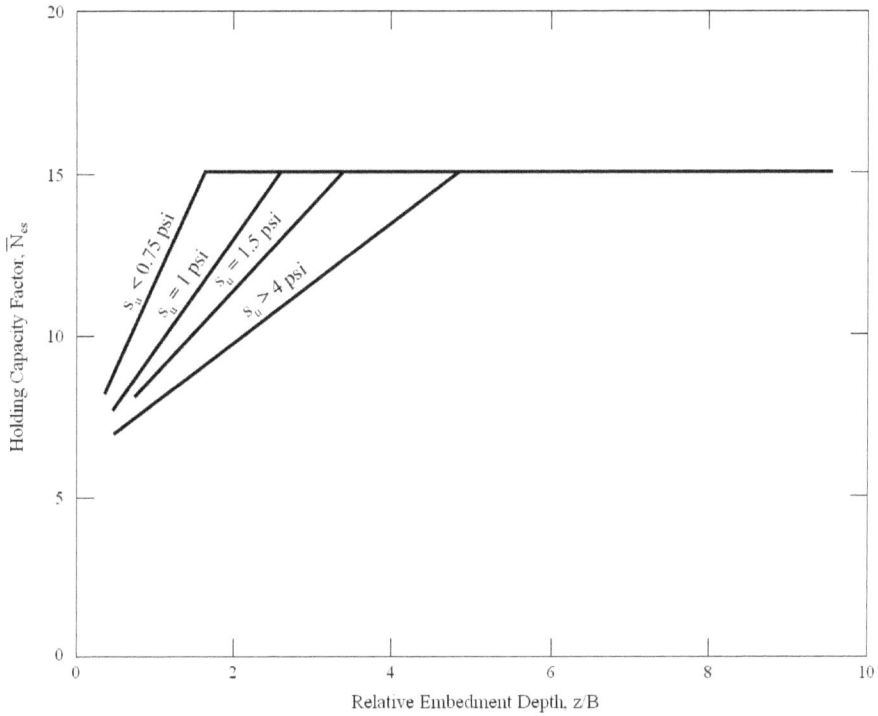

Figure 6.5-2. Short-term holding capacity factors for cohesive soil where full suction develops beneath the plate.

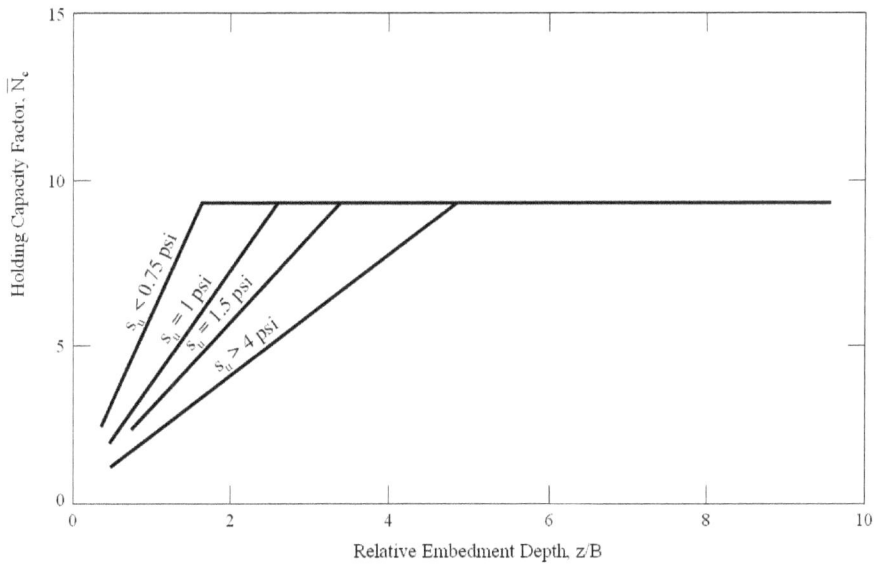

Figure 6.5-3. Long-term holding capacity factors and short-term no-suction factors for cohesive soils.

6-13

6.5.4 Long-Term Capacity in Cohesive Soils

The long-term condition exists when a static load is applied to the anchor over a time sufficiently long to allow near-complete dissipation of excess pore water pressures. This time duration may be a day for silts, a week for silty clays, and considerably longer for clays. In cohesive soils, the long-term holding capacity is governed by the effective soil drained strength parameters: the drained cohesion intercept, $\bar{c}$; and the drained friction angle, $\bar{\phi}$. The long-term static holding capacity, F_{lt}, is:

$$F_{lt} = A(\bar{c}\overline{N}_c + \gamma_b z\overline{N}_q)[0.84 + 0.16(B/L)] \tag{6-4}$$

where:

$\bar{c}$ = drained soil cohesion [F/L^2]

$\overline{N}_c$ = long-term holding capacity factor in cohesive soil (from Figure 6.5-3)

γ_b = buoyant unit weight of the soil [F/L^3]

$\overline{N}_q$ = holding capacity factor for a drained soil condition (from Figure 6.5-4)

Note that the limiting value of F_{lt} is F_{st} (the applied long-term load cannot exceed the short-term load without initiating a failure), i.e. F_{lt} (max) = F_{st}.

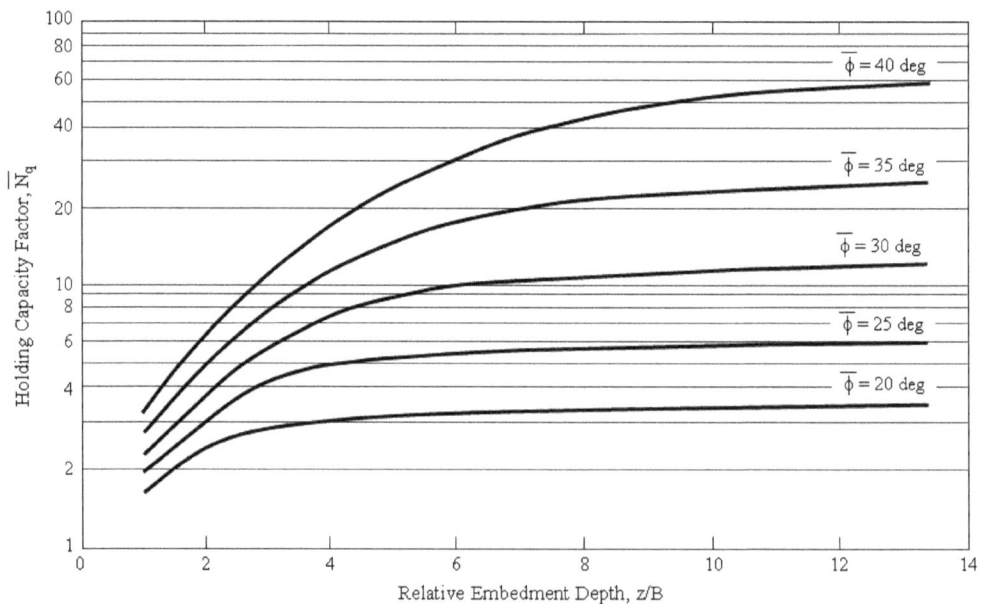

Figure 6.5-4. Holding capacity factors for cohesionless soils.

For very soft underconsolidated sediments, such as delta muds, the shear failure mode may be different than with normally consolidated sediments. Section 3.5.2 of Chapter 3 gives a brief discussion of underconsolidated soils. This condition can result in a lower holding capacity than would be calculated by Equation 6-4. The reduced holding capacities in these very soft soils can be conservatively predicted by reducing the $\bar{c}$ and $\bar{\phi}$ values before entering Figure 6.5-3 and Figure 6.5-4 to obtain the holding capacity factors. The new $\bar{c}$ and $\bar{\phi}$ values are calculated as follows:

$$\bar{c}' = \left(\frac{2}{3}\right)\bar{c} \tag{6-5}$$

$$\bar{\phi}' = \arctan\left(\frac{2}{3}\tan\bar{\phi}\right) \tag{6-6}$$

6.5.5 Short- and Long-Term Capacity in Cohesionless Soils

The short-term loading condition in cohesionless soils is also the drained or long-term failure condition. In sands and gravels, virtually all of the excess pore water pressures resulting from a static loading dissipate as the load is applied. Thus, for cohesionless soils, the soil failure is assumed to be drained for both static short- and long-term conditions.

The static holding capacity in cohesionless soils (sands and gravels) for both short- and long-term conditions, F, is:

$$F = A\gamma_b z\bar{N}_q[0.84 + 0.16(B/L)] \tag{6-7}$$

where:

A = projected maximum fluke area perpendicular to direction of pullout [L²]

γ_b = buoyant unit weight of the soil [F/L³]

z = depth of the fluke after keying [L]

$\bar{N}_q$ = holding capacity factor in cohesive soil

B = plate minimum dimension, usually width [L]

L = plate maximum dimension, usually length [L]

The holding capacity factor $\bar{N}_q$ is obtained from Figure 6.5-4 using the relative embedment depth and the soil friction angle. When dealing with very loose sands (i.e., relative density less than 40%), the soil friction angle, ϕ, should be reduced in the same manner as for the drained cohesive soil case (i.e., by Equation 6-6) before Figure 6.5-4 is entered.

6.5.6 Disturbance Corrections

Equation 6-3 contains a correction factor for soil disturbance to correct for soil remolding during fluke penetration and keying. Values for the factor, h, were determined for the four soil types listed in Table 6.5-1 by anchor tests. These values are recommended for application to similar soil types in calculating plate anchor holding capacity (Ref. 6-9).

Soil sensitivity, S_t, (the ratio of undisturbed to remolded shear strength), is an important indicator of the amount of disturbance likely. For soils with considerably different S_t values than those reported in Table 6.5-1, an estimate must be made for the h value. More sensitive soils should display greater reductions in strength and be assigned a lower value of h.

Table 6.5-1. Values for Strength Reduction Factor for Use in Equation 6-3

Soil Type	Strength Reduction Factor, h
Very soft, moderately sensitive clayey silt: $s_u \cong 1$ psi, $S_t \cong 3$	0.8-0.9
Soft, normally consolidated, silty clay: $s_u \cong 2$ psi, $S_t \cong 3$	0.8
Pelagic clay: $s_u \cong 1.2$ psi, $S_t \cong 3$	0.7
Foraminiferal sand-silt, 77-86% carbonate: $s_u \cong 2.2$ psi, $S_t \cong 10$	0.25

6.5.7 Factors of Safety

The factor of safety to be applied to anchor holding capacity determined from Equations 6-3, 6-4, and 6-7 varies with the type and purpose of the mooring and with the level of environmental data on the site. For those applications where little is known about the soil conditions at the site, or for critical installations, a safety factor of 3 is recommended. When adequate site data permit a high level of design confidence, or when that mooring element is noncritical, the factor of safety may be reduced to 2.

6.6 DYNAMIC HOLDING CAPACITY

6.6.1 Loading Conditions

Dynamic loads are defined as those rapidly applied but of short duration (< 1 minute). They are divided into two categories: (1) cyclic or repetitive loadings and (2) impulse loading (basically a single event). Both types can alter plate anchor holding capacity by changing the existing conditions in the soil surrounding the anchor. These loading types are illustrated by the anchor line load history in Figure 6.6-1 and are discussed in this section. The calculation procedures in this section provide rough but conservative estimates of the effect of dynamic loadings on anchor capacity. A more complete discussion on these loadings and their causes is given in Reference 6-10.

6.6.2   Cyclic Loading

6.6.2.1 Definitions

Cyclic loading can be considered an impulse loading that occurs in a repetitive manner rather than as a single event. For design purposes cyclic loadings are separated into three categories: (1) cyclic line loading of the anchor that may lead to a soil strength loss in the vicinity of the anchor and subsequent anchor failure; (2) cyclic line loading that may cause anchor upward movement (creep), which could accumulate to move the anchor into more shallow soil and thereby lower its short-term static holding capacity; and (3) earthquake-caused cyclic loading of the soil mass with resulting near-complete loss of strength in the entire soil mass and a sudden anchor failure. Cyclic loads are characterized by a pure cyclic "double-amplitude" loading component, P_c, superimposed on a basically static loading component, P_s. Cyclic and static load magnitudes are expressed as a percentage of the static short-term anchor holding capacity (as   determined by Equations 6-3 or 6-7). Figure 6.6-1 illustrates this nomenclature with an example where P_s is approximately 18% of the short-term holding capacity, P_c is approximately 33% of that capacity, and four load cycles occur within approximately 0.55 minute. In design, a cyclic load must have a double amplitude greater than 5% of the static short-term holding capacity. Smaller cyclic loads are difficult to measure or predict and can be ignored in the design.

Two additional parameters are required to describe a cyclic loading condition. The first is the total number of load cycles expected in the anchor's lifetime, n_T. This parameter is needed to evaluate the potential for anchor creep. The second is the number of cycles, n_c, that occurs in a limited time period required for dissipation of excess pore pressure, t_{cd}. The parameter n_c is used to evaluate soil strength loss and potential for liquefaction.

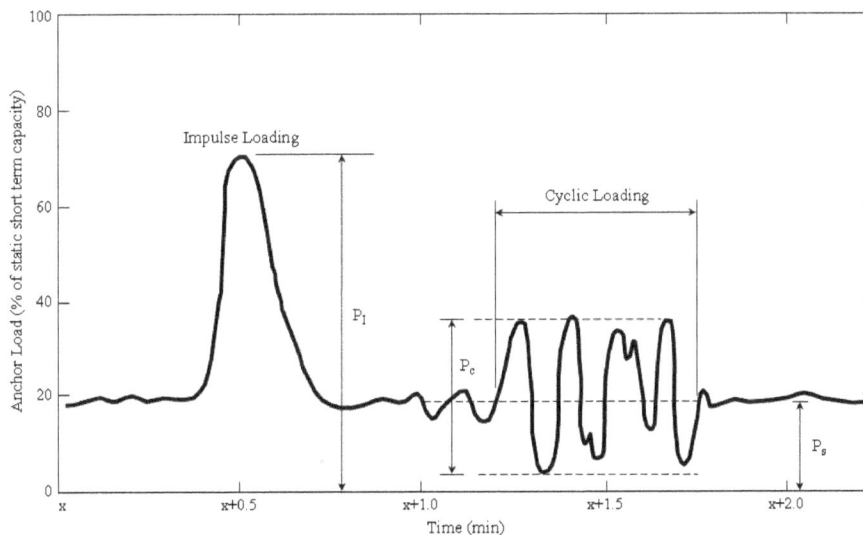

Figure 6.6-1. Nomenclature for types of non-steady loading (Ref. 6-10).

6-17

6.6.2.2 Strength Loss

Virtually all soils are subject to some strength loss from extended cyclic loading. The amount of strength loss, however, varies considerably depending on soil type, state, and the nature of the cyclic loading. In general, the following lower the soil's susceptibility to strength loss: a denser soil, a more plastic soil, a lower cyclic loading magnitude, a smaller number of load cycles, and a longer time period over which the cycles occur.

Some low relative density cohesionless soils are susceptible to complete liquefaction. Sediments of this type (such as uniform fine sands, coarse silts, and some clean deep-sea oozes) can experience a near-total strength loss under cyclic loading. Use of plate anchors in these soils is not recommended if significant cyclic loading is expected.

For other soils, the susceptibility of a given plate anchor to cyclic-load-caused strength reductions can be evaluated by estimating the maximum cyclic load that can be sustained by the anchor without pore pressure dissipation, this is done using the soil permeability, k, whose value must be determined from testing of undisturbed soil from the vicinity of where the anchor will lie in the soil mass. Table 6.6-1 shows typical values of k, which are used with Figure 6.6-2 to estimate t_{cd}, the time required for dissipation of excess pore pressures. The maximum number of double amplitude cyclic loadings that can occur within the time period t_{cd} is estimated from the known or expected loading conditions. Figure 6.6-3 is then used to determine the maximum value of cyclic load, P_c, that can be sustained without a significant soil strength loss.

Figure 6.6-3 can also be used to determine the maximum number of cycles that can be sustained without soil strength loss for a given cyclic load level. The prediction curves of Figure 6.6-3 apply where the average static load is less than 33% of the anchor's static short-term holding capacity. When the average static load is greater than 33% of the static holding capacity, an adjustment must be made to the cyclic and static loads before using Figure 6.6-3. The amount of the static load above 33% (P_s - 33) is added to the value of P_c. This new value of P_c is then used to determine the maximum number of cycles from Figure 6.6-3.

Table 6.6-1. Average Values of Soil Permeability (Ref. 6-10).

Soil Type	Permeability, k (fps)
Uniform Coarse Sand	1×10^{-2}
Uniform Medium Sand	3×10^{-3}
Well-Graded Clean Sand	3×10^{-3}
Uniform Fine Sand	1×10^{-4}
Well-Graded Silty (Dirty) Sand	1×10^{-5}
Uniform Silt	2×10^{-6}
Silty Clay	3×10^{-8}
Low Plasticity Clay (Kaolinite), $PI < 20$	3×10^{-8}
Medium Plasticity Clay (Illite), $PI = 20\text{-}60$	3×10^{-9}
High Plasticity Clay, $PI = 60\text{-}200$	3×10^{-10}
Very High Plasticity Clay, $PI > 200$	3×10^{-11}

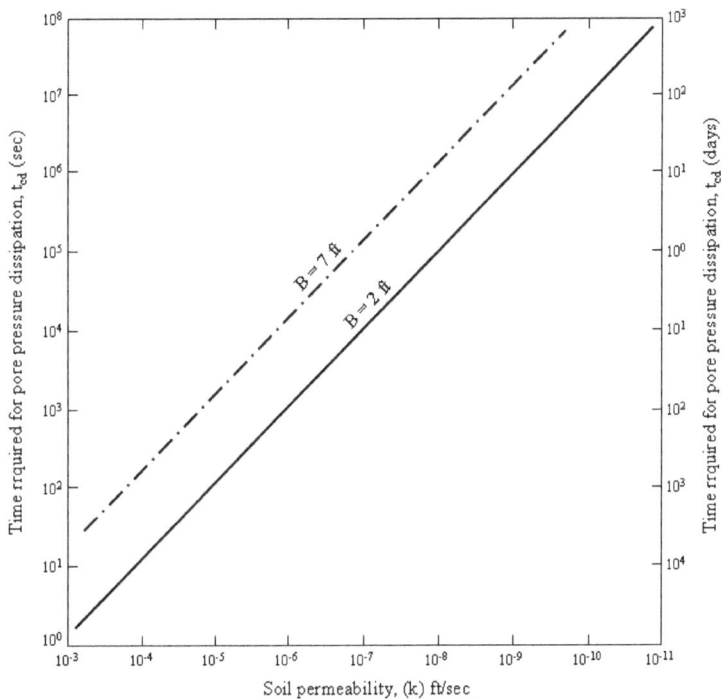

Figure 6.6-2. Time required for dissipation of stress-induced excess pore pressure (Ref. 6-10).

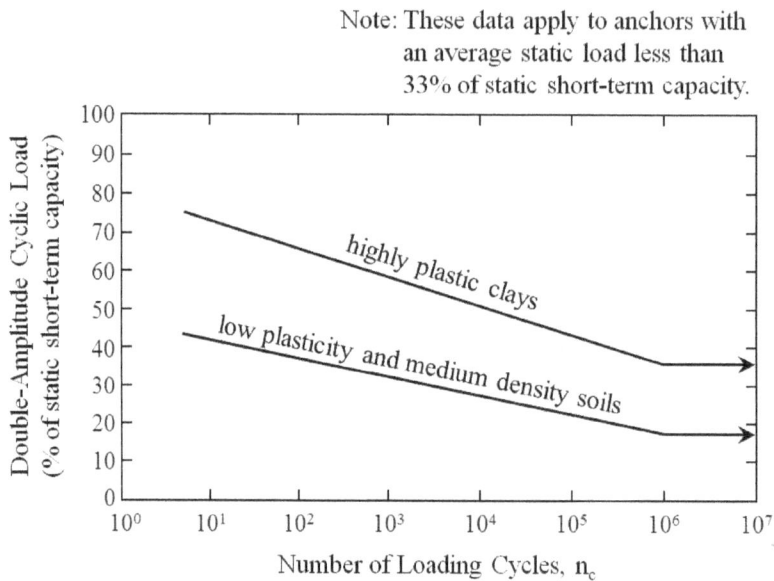

Note: These data apply to anchors with an average static load less than 33% of static short-term capacity.

Figure 6.6-3. Maximum cyclic load capacity without soil strength loss (Ref. 6-10).

6.6.2.3 Cyclic Creep

Cyclic creep of an embedment anchor can occur under loading conditions that appear quite safe relative to the above criteria for cyclic strength loss. To evaluate cyclic creep potential, the number and magnitude of significant loading cycles occurring during the lifetime of an anchor are the controlling load factors. This loading usually must be summarized by statistical techniques, in a spectral or quasi-spectral format, to estimate total number of uniform loading cycles that can occur over the anchor's lifetime (Ref. 6-10).

The maximum number of cyclic loads that can occur over the lifetime of an anchor without significant upward creep is shown by the curves presented in Figure 6.6-4. These criteria are applicable to cases where the average static load, P_s, is less than 20% of static short-term anchor capacity. For cases where the 20% static load criterion is exceeded, the double-amplitude cyclic load, P_c, should be adjusted by an amount equal to that portion of P_s above 20% (i.e., by P_s - 20). For example, if the static load component of a mooring is 28% of the static short-term holding capacity, and the appropriate curve on Figure 6.6-4 indicates an allowable maximum double-amplitude cyclic load of 30% over the lifetime of the structure, then that allowable double-amplitude cyclic load should be reduced by 8% (28% minus 20%). Therefore, the allowable cyclic load becomes 22% (30% minus 8%). This procedure is very conservative for anchor systems expected to have a long service life and be subjected to many cycles of significant cyclic loading. It was designed to be conservative because cyclic creep of anchors is not well-understood.

6-20

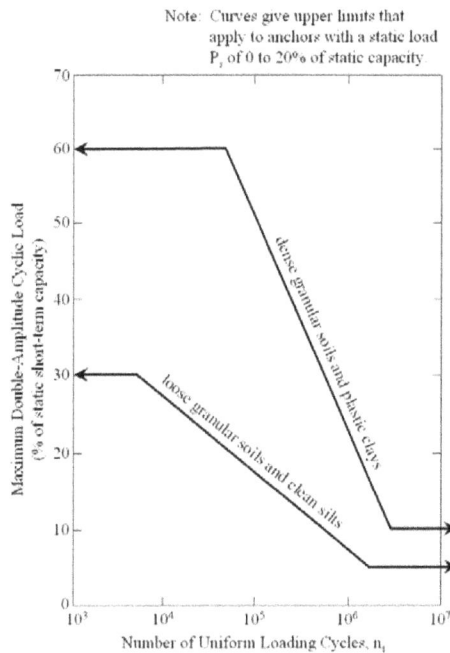

Figure 6.6-4. Maximum (lifetime) cyclic load capacity without development of cyclic creep (Ref. 6-10).

6.6.2.4 Factors of Safety

Because the above approaches to design for cyclic loadings are quite conservative, a lower safety factor is recommended for use with the cyclic loading aspects of anchor design. It is recommended that the following factors of safety be used: 1.75 for critical installations or where the soils data are not well known and 1.25 for noncritical installations or where the soils data are very well known.

6.6.3   Earthquake Loading

In contrast to line-applied cyclic loading of an anchor, earthquakes load the entire soil mass. The loading, however, occurs for only some 10 to 30 significant and rapidly applied loading cycles. Cohesive soils and most medium to dense sands (materials that are less susceptible to liquefaction) subjected to such loading are treated by the techniques outlined in Section 6.6.2. Relatively clean cohesionless soils, of medium to low density, are considerably more susceptible to liquefaction under earthquake loadings. Anchors embedded in these soils, even when under very low static loads, will completely fail and pull out if the soil liquefies. The prediction of such events is not treated here, as it is not well-understood and is very complex;

6-21

however, techniques for evaluating the liquefaction potential of soils are presented in References 6-7 and 6-10.

6.6.4 Impulse Loading

6.6.4.1 Definitions

Procedures for predicting the holding capacity under impulse loading are presented below. They are appropriate for use with circular, square, or rectangular ($L/B \leq 2$) anchor flukes only. The procedures consist of applying a number of influence factors to Equations 6-3 and 6-7, the basic equations for calculating static short-term anchor holding capacity. The influence factors yield conservative results in all cases.

An impulse load, P_I is defined as basically a single load that is applied quickly but does not remain for a long time (less than 10 minutes for clays and less than 10 seconds for sands). An example is shown in Figure 6.6-1. These loads are considered single events only where enough time elapses between similar events to allow the soil to return to its "normal" state without any residual effect from other impulse loads. In the absence of other dynamic loads, impulse loading will result in a higher anchor holding capacity during the loading event than that computed for the static short-term holding capacity. That is, the anchor will be able to resist an impulse loading higher than its static short-term capacity.

6.6.4.2 Cohesive Soil

The holding capacity under impulse loading, F_I, is the maximum load that can be applied to an anchor under impulse conditions. It is determined for cohesive soils by:

$$F_I = I R_c R_I R_f I_f (F_{st})$$

(6-8)

where:

F_{st} = static short-term anchor holding capacity (Equation 6-3 for cohesive soil) [F]

I = influence factor for adjusting the soil strength for strain rate

R_c = reduction factor for cyclic loading

R_I = reduction factor for repeated impulse loading

I_f = inertial factor for capacity increase under very rapid and short duration loading (i.e., for impulse duration less than 0.01 second)

The factor I is a strain rate used to adjust for an increase in the soil undrained shear strength during impulse loading. The value of I is obtained from Figure 6.6-5 on the basis of impulse load duration and a general description of soil type.

6-22

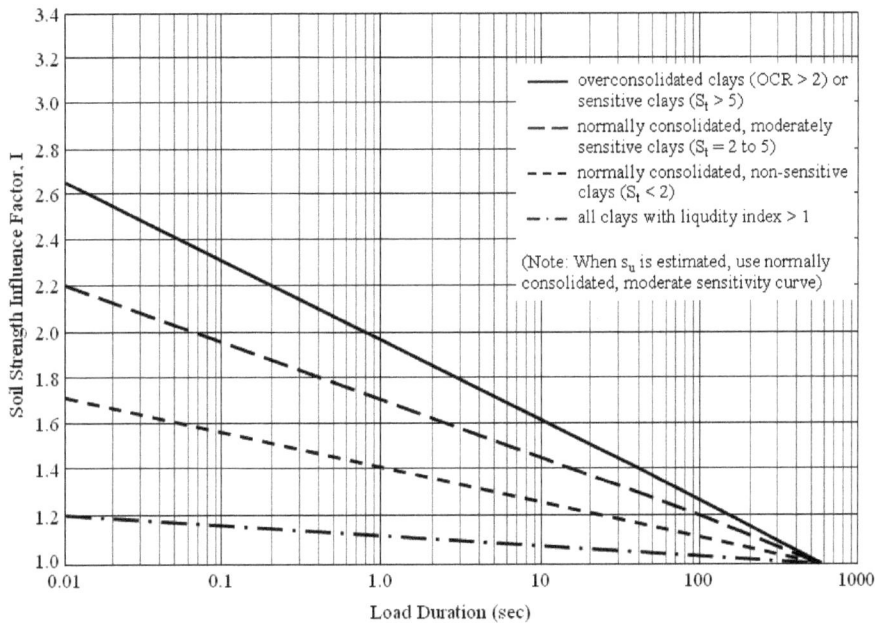

Figure 6.6-5. Strain rate factor, I, for cohesive soil (Ref. 6-11).

The factor R_c depends on the loading history prior to the impulse loading. It is used to adjust F_I for the influence of other nonstatic loads that are occurring at about the same time and is determined as follows: (1) if the impulse load is the first event, then $R_c = 1.0$; or (2) if cyclic loads immediately precede the impulse event, then:

$$R_c = P_c / F_{st} \qquad (6\text{-}9)$$

The factor R_I adjusts F_I for repeated impulse loadings. It is determined as follows: (1) if there is only one impulse load in a 4-hour period, then $R_I = 1$; or (2) if there is more than one impulse load in a 4-hour period, then:

$$R_I = 1.33 e^{-1.15 \bar{f}_c} \qquad (6\text{-}10)$$

where $\bar{f}_c$ is the average frequency, in impulses per hour, over a 4-hour period.

The factor I_f, is used to adjust (increase) F_I for the inertia of the soil mass at very short duration loadings (i.e., where the loading is known to be applied for less than 0.1 second). It is determined from Figure 6.6-6.

6-23

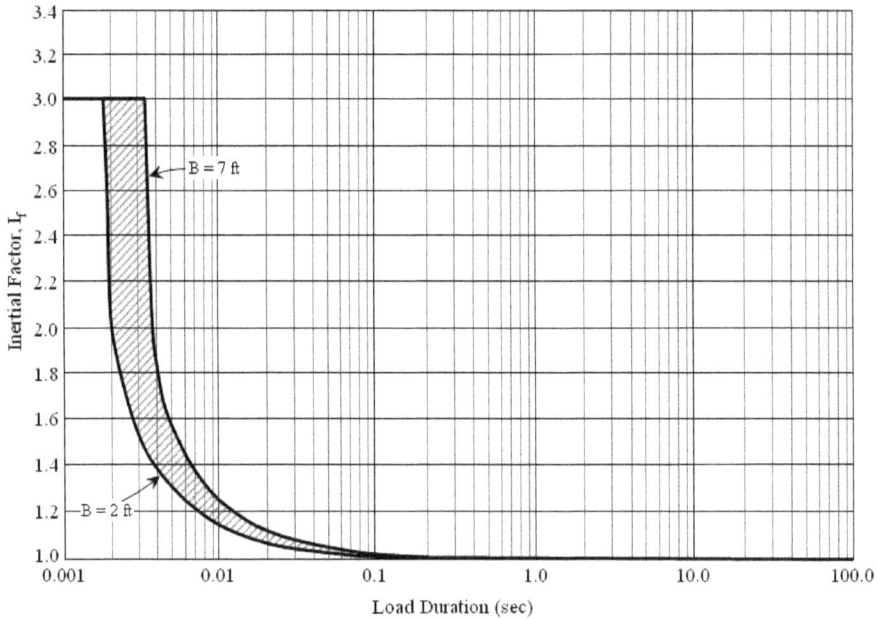

Figure 6.6-6. Inertial factor, I_f, for cohesive and cohesionless soils (Ref. 6-11).

6.6.4.3 Cohesionless Soil

The holding capacity under impulse loading in cohesionless soils is also derived by applying a series of influence factors to the calculated short-term static holding capacity. The maximum anchor capacity impulse loading is given by:

$$f_I = \left(\frac{\overline{N}_{qI}}{\overline{N}_q} \right) R_c R_I I_f (F) \tag{6-11}$$

where:

F = static short-term anchor holding capacity (Equation 6-7 for cohesionless soil) [F]

$\overline{N}_{qI}$ = cohesionless soil holding capacity factor adjusted for impulse loading

$\overline{N}_q$ = cohesionless soil holding capacity factor (from Figure 6.5-4)

R_c = reduction factor for cyclic loading

R_I = reduction factor for repeated impulse loading

6-24

I_f = inertial factor for capacity increase under very rapid and short-duration loading (i.e., for impulse duration less than 0.01 second)

The adjusted holding capacity factor, $\overline{N}_{ql}$, accounts for the effect of the impulse loading on the soil friction angle, ϕ. In order to obtain $\overline{N}_{ql}$, first a new friction angle adjusted for the impulse loading effect, ϕ_I, is calculated as shown below (Ref. 6-11). Then, $\overline{N}_{ql}$ is determined from the curves in Figure 6.5-4.

$$\phi_I = \sin^{-1}\left[\frac{I\sin\phi}{1+(I-1)\sin\phi}\right] \tag{6-12}$$

where I is the influence factor for adjusting the soil strength (from Figure 6.6-7 for cohesionless soil).

The factor R_c is determined in the same manner as for cohesive soil. This is described in Section 6.6.4.2.

The factor for repeated impulse loadings, R_I, is dependent on the frequency of those impulse loads, $\overline{f_s}$: (1) if $\overline{f_s}$ ≤ one impulse/10 min, then R_I = 1.0; or (2) if $\overline{f_s}$ > one impulse/10 min, then R_I is obtained from:

$$R_I 2e^{-0.116\overline{f_s}} \tag{6-13}$$

where $\overline{f_s}$ is the average number of impulses per 10 minutes.

The inertial factor I_f is determined in the same manner as for cohesive soil (described in Section 6.6.4.2). I_f is determined from Figure 6.6-6.

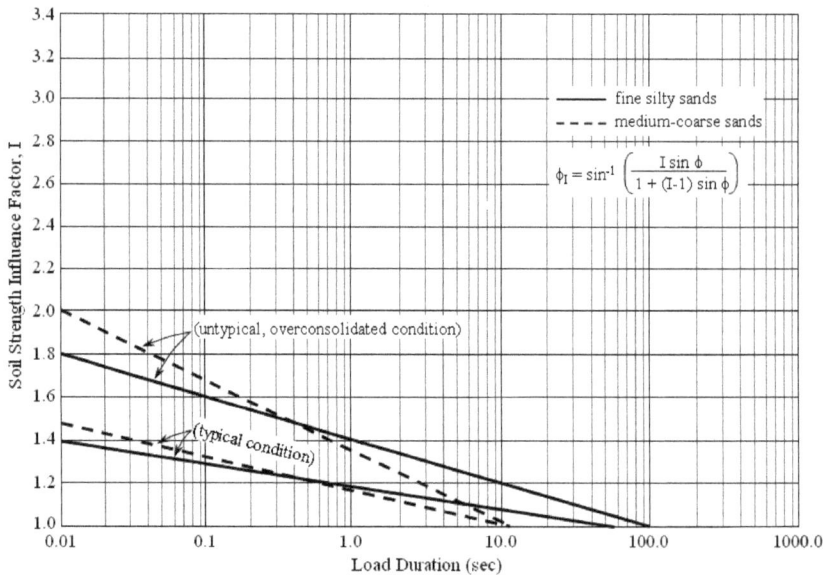

Figure 6.6-7. Strain-rate factor, *I*, for cohesionless soil (Ref. 6-11).

6.6.4.4 Factors of Safety

To calculate allowable design loadings from the maximum impulse loadings calculated from this section, the factors of safety recommended in Section 6.5.7 should be applied.

6.7 OTHER INFLUENCES ON HOLDING CAPACITY

6.7.1 Holding Capacity on Slopes

Two major items should be considered in predicting the holding capacity of embedded plate anchors on slopes: first, stability of the slope itself under the additional influence of the embedment anchor; and second, the influence of the inclined seafloor on the soil resistance mobilized (i.e., holding capacity) by the loaded anchor.

The effect of an anchor on slope stability is extremely complex, involving the effects of anchor installation and anchor loading on the slope. Table 6.7-1 presents a list of factors associated with direct-embedment anchor installation and loading which influence submarine slope stability (Ref. 6-12). All factors lead to a lower resistance to soil mass down-slope sliding and, therefore, greater slope instability. It is important to note that the influence of plate anchors on slope stability depends to a high degree on the type and sensitivity of the sediment. Slope angle itself is not a clear indicator of potential problems.

The inclined seafloor also influences the anchor by lowering the resistance to soil failure and, therefore, lowering the holding capacity that can be developed. A significant portion of the soil shear strength is mobilized to support the soil slope against gravitational forces. For downslope loading, most of the soil shear stresses developed to resist anchor pull-out will be in addition to those resisting slope failures. The result is that a smaller amount of the soil's shear strength is available to resist anchor pullout than is available with a horizontal seafloor. This becomes less important as the anchor is loaded in a more vertical or more upslope direction.

Methods for calculating the reduced holding capacity of anchors on slopes have not been developed. However, as a conservative approach, the holding capacity of a direct-embedment anchor on a specific slope can be estimated by multiplying the holding capacity calculated for a horizontal seafloor by a reduction factor, R_s, determined by:

$$R_s = \frac{F_s - 1}{F_s} \tag{6-14}$$

where F_s is the factor of safety against a slope failure without the anchor. R_s represents the amount of soil strength remaining, or the amount not mobilized in maintaining the slope stability. Note that in computing the anchor holding capacity, both anchor depth and the holding capacity factors $\overline{N_c}$ and $\overline{N_q}$ are based on a depth of embedment measured perpendicular to the seafloor surface.

Table 6.7-1. Factors Associated With Direct-Embedment Anchors Which Can Influence Submarine Slope Stability (Ref. 6-12)

Factor	Reasons for Consideration
Impact Loading During Embedment	Effect similar to earthquake loading but with greater local influence; more critical problem in loose soils.
Remolding or Disturbance of Soils During Installation	Needs consideration in all anchor installations; effect varies significantly from one soil to another.
Cyclic Loading by Anchor	Important consideration in all anchor installations.
Local Instability After Anchor Pullout	Potentially can progress to major slope failure.
Direct Application of Anchor Load to Slope	Probably not more significant than a local instability problem but can progress into a large slide.

6.7.2 Creep Under Static Loading

Creep failure of direct-embedment anchors under static loading conditions is possible. In some onshore soils, data have been taken showing soil strength reductions of 60% for some soft cohesive soils under only static loading (Ref. 6-13). However, tests on two pelagic clays (Ref. 6-14) and a calcareous ooze (Ref. 6-15) indicate that, for the undrained condition, shear strength reductions may not be nearly so great for seafloor sediments. Further, for soil overconsolidation ratios (OCR) reasonable in seafloor soils, negative pore pressures will not be generated above the anchor plate, and the shear strength above the anchor plate will always increase with time (Ref. 6-14). Therefore, the creep failure potential for plate anchors under only static loading is minimal. A safety factor of 2 against creep failure is recommended on the maximum long-term holding capacity (Ref 6-3).

6.8 HOLDING CAPACITY IN CORAL AND ROCK

6.8.1 Coral

A large number of plate anchors have been driven into coral, such as in Apra Harbor, Guam (see Figure 6.2-1b). This anchor design has a taper ram on the end of the beam to help break up the coral locally. The plate is then tapered and sharpened. Large impact hammers and heavy follower beams are then use the drive the anchor into the coral. Experience shows 300 kips (1,300 kN) working capacity can easily be achieved by driving the anchor through a hard layer and into a softer layer. The anchor then keys up against the layer of rock.

6.8.2 Rock

There is little experience with plate anchors in rock and other anchors, such as a drilled and grouted pile, may be more suitable.

6.9 DESIGN OF PILE-DRIVEN PLATE ANCHORS

6.9.1 General

A pile-driven plate anchor (PDPA) is constructed from a steel plate, an I-beam section, and a pad eye (Ref. 6-1). The PDPA is driven into the seafloor sediments and keyed to provide a fixed-point anchor mooring system, as illustrated in Figure 6.9-1. For use in soft clays or mud, a keying flap is added to accelerate the keying action. The procedure for installing a PDPA is described below:

- Attach a PDPA with chains or cable to a follower (generally a section of I-beam).

- Use a crane to lift the PDPA and follower, and place the anchor tip on the seafloor.

- Drive the PDPA into seafloor to a pre-determined depth with either an impact or vibratory hammer.

- Retrieve the anchor follower.

- Pull the chain or cable to key the plate anchor and lock it in position in the bottom.

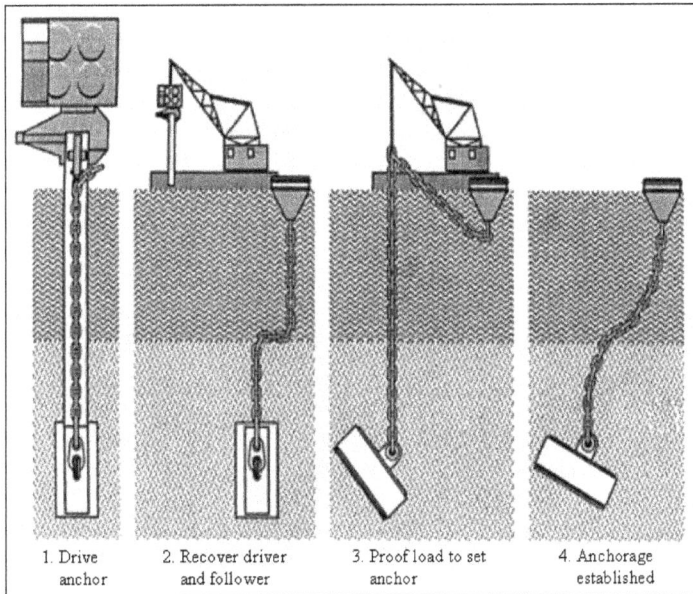

Figure 6.9-1. Installation of Pile-Driven Plate Anchor.

6.9.1.1 Site Investigation

A site investigation should be performed to determine bathymetry, topography, soil strata, soil strengths, and engineering properties as discussed in Section 6.3.3. Soil samples obtained from field investigation may be sent to a laboratory to determine soil engineering properties, including soil shear strength, cohesive, angle of internal fiction, relative density and submerged unit weight. In addition to the soil investigation, seafloor soundings may be conducted, and it is necessary to determine the locations of any utility lines, cables, pipes, unexploded ordinance or other debris prior to installation.

6.9.1.2 Factors of Safety

A factor of safety of 2 is generally recommended for most applications of pile-driven plate anchors.

6.9.1.3 Shape of Plate Anchor

A typical plate anchor is made of A36 steel with recommended length to width ratio of 1.5 to 2.0. The plate thickness and other steel components vary with the design load in accordance of AISC Steel Construction Manual (Ref. 6-16). For a plate anchor driven in soft soils and mud, a key flap is needed to ensure its key-in, as illustrated in Figure 6.9-2.

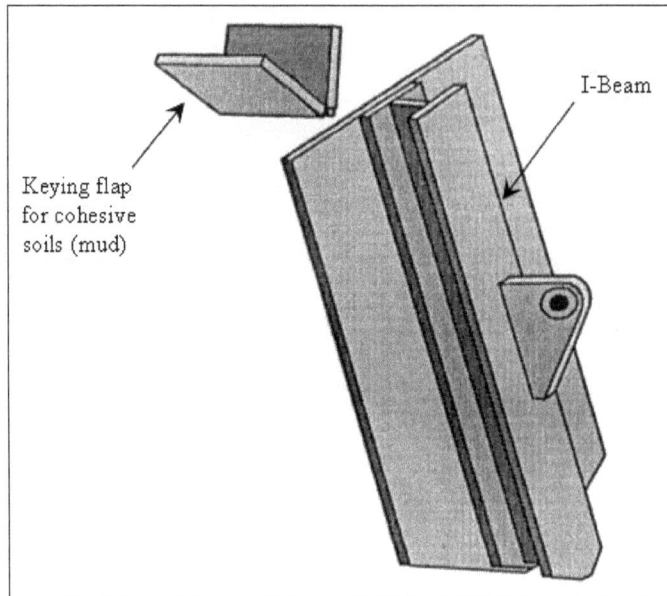

Figure 6.9-2. Typical shape and components of a pile-driven plate anchor (plate length to width ratio: 1.5 to 2.0).

6.9.2   Ultimate Holding Capacity of Plate Anchor in Soft Clays and Mud

The ultimate holding capacity can be predicted by Equations 6-3 and 6-4 for short- and long-term anchor holding capacities respectively. The holding capacity factor, $\overline{N_c}$ varies with soil shear strength and relative embedment depth as shown in Figure 6.5-2 and Figure 6.5-3 for both short- and long-terms conditions respectively. For keyed depths more than five times the plate width, the holding capacity factors are considered to be constant as:

$\overline{N_c}$ =   15, if the undrained shear strength is known and the loading is short term;

=   12, recommended relationship for all cases;

=   9, non-saturated conditions or long-term loading with consolidated-drained shear strength parameters

6.9.3   Holding Capacity of Plate Anchor in Sands

The anchor holding capacity for a driven plate anchor in sands can be estimated by Equation 6-7 and the bearing capacity factor, $\overline{N_q}$ in sands is shown in Figure 6.5-4. Since $\overline{N_q}$ varies with the angle of internal friction and it is difficult to obtain undisturbed sand samples underwater, it is recommended that standard penetration test (SPT) be performed at the anchor site. Based upon the blow counts of the SPT, the angle of internal friction can be estimated as shown in Chapter 5, Table 5.3-1.

6.10 EXAMPLE PROBLEMS

6.10.1 Problem 1 – A Pile-Driven Plate Anchor Used in Cohesive Soil

6.10.1.1 Problem Statement

Design a pile-driven plate anchor for specified required mooring loads. Also, determine the depth to which the anchor is driven prior to keying.

Data: Due to limited mooring space in a harbor, it is decided to install a pile-driven plate anchor to provide a short-term, horizontal mooring load of 200 kips. The water depth is 25 feet, and sediments in the harbor are mud of soft silty clay with a shear strength increasing at a rate of 10 psf per foot of depth. The depth of bed rock is more than 100 feet. The contractor has pile-diving equipment and a section of 90-foot pile follower for the installation. Assume the embedded anchor chains contribute 25 kips to the holding capacity of the anchor.

6.10.1.2 Problem Solution

The trial-and-error computational procedures for the problem's solution are presented below. These follow the procedures outlined in this chapter.

Problem 6.10-1

ANALYTICAL PROCEDURES	COMPUTATIONS
1. Check the maximum allowable penetration depth of the plate anchor.	Given a 90-ft pile follower, the maximum allowable penetration depth is $= 90\,ft - 25\,ft$ of water depth $\qquad$ + plate anchor length, L $= 65\,ft + L$

Problem 6.10-1

ANALYTICAL PROCEDURES	COMPUTATIONS
2. Estimate the ultimate anchor holding capacity using an appropriate factor of safety.	Required horizontal mooring load = 200 *kips* Chain holding capacity = 25 *kips* (given) $F_s = 2.0$ Ultimate horizontal anchor holding load is = (200 *kips* − 25 *kips*)(2.0) = 350 *kips*
3. Select a trial key-in depth, z, and estimate the soil shear strength, s_u, at the trial depth.	Say $z = 60\,ft$ s_u @ 60 *ft* = (60 *ft*)(10 *psf/ft*) = 600 *psf*
4. Determine the short-term holding capacity factor, $\overline{N}_{cs}$ (from Figure 6.5-2 with the z/B ratio).	Assume $B = 6\,ft$ z/B = (60 *ft*)/(6 *ft*) = 10 $\overline{N}_{cs}$ = 15
5. Determine the strength reduction factor, h (from Table 6.5-1 for silty clay).	$h = 0.8$
6. Select the plate anchor size.	Try $L = 10\,ft$ with $B = 6\,ft$ $A = L{\cdot}B$ = (10 *ft*)(6 *ft*) = 60 *ft²* B/L = (6 *ft*)/(10 *ft*) = 0.6

Problem 6.10-1

ANALYTICAL PROCEDURES	COMPUTATIONS
7. Compute the anchor holding capacity, F_{st}, based on the plate anchor size and key-in depth. From Equation 6-3, $F_{st} = A\ s_u\ \overline{h}\ \overline{N_{cs}}\ [0.84 + 0.16(B/L)]$	$F_{st} = (60\ ft^2)(600\ psf)(0.8)(15)[0.84 + 0.16(0.6)]$ $= 404{,}352\ lb = 404\ kips$ OK; F_{st} > required horizontal mooring load of 350 *kips*. Try to reduce plate anchor size. Try $L = 10\ ft$ with $B = 5\ ft$ $A = 50\ ft^2$ and $B/L = 0.5$ For $z/B = (60\ ft)/(5\ ft) = 12$, $\overline{N_{cs}} = 15$ $F_{st} = (50\ ft^2)(600\ psf)(0.8)(15)[0.84 + 0.16(0.5)]$ $= 331{,}200\ lb = 331\ kips$ NO GOOD; F_{st} < 350 kips. Use $L = 10\ ft$ with $B = 6\ ft$.
8. Check the plate length to width ratio. Is $1.5 \le (L/B) \le 2.0$?	$(L/B) = (10\ ft)/(6\ ft) = 1.67$ YES, the length to width ratio is acceptable.
9. Calculate the maximum penetration depth from Equation 6-1 and check against the maximum allowable penetration depth. $z = z_p - 2L$	From STEP 1, max allowable penetration depth is $(65\ ft + 10\ ft) = 75\ ft$ From Equation 6-1, $z_p = z + 2L = 60\ ft + 2(10\ ft) = 80\ ft > 75\ ft$ Unacceptable. Use a 5-ft follower extension or try a larger anchor key-in at a shallower depth.

Problem 6.10-1

ANALYTICAL PROCEDURES	COMPUTATIONS
10. Try another (shallower) key-in depth and larger plate anchor.	Try $z = 50\,ft$ s_u @ $50\,ft = (10\,psf/ft)(50\,ft) = 500\,psf$ Try $L = 12\,ft$ with $B = 6\,ft$ $z/B = (50\,ft)/(6\,ft) = 8.33$ $\overline{N_{cs}} = 15$ and $h = 0.8$ $A = (12\,ft)(6\,ft) = 72\,ft^2$ $B/L = (6\,ft)/(12\,ft) = 0.5$
11. Calculate the anchor holding capacity from Equation 6-3.	$F_{st} = (72\,ft^2)(500\,psf)(0.8)(15)[0.84 + 0.16(0.5)]$ $\quad = 397,440\,lb = 397\,kips$ OK; $F_{st} >$ required horizontal mooring load of 350 $kips$.
12. Check the plate length to width ratio. Is $1.5 \le (L/B) \le 2.0$?	$(L/B) = (12\,ft)/(6\,ft) = 2.0$ YES, the length to width ratio is acceptable.
13. Calculate the maximum penetration depth from Equation 6-1 and check against the maximum allowable penetration depth.	From STEP 1, max allowable penetration depth is $(65\,ft + 12\,ft) = 77\,ft$ From Equation 6-1, $\quad z_p = z + 2L = 50\,ft + 2(12\,ft) = 74\,ft < 77\,ft$ OK; the maximum penetration depth does not exceed the maximum allowable penetration depth.

SUMMARY

1. The pile-driven plate anchor to be used has a length, $L = 12$ ft, and a width, $B = 6$ ft. The plate anchor is driven to a depth of 74 ft below the seafloor prior to keying.

2. The anchor holding capacity is 397 kips.

6.10.2 Problem 2 - A Pile-Driven Plate Anchor Used in Cohesionless Soil

6.10.2.1 Problem Statement

Design a pile-driven plate anchor for specified required mooring loads. Also, determine the depth to which the anchor is driven prior to keying.

Data: Due to the limited mooring space in a harbor, a pile-driven plate anchor is needed to provide horizontal mooring load of 200 kips in the harbor. The water depth in the harbor is 35 feet and the harbor sediments are mainly dense sand. The contractor has pile-diving equipment and a section of 90-foot pile follower for the installation. Assume embedded chains in sand contribute 20 percent of anchor holding capacity.

6.10.2.2 Problem Solution

The trial-and-error computational procedures for the problem's solution are presented below. These follow the procedures outlined in this chapter.

Problem 6.10-2

ANALYTICAL PROCEDURES	COMPUTATIONS
1. Check the maximum allowable penetration depth of the plate anchor.	Given a 90-ft pile follower, the maximum allowable penetration depth is $= 90\,ft - 35\,ft$ of water depth $\qquad +$ plate anchor length, L $= 55\,ft + L$ Note: In dense sand, the required anchor penetration depth is usually shallower than that in soft clays or mud.
2. Estimate the ultimate anchor holding capacity using an appropriate factor of safety.	Required horizontal mooring load = 200 $kips$ Chain holding capacity = 20% = 0.2 (given) $F_s = 2.0$ Ultimate horizontal anchor holding load is $= (200\ kips)(1-0.2)(2.0) = 320\ kips$

Problem 6.10-2

ANALYTICAL PROCEDURES	COMPUTATIONS
3. Determine soil properties ϕ, and γ_b (cohesionless soil). Because the properties have not been measured, they must be estimated (Chapter 5, Table 5.3-1).	For a dense cohesionless sand, $\phi \approx 40 \ deg$ $\gamma_b \approx 60 \ pcf$
4. Select a trial key-in depth, z.	Say $z = 10 \ ft$
5. Determine the holding capacity factor, $\overline{N_q}$ (from Figure 6.5-4 with ϕ and the z/B ratio).	Let $B = 2 \ ft$ $z/B = (10 \ ft)/(2 \ ft) = 5$ $\overline{N_q} \cong 25$
6. Select the plate anchor size.	Try $L = 4 \ ft$ with $B = 2 \ ft$ $A = L \cdot B = (4 \ ft)(2 \ ft) = 8 \ ft^2$ $B/L = (2 \ ft)/(4 \ ft) = 0.5$
7. Compute the anchor static holding capacity, F, based on the plate anchor size and key-in depth. From Equation 6-7, $F = A \ \gamma_b \ z \ \overline{N_q} \ [0.84 + 0.16(B/L)]$	$F = (8 \ ft^2)(60 \ pcf)(10 \ ft)(25)[0.84 + 0.16(0.5)]$ $= 110,400 \ lb = 110 \ kips$ NO GOOD; F < required horizontal mooring load of 320 $kips$. Use a deeper key-in depth. Try $z = 20 \ ft$ For $z/B = (20 \ ft)/(2 \ ft) = 10$, $\overline{N_q} \cong 50$ $F = (8 \ ft^2)(60 \ pcf)(20 \ ft)(50)[0.84 + 0.16(0.5)]$ $= 441,600 \ lb = 442 \ kips$ OK; $F > 350 \ kips$. Use $z = 20 \ ft$ with $L = 4 \ ft$ and $B = 2 \ ft$.
8. Check the plate length to width ratio. Is $1.5 \le (L/B) \le 2.0$?	$(L/B) = (4 \ ft)/(2 \ ft) = 2.0$ YES, the length to width ratio is acceptable.

Problem 6.10-2

ANALYTICAL PROCEDURES	COMPUTATIONS
9. Calculate the maximum penetration depth from Equation 6-2 and check against the maximum allowable penetration depth. $z = z_p - 1.5L$	From STEP 1, max allowable penetration depth is $(55\ ft + 4\ ft) = 59\ ft$ From Equation 6-1, $z_p = z + 2L = 20\ ft + 1.5(4\ ft) = 26\ ft < 59\ ft$ OK; the maximum penetration depth does not exceed the maximum allowable penetration depth.

SUMMARY

1. The pile-driven plate anchor to be used has a length, L = 4 ft, and a width, B = 2 ft. The plate anchor is driven to a depth of 20 ft below the seafloor prior to keying.

2. The anchor holding capacity is 442 kips.

3. The plate anchor design is based on an estimated value for the soil friction angle. Since the friction angle greatly affects the value for $\overline{N_q}$, it is recommended that a Standard Penetration Test (SPT) be performed to validate the estimated value.

6.11 REFERENCES

6-1.   Forrest, J; Taylor, R and Brown, L (1995) "Design Guide for Pile-driven Plate Anchors" Technical Report TR-2039-OCN, Naval Facilities Engineering Service Center, Port Hueneme, CA   March 1995.

6-2.   H.S. Stevenson and W.A. Venezia. Jetted-In Marine Anchors, Naval Civil Engineering Laboratory, Technical Note N-1082. Port Hueneme, CA, Feb 1970.

6-3.   R.J. Taylor. "Interaction of Anchors with Soil and Anchor Design," presented at short course on Recent Developments in Ocean Engineering, University of California, Berkeley, CA, Jan 1981.

6-4.   N. Kerr. "The Hydropin: A New Concept in Mooring," Transactions, North East Coast Institution of Engineers and Shipbuilders, Vol. 92, No. 2, Nov 1975, pp. 39-44.

6-5.   R.J. Taylor, D. Jones, and R.M. Beard. "Uplift-Resisting Anchors," in Ocean Engineering, Vol 6. Pergamon Press, 1979, pp. 3-137.

6-6.   R.M. Beard. Expendable Doppler Penetrometer: A Performance Evaluation, Civil Engineering Laboratory, Technical Report R-855. Port Hueneme, CA, Jul 1977.

6-7.   R.M. Beard. Holding Capacity of Plate Anchors, Civil Engineering Laboratory, Technical Report R-882. Port Hueneme, CA, Oct 1980.

6-8.   A.S. Vesic. "Breakout Resistance of Objects Embedded in Ocean Bottom," in Civil Engineering in the Oceans II. New York, NY, American Society of Civil Engineers, 1970, pp. 137-165.

6-9.   P.J. Valent. Results of Some Uplift Capacity Tests on Direct Embedment Anchors, Civil Engineering Laboratory, Technical Note N-1522. Port Hueneme, CA, Jun 1978.

6-10.   H.G. Herrmann. Design Procedures for Embedment Anchors Subjected to Dynamic Loading Conditions, Naval Civil Engineering Laboratory, Technical Report R-888. Port Hueneme, CA, Nov 1981.

6-11.   B.J. Douglas. Effects of Rapid Loading Rates on the Holding Capacity of Direct Embedment Anchors, Civil Engineering Laboratory, P.O. No. M-R420. Port Hueneme, CA, Oct 1978.

6-12.   F.H. Kulhawy, D.A. Sangrey, and S.P. Clemence. Direct Embedment Anchors on Sloping Seafloors, State-of-the-Art, Civil Engineering Laboratory, P.O. No. M-R510. Port Hueneme, CA, Oct 1978.

6-13. A. Singh and J.K. Mitchell. "General Stress-Strain-Time Function for Soils," Journal of the Soil Mechanics and Foundations Division, American Society of Civil Engineers, vol 94, no. SMI, Jan 1968, pp 21-46.

6-14. R.M. Beard. Long-Term Holding Capacity of Statically Loaded Anchors in Cohesive Soils, Civil Engineering Laboratory, Technical Note N-1545. Port Hueneme, CA, Jan 1979.

6-15. P.J. Valent. Long-Term Stress-Strain Behavior of a Seafloor Soil, Civil Engineering Laboratory, Technical Note N-1515. Port Hueneme, CA, Feb 1978.

6-16. American Institute of Steel Construction. AISC Steel Construction Manual, 13th Edition, May 2005.

6.12 SYMBOLS

A	Projected fluke area perpendicular to direction of pull out [L^2]
B	Fluke or plate minimum dimension, usually width [L]
$\bar{c}$	Drained soil cohesion [F/L^2]
$\bar{c}'$	Reduced drained soil cohesion for very soft underconsolidated soils [F/L^2]
e	Exponential base for natural logarithms
F	Holding capacity in cohesionless soils [F]
F_I	Holding capacity under impulse loading [F]
F_{lt}	Long-term static holding capacity [F]
F_s	Factor of safety
F_{st}	Short-term static holding capacity [F]
$\bar{f_c}$	Average number of impulse loadings per hour, in clay, over a 4-hour period [I/T]
$\bar{f_s}$	Average number of impulse loadings per 10 minutes, in sand [1/T]
h	Correction factor for soil disturbance due to penetration and keying
I	Influence factor for adjusting soil strength for impulse loading
I_f	Inertial factor for holding capacity increase under impulse loading
k	Soil permeability [L/T]
L	Fluke or plate maximum dimension, usually length [L]
$\overline{N_c}$	Long-term holding capacity factor in cohesive soil
$\overline{N_{cs}}$	Short-term holding capacity factor in cohesive soil
$\overline{N_q}$	Holding capacity factor for a drained soil condition
$\overline{N_{ql}}$	Cohesionless soil holding capacity factor adjusted for impulse loading
n_c	Actual number of cycles occurring during time period, t_{cd}
n_T	Total number of cycles occurring during anchor lifetime
PI	Plasticity index
P_c	Double-amplitude cyclic load component [F]
P_I	Magnitude of the impulse line load [F]
P_s	Static or nearly static line load component [F]
$\bar{p}_{vo}$	Effective overburden stress in soil [F/L^2]
R_c	Reduction factor for holding capacity under cyclic loading
R_I	Reduction factor for holding capacity under repeated impulse loading
R_s	Reduction factor applied to anchor holding capacity to account for slope instability
S_t	Soil sensitivity

s_u	Soil undrained shear strength [F/L^2]
t_{cd}	Time duration required for dissipation of excess pore-pressure [T]
z	Embedded depth of fluke after keying [L]
z_k	Distance required for fluke keying [L]
z_p	Maximum penetration depth of keying-type fluke [L]
γ_b	Soil buoyant unit weight [F/L^3]
ϕ	Soil friction angle [deg]
$\bar{\phi}$	Drained cohesive soil friction angle [deg]
$\bar{\phi}'$	Reduced drained cohesive soil friction angle for very soft underconsolidated soils [deg]
ϕ_I	Soil friction angle adjusted for impulse loading [deg]

[This page intentionally left blank]

7 DRAG-EMBEDMENT ANCHORS

7.1 INTRODUCTION

7.1.1 Purpose and Scope

This chapter consolidates the performance, selection, and design information for drag-embedment anchors. The functioning of the drag anchor is described, and a brief description of geotechnical information necessary for design follows. Criteria for selection of an anchor are listed; techniques for designing or sizing are then presented. A section on field solutions for anchor performance problems follows.

This chapter deals only with geotechnical aspects of mooring system design. Prediction of mooring system loads, selection of mooring line materials, and maintenance and inspection of mooring system components are not discussed. For prediction of environmental loadings on moored vessels, the reader is referred to the Department of Defense UFC 4-150-06 "Military Harbors and Coastal Facilities" (Ref. 7-1), the Det Norske Veritas "Rules for the Design, Construction and Inspection of Offshore Structures" (Ref. 7-2), the rules of the American Bureau of Shipping (Ref. 7-3), and the API "Recommended Practice for the Analysis of Mooring Systems for Floating Drilling Units" (Ref. 7-4). Information on mooring system component selection and system design and analysis are generally available (Refs. 7-5 and 7-6). Mooring inspection and maintenance rules have been drawn up by Det Norske Veritas (Ref. 7-2).

7.1.2 Drag Anchor Description

Most drag anchors are referred to by their manufacturer trade name. While specialty anchors may have a very different shape, many widely used anchors share common features. All have a shank, through which the mooring line load is applied, and a fluke or flukes, which are the digging parts of the anchor and provide the bearing area to mobilize sediment resistance. These and other components common to most drag anchors are shown in Figure 7.1-1. Tripping palms (often called "mud palms" or "palms") improve the capability of the anchor fluke to open and dig into the seafloor. Stocks or stabilizers are used on many anchors to improve their biting into the seafloor and their rotational stability. The biting or leading edge of the fluke is the fluke tip. The area where the fluke connects to the shank is the crown or head of the anchor.

The chain or wire rope mooring line attached to the anchor shank is an integral part of a drag anchor system. One section of this mooring line is dragged below the seafloor surface by the anchor, while a variable length of the mooring line usually rests on and is dragged along the seafloor surface. Both mooring line sections develop resistance to horizontal movement and contribute to the holding capacity of the anchor system.

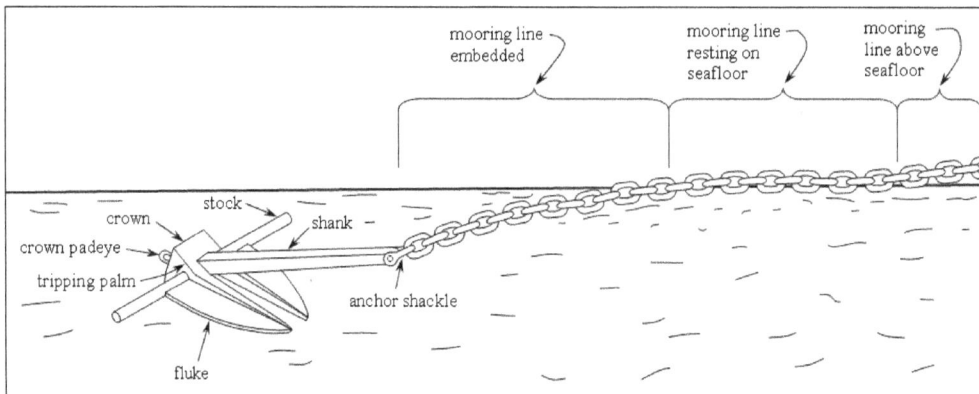

Figure 7.1-1. Features of a drag anchor (Ref. 7-5).

7.1.3 Types of Drag Anchors

Anchors can be classified by their general physical characteristics, their intended function or performance in different types of soils. Types of representative drag embedment anchors are shown in Figure 7.1-2[2] (Ref. 7-7). Detailed discussions of these anchors are presented in the following sections.

Additionally, drag anchors may be classified as:

- Movable or fixed fluke
- Bilateral or unilateral fluked
- Hard or soft seafloor anchors
- Standard or high holding power anchors

Movable fluke anchors are hinged at the crown so that the anchor can lie flat, with the plane of the flukes parallel to the axis of the shank when on the deck of a work boat (Figure 7.1-3a). The flukes can still open fully for digging in when the anchor is placed on the seabed (Figure 7.1-3b). Fixed fluke anchors are those with either movable flukes which have been welded or blocked in the open position, or those where the fluke and shank have been cast or fabricated as a single piece (Figure 7.1-4).

---

[2] Most of the anchors in this chapter are still available for purchase. The HOOK, STEVFIX, and STEVMUD have been updated with the STEVIN or STEVPRIS. The ADMIRALTY AC-12, BOSS, STOKES, EELLS, BEIJERS, and all the STOCKED anchors are not available for purchase based on a quick market search. The data from these unavailable anchors was left in this chapter as it shows the performance improvement over the years and these anchors also have a significant amount of qulailty test data.

Bilateral fluked anchors are those constructed so the flukes can open freely to either side of the anchor shank (Figure 7.1-5a). Unilateral fluked anchors are constructed so the fluke must remain only on one side of the shank (Figure 7.1-5b). Unilateral fluked anchors require greater care in handling and placement on the seafloor to ensure that the flukes are down for digging; improper placement could result in the anchor being on its back and unable to dig into the seafloor.

Anchors which work best in hard soil seafloors (i.e., hard clays and most sands and gravels) are sharp-fluked with close-set fluke tips to initiate penetration. They have long stabilizer bars to counter roll instability tendencies (Figure 7.1-5a). Anchors for soft seafloors, on the other hand, maximize fluke area and emphasize streamlining to achieve deep seafloor penetration (Figure 7.1-6).

7.1.4   Application of Drag Anchors

Drag anchors are standard equipment for the temporary mooring of all mobile craft of substantial size and are often selected for the permanent mooring of floating platforms. They are the leading contenders for temporary moorings because they are efficient, require a minimum of specialized support, and are reuseable. Often, catenary system motion (in reaction to an increase in line tension) and load characteristics of the drag anchor system are required in a mooring system to cope with dynamic load components.

On the disadvantage side, the drag anchor is often a poor performer on very hard seafloors. Further, the drag anchor system is not a constructed anchor system, like the pile or gravity anchor system. After placement, the drag anchor must trip, dig-in, and remain stable with drag. All of these are statistical performance functions and have a probability of occurrence lower than 100%, particularly in hard soils.

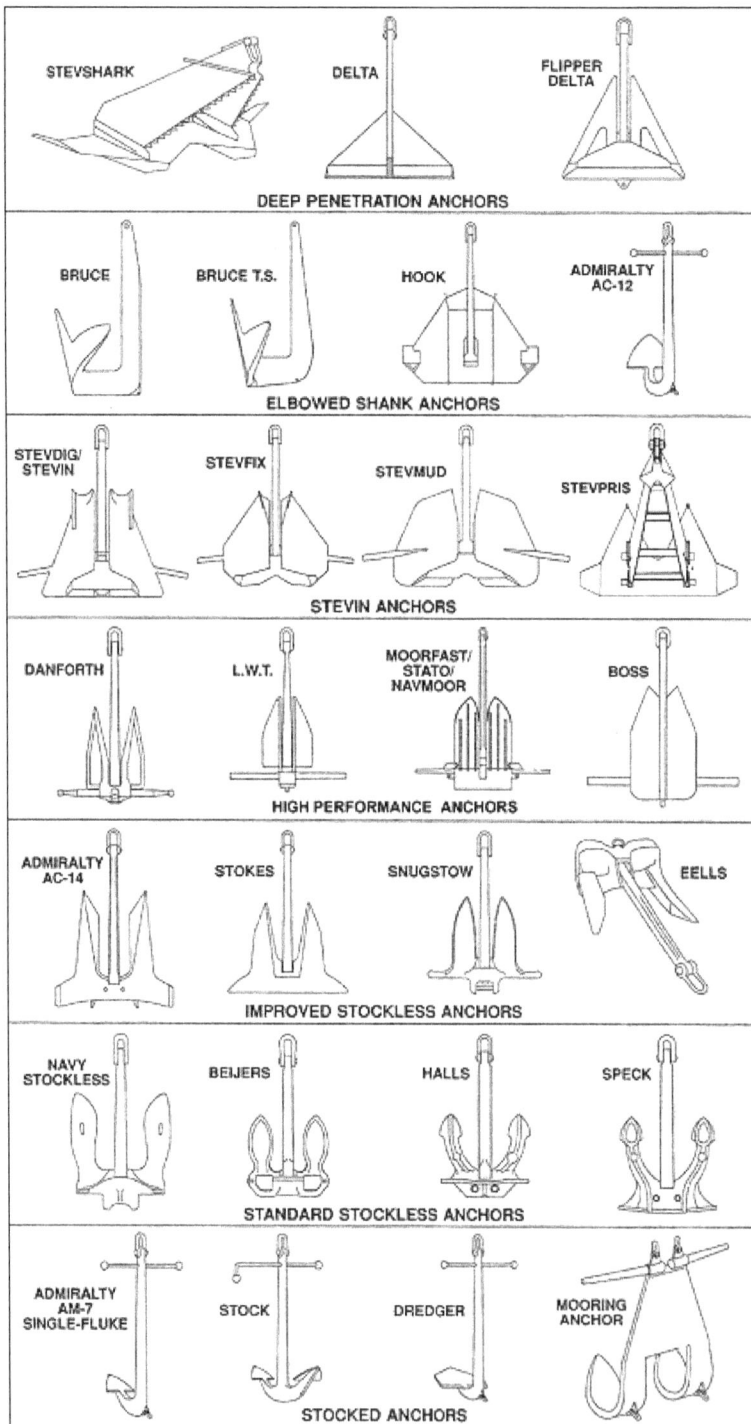

STEVSHARK · **DELTA** · **FLIPPER DELTA**

DEEP PENETRATION ANCHORS

BRUCE · **BRUCE T.S.** · **HOOK** · **ADMIRALTY AC-12**

ELBOWED SHANK ANCHORS

STEVDIG/ STEVIN · **STEVFIX** · **STEVMUD** · **STEVPRIS**

STEVIN ANCHORS

DANFORTH · **L.W.T.** · **MOORFAST/ STATO/ NAVMOOR** · **BOSS**

HIGH PERFORMANCE ANCHORS

ADMIRALTY AC-14 · **STOKES** · **SNUGSTOW** · **EELLS**

IMPROVED STOCKLESS ANCHORS

NAVY STOCKLESS · **BEIJERS** · **HALLS** · **SPECK**

STANDARD STOCKLESS ANCHORS

ADMIRALTY AM-7 SINGLE-FLUKE · **STOCK** · **DREDGER** · **MOORING ANCHOR**

STOCKED ANCHORS

Figure 7.1-2. Examples of different drag embedment anchors (Ref. 7-7).

7-4

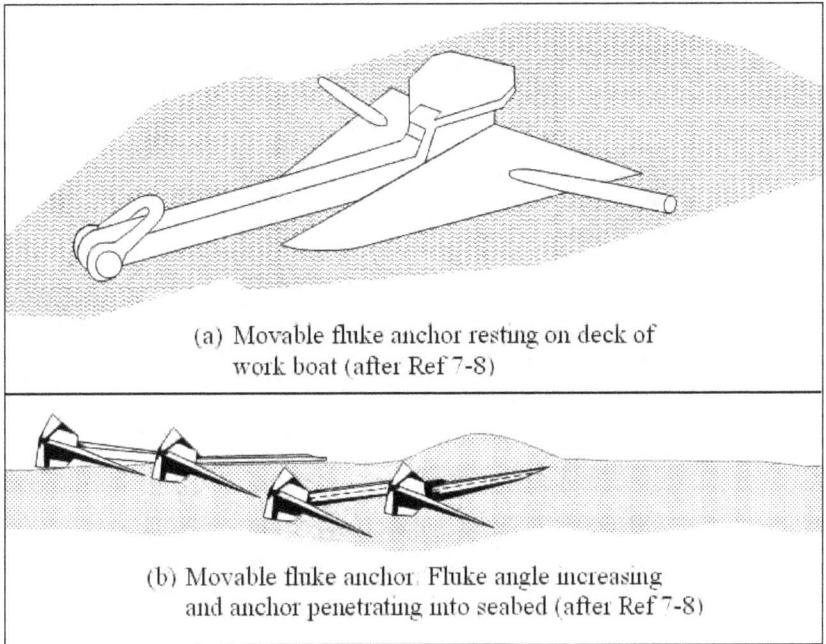

(a) Movable fluke anchor resting on deck of work boat (after Ref 7-8)

(b) Movable fluke anchor. Fluke angle increasing and anchor penetrating into seabed (after Ref 7-8)

Figure 7.1-3. Example of a movable fluke anchor: STEVIN cast.

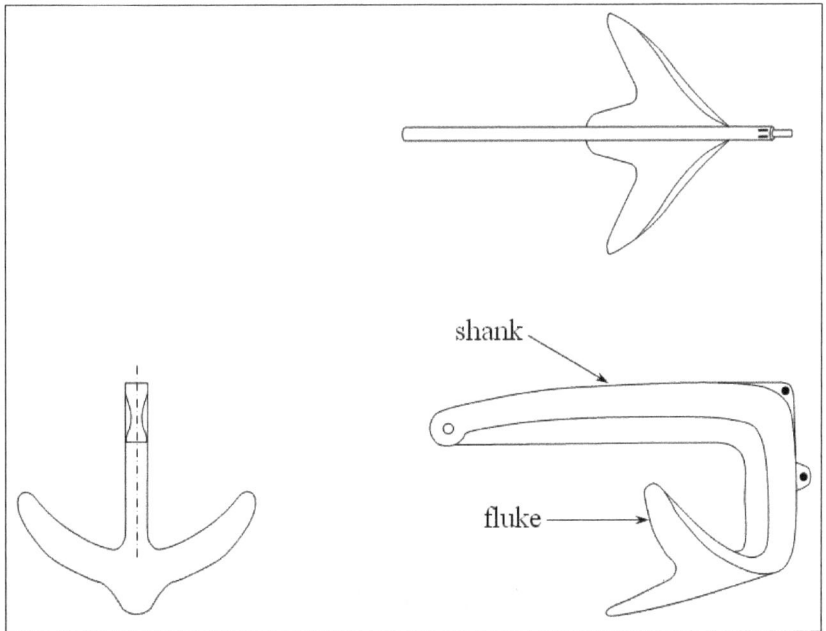

shank

fluke

Figure 7.1-4. Example of a fixed fluke anchor: BRUCE cast (Ref. 7-9).

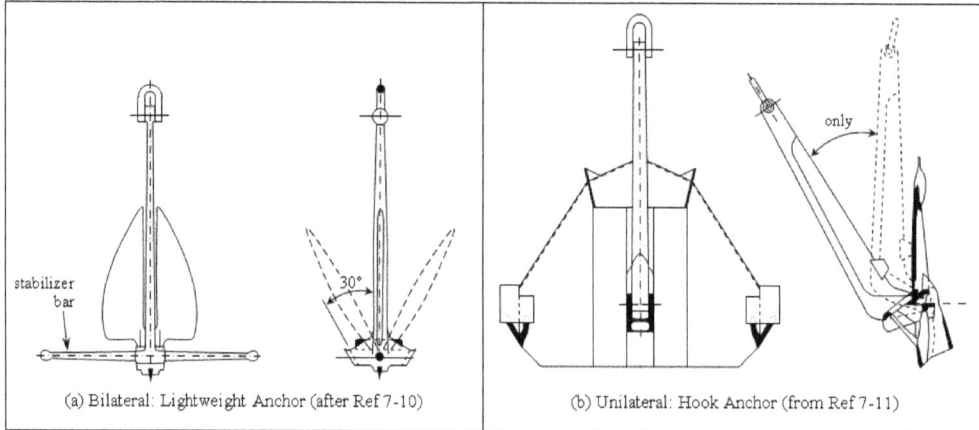

Figure 7.1-5. Example of bilateral fluke anchors.

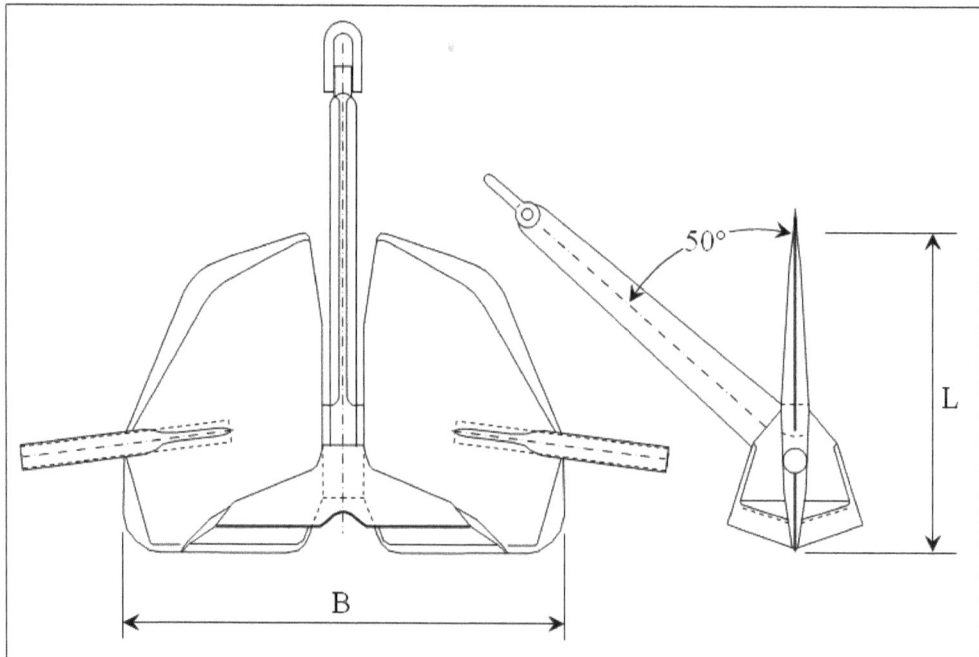

Figure 7.1-6. Example of a soft soil anchor: STEVMUD (Ref. 7-11).

7.2 FUNCTIONING OF A DRAG ANCHOR

7.2.1 General

The drag anchor is analogous to an inverted kite made to "fly downward" into the soil. The shank acts as the kite bridle, maintaining the angle of attack of the fluke to the soil to maximize the developed mooring line tension in the horizontal direction at the seafloor. The anchor is pulled along the seafloor until it digs in or penetrates to its place of maximum holding capacity. The tripping palms assist by causing the movable fluke anchors, when placed on the seafloor, to begin digging in when dragged. The stabilizers assist, especially on harder seafloors, in orienting and maintaining the anchor in the horizontal position. When working properly, the drag anchor will embed itself into the soil to some equilibrium depth dependent on existing mooring line, anchor, and soil conditions. When not working properly, the anchor will not embed as deeply nor develop as high a holding capacity, or it may not embed at all.

7.2.2 Tripping

Most drag anchors are the movable, bilateral fluke variety, where the flukes are free to move to either side of the shank and where the anchor can lie flat on the deck of a work boat with flukes nearly parallel to the shank and deck. Generally, the movable fluke anchor is easier to handle, deploy, and recover. Occasionally, a movable fluke anchor on the seafloor may not trip.

On soft seafloors (i.e., soft clays and muds) those anchors with very heavy crowns, small or nonexistent tripping palms, or with the shank-to-fluke hinge far back on the fluke tend to have tripping problems. This holds true especially when the anchor is lowered and placed crown first by the--mooring line (Figure 7.2-1a). Anchors of this type (e.g., Stockless, Lightweight (LWT)), when dragged to set them, will often not dig into the seafloor but will instead slide at mudline level with the movable flukes oriented parallel to the shank or pointing slightly upward, serving only as a deadweight anchor at the mud surface (Figure 7.2-1b).

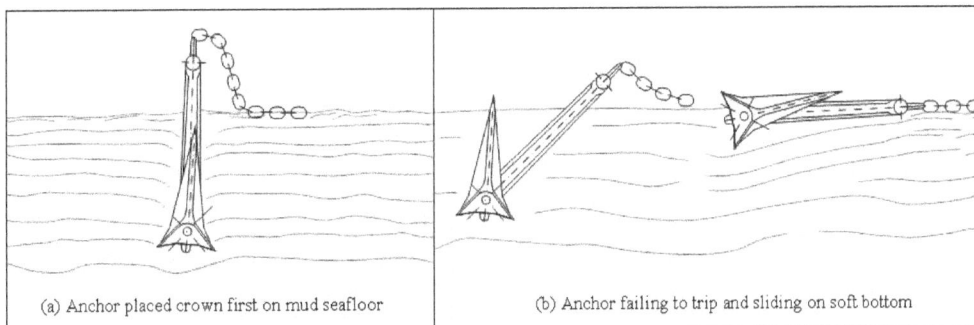

(a) Anchor placed crown first on mud seafloor   (b) Anchor failing to trip and sliding on soft bottom

Figure 7.2-1. Development of a tripping problem in soft seafloors with an improperly set anchor.

This problem in soft soil can be largely eliminated by proper anchor setting procedures (depicted in Figure 7.2-2). Generally, implementation of this procedure requires two vessels to lay a mooring leg: one paying out the mooring line and the second handling, lowering, and positioning the anchor for digging in when the mooring line is pulled.

Anchor tripping difficulties may also occur in dense and hard soils. Even with proper anchor handling procedures, the fluke tips may not be able to develop sufficient local stress to initiate digging in. Under these conditions, the anchor may simply slide without tripping (Figure 7.2-3a) or may dig-in slightly to effect standing up (Figure 7.2-3b), and then fall on its side and drag (Figure 7.2-3c). Those anchors having a relatively heavy crown and a shank connection well back of the center of fluke area (Moorfast, Offdrill, etc.) appear susceptible to this penetration problem in hard soils. The hard soil tripping problem has been corrected by sharpening the fluke tips to improve digging capability, by welding barbs on the tripping palms to increase the tripping moment, and by reducing fluke angle several degrees below the sand setting. Most drag anchor types incorporate stocks or stabilizers to prevent the standing anchor from falling completely on its side (Figure 7.2-3c). Thus, the stocks serve to hold the fluke tips in a digging position.

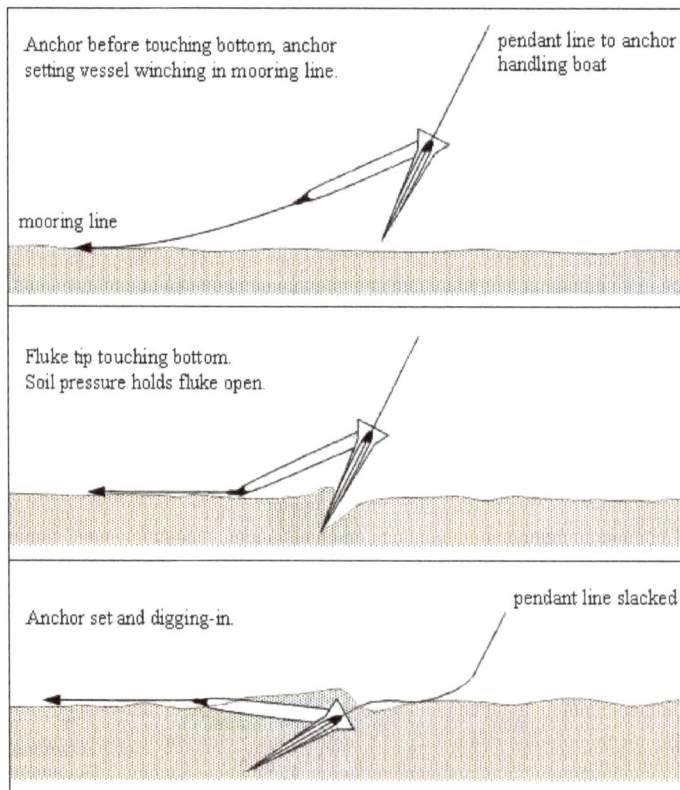

Figure 7.2-2. Proper anchor setting sequence using two floating platforms.

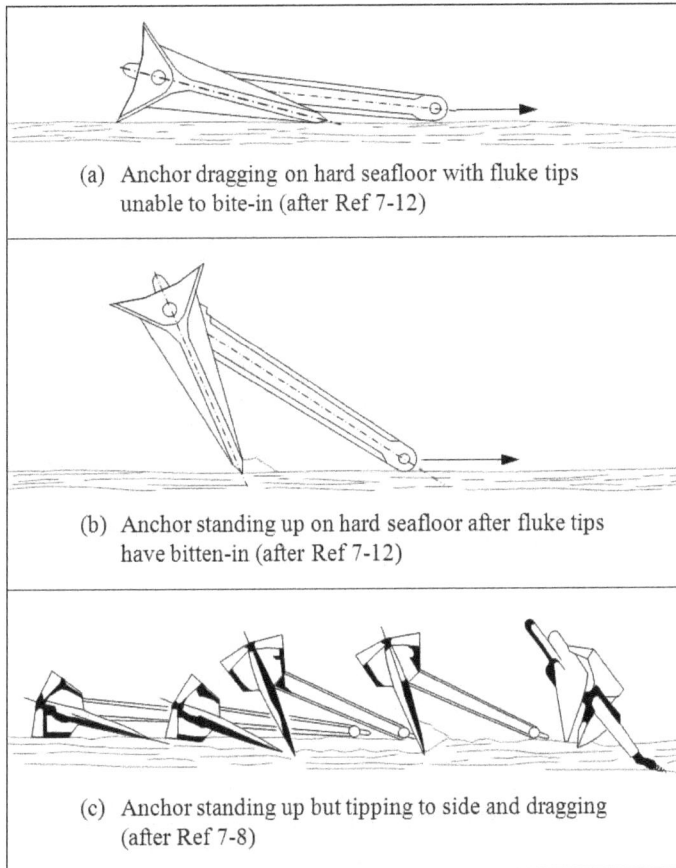

(a) Anchor dragging on hard seafloor with fluke tips unable to bite-in (after Ref 7-12)

(b) Anchor standing up on hard seafloor after fluke tips have bitten-in (after Ref 7-12)

(c) Anchor standing up but tipping to side and dragging (after Ref 7-8)

Figure 7.2-3. Development of a tripping problem in hard seafloors.

7.2.3 Embedment

Embedment, or penetration of the fluke into the seafloor once digging-in is initiated, is governed by the fluke angle, the soil type, the degree of anchor streamlining, and the smoothness of the fluke surfaces. In addition, the mooring line angle to the seafloor must be zero to ensure that even a properly selected drag anchor will embed properly.

There is a critical fluke-to-shank angle (termed the "fluke angle," β) at which the anchor holding capacity is maximized. This critical angle decreases as:

- Fluke length increases
- Shank length decreases
- Fluke surface becomes rougher

For fluke angles greater than critical, the standing anchor will penetrate only slightly into the sediments and will slide in a standing or tipped orientation (Figure 7.2-3b and c) at a very small holding capacity. For fluke angles less than critical, anchor embedment depth will be reduced, and the anchor will develop less than its maximum potential holding capacity in that particular environment.

Data gathered for STATO anchors in the 1950s (Ref. 7-13) showed an optimum fluke angle of 50° for soft mud (clayey silt) and 34° for sand. Tests performed in the 1980s (Ref. 7-11) in denser sand showed that, for the STATO, a fluke angle of 30° is more appropriate for use in sands. Most anchors are manufactured with a fluke angle of about 50° for soft soils, with bolted or welded wedges or inserts providing an option for fluke angle reduction for hard soils. Changes to standard fluke angle settings may be advised to improve performance, especially on hard, dense sands and gravels. For instance, where a fluke angle change to 28° may be necessary to initiate the biting and digging-in of large Moorfast- and Offdrill-type anchors.

Anchor penetration is influenced by the anchor's degree of streamlining. The newer anchor designs with tapered and sharpened flukes, narrowed and chamfered shanks, and open tripping palms have good penetration capability and can reach stronger soils deep in the soil profile.

An increase in fluke roughness limits anchor penetration and therefore influences holding capacity. Flukes with smooth surfaces mobilize less soil resistance to penetration in the plane of the fluke. Thus, smooth flukes penetrate deeper and reach the stronger soils which usually occur at greater depths.

At optimum fluke angles, penetration behavior of an anchor is vastly different for hard and for soft soils. In hard soils (stiff clays and the denser sands), the drag anchor does not penetrate deeply. Rather, the crown of the anchor may remain above the surrounding seafloor surface, as with a STATO (Figure 7.2-4a), or it may penetrate only a few feet, as with the more streamlined anchors such as the STEVFIX. In stiff clays and dense sands, anchor penetration depth is typically less than its fluke length since the anchor crown and shank have a significant effect on the penetration process (Table 7.3-1). In soft soils (soft clayey silts and clays) the anchors will penetrate more deeply (Figure 7.2-4b), from 45 feet for large STATO-type anchors to 60 feet for a 20,000-pound STEVIN (as reported in Ref. 7-8).

Based upon anchor field tests anchor performance data including anchor penetration (d_T) and drag distance (D) can be normalized with respect to its fluke length (L). The normalized drag distance (D/L) and anchor penetration depth (d_T/L) for 10 anchors tested by the Navy are shown in Figure 7.2-5, which shows that each anchor has its own distinctive dragging pattern (Refs. 7-14 and 7-15). The estimated maximum fluke tip penetration in soft clays and mud are summarized in Table 7.3-1.

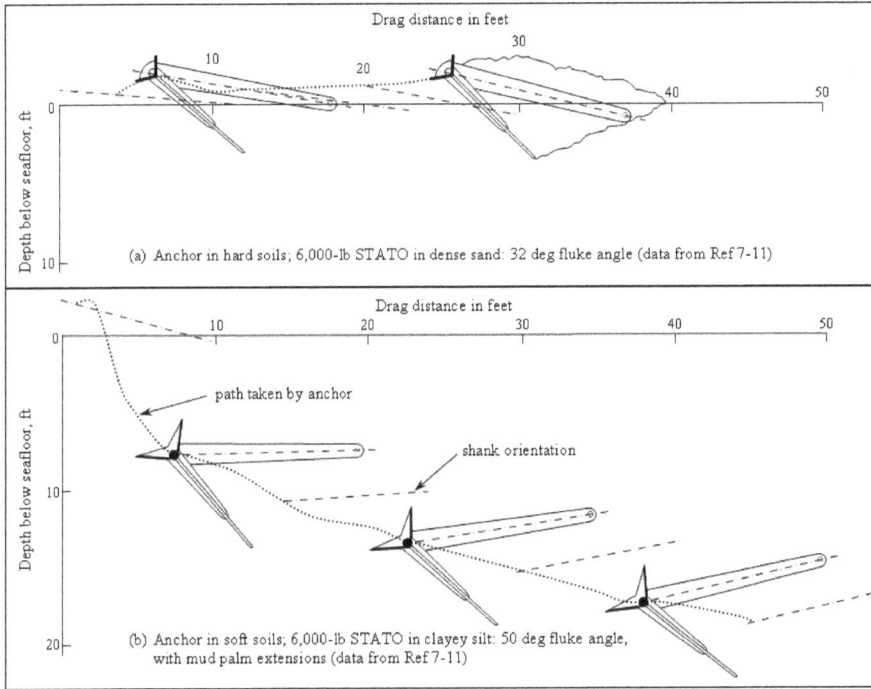

Figure 7.2-4. Penetration and orientation behavior of an anchor in hard and soft seafloors.

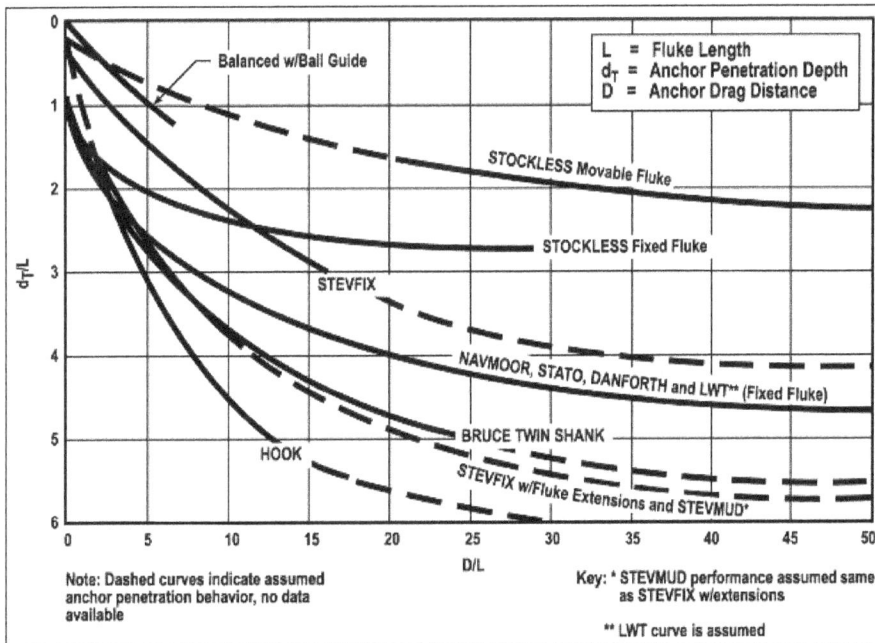

Figure 7.2-5. Normalized anchor penetration depth and drag distance.

7-11

7.2.4 Stability

A drag anchor may exhibit instability after the initial biting and digging-in due to differences in the soil resistance encountered by the two flukes, initial differences in the fluke penetration depths, a slight change in the direction of mooring line pull, or some other source of asymmetry. Drag anchors in sand will often become unstable and roll after only a few feet of drag. Once the anchor begins to roll, the soil pressure on the rising fluke becomes much reduced, and the force couple in the direction of the roll is increased, thus speeding the rolling of the anchor on its side (Figure 7.2-6a). Stocks or stabilizers are designed to develop a countering force couple to resist the roll motion (Figure 7.2-6b). In soft clays and silts (mud), the stabilizers probably provide a significant stabilizing influence only for the first few feet of drag, up to the point where both stabilizers pass beneath the mudline. Even when fully embedded, every drag anchor is potentially unstable. Once beneath a soft seafloor surface, anchor stability is primarily a function of the symmetry of the anchor, the variation in the direction of mooring line pull, and the homogeneity of the soil. At present, insufficient data exist to qualify the probability of an instability developing in either hard or soft soils.

Instability after anchor dig-in also includes the phenomenon of anchor "balling up" and pulling out (Ref. 7-16). This balling up phenomenon refers to formation of a large ball of soil on the entire fluke and crown assembly (Figure 7.2-7), which can occur after dragging 50 to 200 feet in soft soils. This ball of soil or "dead zone" travels with the anchor, distorting its shape and significantly limiting anchor penetration capability and its stability. A balled-up anchor that rises to the surface due to rotational instability will not re-embed with further dragging. It must be recovered and cleaned before it can be reset.

The process for development of this balling up is hypothesized as follows. The soil in the dead zone is subjected to a high total stress and the soil develops large positive excess pore water pressures. In silt to clayey silt soils (muds), some of this excess pore water pressure dissipates to the surrounding soils as the anchor drags. The soil immediately in front of the flukes, therefore, becomes stronger than the undisturbed soil and is able to adhere to and build up on the flukes to become a large, compact lump.

The balling-up phenomenon has been reported (Ref. 7-12) with LWT, STATO, and Moorfast anchors, but not with the more streamlined BOSS anchor. Performance differences lie in the orientation of the fluke with respect to the anchor trajectory (Refs. 7-11 and 7-17). The STATO fluke surface, for instance, is oriented much more obliquely to its trajectory in the soil than the fluke surface of the more streamlined Hook anchor. This more oblique orientation and the STATO's large, tripping palms are believed responsible for a tendency to ball up in some soft soils. The fluke, tripping palm, and shank streamlining of the newer anchors (e.g., BOSS, Hook, STEVMUD, and Flipper Delta) are responsible for reducing the potential for this happening.

A positive feature associated with balling up is that anchors with this characteristic achieve their maximum capacity in a much shorter drag distance because the soil shear surfaces developed by these anchors are substantially greater per unit fluke area than those developed by the newer streamlined anchors. This will be apparent in Section 7.5.3.1.

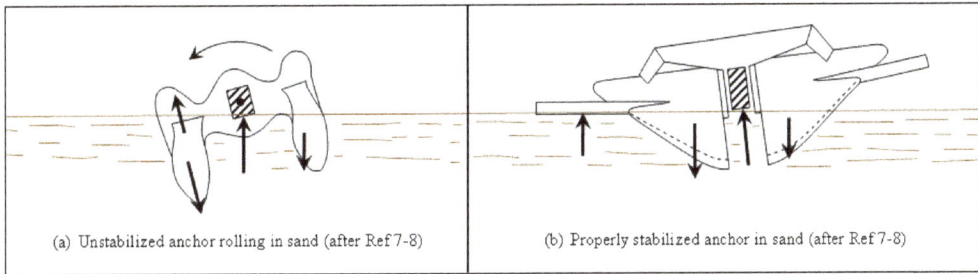

(a) Unstabilized anchor rolling in sand (after Ref 7-8)   (b) Properly stabilized anchor in sand (after Ref 7-8)

Figure 7.2-6. Forces on unstabilized and stabilized anchors in sand.

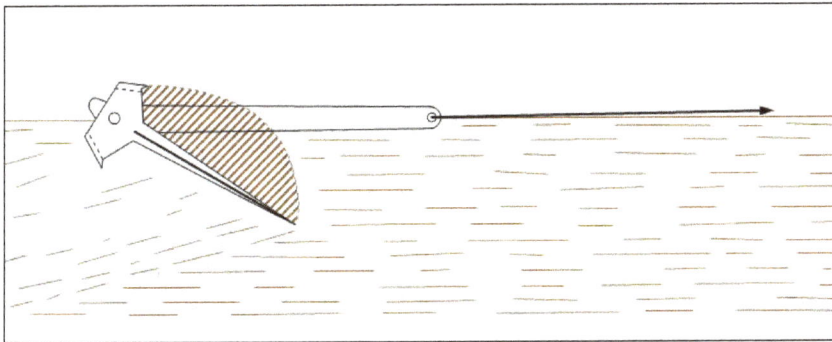

Figure 7.2-7. Anchor in soft soil, after balling-up and pulling-out (Ref. 7-16).

7.2.5   Soaking

"Soaking" of an anchor is the practice of allowing a newly embedded anchor to rest for a period of time, typically 24 hours, before applying the required proof load. The mechanism that makes soaking work is similar to that causing balling up (i.e., consolidation of the silt and clay soil around the fluke, causing strengthening of that soil and increasing the anchor's holding capacity with time).

7.3   SITE INVESTIGATION

7.3.1   Site Data Needed

Although this chapter does present the results of initial work toward a rational methodology for determining the capacity of drag anchors, present technology cannot utilize detailed soil strength and behavior information for drag anchor design.

Drag anchor design usually consists of selecting an anchor type and size. This selection and sizing requires the following knowledge of site conditions:

- Topography and thickness of significant soil layers
- Sediment type (cohesive versus cohesionless)
- Undrained shear strength for cohesive soils

7.3.2   Topography and Layer Thickness

Knowledge of topography and sediment thickness at the proposed anchor locations is necessary for determining whether or not drag anchors can work at a given site. First, topography is an indicator of seafloor material type. Irregular or rugged topography may indicate outcropping of rock or hard strata or the existence of boulders or talus. Drag anchors should not be expected to function well, if at all, at such sites. Second, topography indicates the bottom slope gradient at the anchor location. An anchor being pulled downslope will have a lower holding capacity than if it were pulled on a horizontal surface in the same soil. In lieu of rational and definitive procedures, it is recommended that the siting of drag anchors be limited to downslopes of no more than 10° and, where it is practical, to slopes of less than 5°.

Topography is also important in the design of the total mooring system. Overall topography may limit anchor placement locations, and affects the required lengths of the mooring legs, the proportions of chain and wire rope in those mooring legs, mooring leg loads, and allowable anchor displacements.

Sediment thickness data are necessary to determine that sufficient depth exists to permit full anchor penetration to equilibrium depth, where the maximum holding capacity will be developed. Drag anchors will penetrate to a depth of about one fluke length in dense sands and to a depth of three to six fluke lengths in soft clays and silts (see Table 7.3-1). For less dense sands penetration will be slightly more deep, and for stronger clays and silts the penetration will be less deep than indicated on Table 7.3-1. This table can be used to gauge the depth to which a site investigation is needed. The ratios for the larger drag anchors, for example, indicate sediment thickness requirements of 15 to 20 feet in dense sands and 60 to 80 feet in the soft clays and silts.

Techniques and equipment for obtaining the necessary soils data are discussed in Chapter 2.

Table 7.3-1. Estimated Maximum Fluke Tip Penetration of Some Drag Anchor Types in Soil

Anchor Type	Normalized Fluke Tip Penetration, $(d_t/L)_{max}$	
	Cohesive Soils (soft clays and silts)	*Cohesionless Soils (dense sands)*
Stockless (movable fluke)	2	1
Stockless (fixed fluke)	3	1
Moorfast Offdrill	4	1
BOSS BRUCE Cast Danforth LWT STATO/NAVMOOR STEVFIX	4-½	1
BRUCE Twin Shank Hook STEVMUD	5	1

7.3.3 Sediment Type and Strength

Data on sediment type (clay, silt, or sand) and consistency (soft or hard) are necessary for selection of anchor type and for sizing. This information can sometimes be interpreted directly from subbottom acoustic records by experienced personnel. But, more often, soil samples are required for visual classification and index property testing (see Chapter 3).

In most instances, samples from short gravity corers will provide the necessary sediment for determining classification and consistency data. Acoustic survey data will usually suffice to characterize the sediments over the expected depth of penetration of the anchor, provided the acoustic data show no significant sediment layering or changes over the planned drag path of the anchor. In homogeneous soils, sediment parameters measured on the surficial samples can be used to develop usable prediction of soil strength over the full anchor penetration depth.

In situations where the sediment stratigraphy is complicated, soil data may be required from samples taken from deeper depths as, for example, in the following instances. Where a thin 5- to 10-foot thick layer of soft soil overlies a hard layer, the anchor may drag in the soft surficial material but be unable to penetrate into the underlying hard material. Drag anchors will also have problems when a, thin 2- to 5-foot layer of sand overlies soft silt or clay. In this case, the anchor fluke angle may have to be set for a sand bottom to initiate dig-in. However, the anchor will not be able to develop a high holding capacity in the underlying soft material.

Soil samples of the entire soil profile to be penetrated will identify potential problems such as these and help in selection of the most suitable drag anchor type and size.

7.3.4 Site Investigation Summary

Figure 7.3-1 presents a flow chart for developing a geotechnical site survey plan for drag anchor design. It may be advantageous or necessary to accomplish some of the on-site tasks according to a different schedule based on known site conditions or on the availability and cost of the survey and positioning equipment and the vessel support.

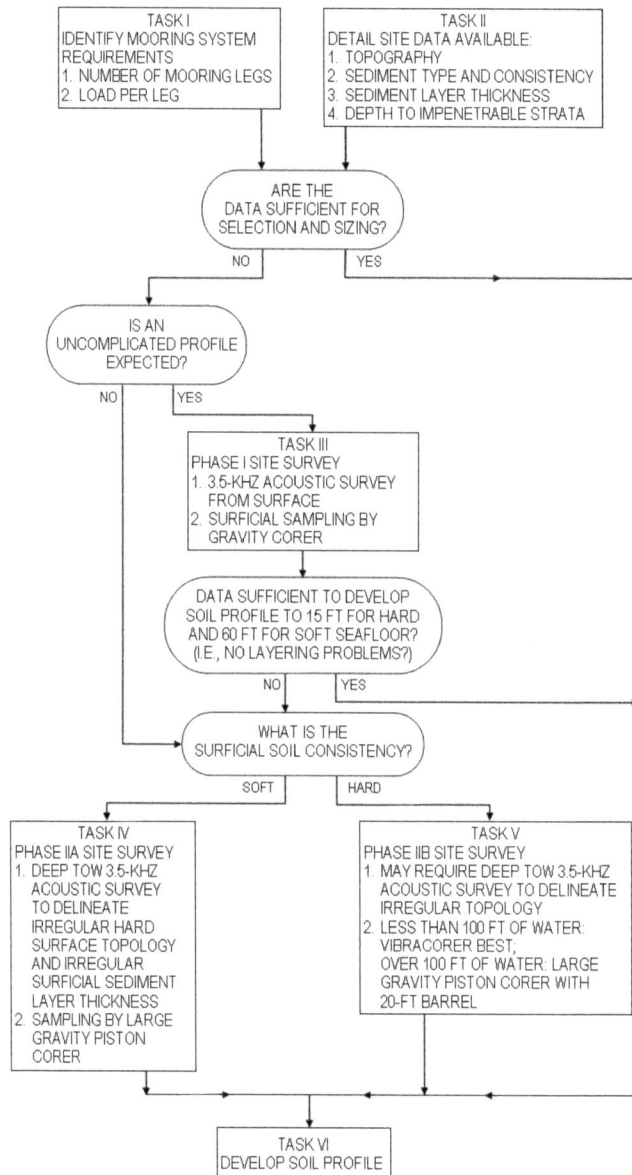

Figure 7.3-1. Site survey plan decision flow chart.

7.4 SELECTING A DRAG ANCHOR

7.4.1 General

Designing an anchor system is a two-step process in which:

- One or more drag anchor types are selected for use based on expected overall performance, availability, and cost.
- The selected anchor type is then sized to develop the required holding capacity.

To aid the anchor selection process, Table 7.4-1 rates common anchor types, based on reports of field experience with prototype anchors and on test findings with small anchors. Separate ratings are given for (1) tripping reliability and dig-in performance, (2) roll stability during setting and dragging, and (3) holding capacity "efficiency" (the ratio of holding capacity to anchor air weight). The anchors are rated as high, medium, or low for each category. Where data on a particular anchor type are not complete, the ratings have been partly based on the performance of geometrically similar anchor types.

7.4.2 Tripping and Penetration Performance

The ratings in Table 7.4-1 on tripping and dig-in performance are for anchors with fluke angles set according to recommendations for cohesive soils (clays and silts) or cohesionless soils (sand), respectively. (Mud palms are used for the STATO on soft clay.)

In soft clays, the tripping/dig-in rating is also a function of the installation method. If the anchor is installed using two work platforms to keep the flukes open before touchdown as shown in Figure 7.2-2, then most anchors will trip and dig-in properly. However, some anchors with limited roll stability may still roll during setting and may pull out. Once back at the seafloor surface the probability of anchor re-embedment would be small.

7.4.3 Stability Performance

The stability performance ratings presented in Table 7.4-1 are based on field tests where the anchors were instrumented to measure roll (Refs. 7-11 and 7-17) or are based on model tests in sand. Since the number of anchor tests used to rate stability is small and the stability of an anchor is statistical in nature, the ratings should be treated as a best estimate.

Table 7.4-1. Rating of Drag Anchor Types Based on Tripping and Dig-In, Roll Stability, and Holding Capacity Efficiency

Anchor Type	Reliability or Holding Capacity in:					
	Cohesive Soils (clays and silts)			Cohesionless Soils[a] (sands)		
	Tripping/ Dig-In	Stability	Holding Capacity	Tripping/ Dig-In	Stability	Holding Capacity
Stockless[b] (movable fluke)	Low	Medium	Low	High	Medium	Low
Stockless[b] (fixed fluke)	High	Medium	Low	High	High	Low
G.S.	c	c	Medium	High	Medium	Medium
Danforth	Medium	Low	Medium	High	Medium	Medium
LWT	Low	Low	Low	High	Medium	Medium
STATO[d]/NAVMOOR	High	Medium	High	High	High	High
Moorfast	Medium	Medium	Medium	Medium	Medium	Medium
Offdrill II	Medium	Medium	Medium	Medium	Medium	Medium
Flipper Delta	c	c	Medium	c	c	Medium
STEVIN	c	c	Medium	c	c	Medium
STEVFIX	Low	Low	High	High	Medium	High
STEVPRIS	c	c	c	High	High	High
STEVDIG	c	c	c	High	Medium	High
STEVMUD	High	c	High	e	e	e
Boss	High	Medium	High	High	c	High
Hook	High	High	Medium	Medium	High	Medium
BRUCE Cast	High	High	Low	High	High	High
BRUCE Twin Shank	High	High	High	c	High	High

[a]Anchor fluke angle set at manufacturer's recommendation for sand.
[b]With stabilizers (ratings not as high without stabilizers).
[c]Insufficient data available for rating.
[d]Anchor fluke angle set at 30 deg for sand.
[e]Anchor not normally used in this seafloor condition.

7.4.4   Holding Capacity Performance

The Table 7.4-1 holding capacity ratings are estimates based on the measured performance of some of the anchor types in comparative field tests (Refs. 7-11 and 7-17) and on the predicted performance for the other anchor types based on their geometric similarity to those which were field tested. This presentation of drag anchor rating based on holding capacity is highly generalized also because it is based on performance in two uncomplicated soil profiles. These holding capacity ratings can be expected to change and become better defined as technical understanding and the predictive capability regarding drag anchor performance improve. These ratings can also be expected to vary for those soil profiles which are more complex.

In soft soils (normally consolidated soft clays and silts), the STEVMUD, a specialized soft soil anchor, has the highest holding-capacity-to-weight ratio. Next highest in developed holding capacity efficiency are the streamlined, deep-digging anchors of substantial fluke area, including the STATO, HOOK, BRUCE Twin Shank, and the STEVFIX anchors. Cast large anchors of the Moorfast and Offdrill II type exhibit lower holding capacity efficiencies because they are less streamlined and do not penetrate as deeply in the sediment profile, and because their fluke areas per pound of steel are smaller.

In sand, the STEVFIX, STEVDIG, STEVPRIS, BRUCE (Cast and Twin Shank), BOSS, and STATO anchors all perform quite well, with no significant differences apparent among them. The cast, bilateral anchors of the Moorfast and Offdrill II type are projected as exhibiting good holding capacities, but not as high as those developed by the anchors of newer design.

7.4.5   Selection of Anchor Type

Selection of the anchor type to be used for a given mooring should be based in large part on the considerations previously discussed. The designer may wish to apply weighted numerical ratings to the anchor types for each of the performance characteristics to provide a "score" for each anchor type. The designer must also consider other very important factors of (1) availability, (2) hardware purchase price, (3) transportation, (4) vessel space requirements, and (5) handling ease.

7.5   SIZING A DRAG ANCHOR

7.5.1   Efficiency Ratio Method

The prediction of drag anchor capacities has traditionally been by empirical approaches. That is, holding capacity is based on field experience with anchors. The most widely used of these approaches is the simple efficiency ratio method. In this method, the anchor's efficiency, e, is defined as a ratio of the horizontal load resistance developed divided by the weight of the anchor (also called the "holding-capacity-to-weight ratio").

$$e = \frac{H_M}{W_A} \qquad\qquad (7\text{-}1)$$

where:

H_M = holding capacity (horizontal load resistance at the seafloor) [F]

W_A = weight of the anchor in air [F]

This method assumes that the drag anchor efficiency is a constant for a given soil type over a wide range of anchor sizes (weights). It also assumes that the mooring leg is properly; designed and the anchor is properly installed and the anchor is dragged the necessary distance to develop its maximum holding capacity.

The efficiency ratio method remains widely used because of its simplicity and familiarity within the user community. However, comprehensive field tests have shown that the method may produce unsafe holding capacity predictions for large anchors. In these tests, the efficiency was shown to decrease with Increasing anchor weight (Ref. 7-1). Therefore, if efficiency constant was used to project anchor holding capacity for a larger anchor, the projection would overpredict that capacity.

To develop the holding capacities predicted by Equation 7-1 or by the two anchor holding capacity predictive techniques given in Sections 7.5.2 and 7.5.3, the mooring line tension must be applied to the anchor system IN A DIRECTION PARALLEL TO THE SEAFLOOR. This is accomplished, in a proper design, by using sufficient chain weight (sometimes accompanied by a concrete sinker weight) in the mooring line leg to keep the chain angle zero even at the highest load condition. In addition to this, an extra shot (90 feet) of heavy chain is often added at the anchor end to ensure that some chain remains on the seafloor.

7.5.2 Power Law Method

The power law method is recommended for use as the best method for predicting the holding capacity of drag-embedment anchors. It accounts for the nonlinear increase in holding capacity with increasing anchor air weight (Ref. 7-1). This method produces a straight-line relationship between anchor holding capacity and anchor weight on a log-log plot. The validity of the power law method has been demonstrated in field tests (Refs. 7-16 and 7-18).

Expressed as an equation, the holding capacity in kips, H_M, for large anchors is determined by:

$$H_M = m \left(W_a \right)^b \tag{7-2}$$

where:

m, b = dimensionless parameters dependent on the anchor and soil type

W_a = anchor weight in air (kips) [F]

7-20

Equation 7-2 is only valid for large anchors weighing 200 pounds or more. Refer to Section 7.5.2.1 for information on determining the holding capacity of anchors weighing less than 200 pounds.

Anchor holding capacity field test data have been used to develop values for m and b used in Equation 7-2. These data were obtained from 1950 NCEL testing of Navy Stockless, STATO, and LWT anchors (Refs. 7-13, 7-19 through 7-22); from Exxon Production Research testing of LWT and BOSS anchors (Refs. 7-16 and 7-18); from 1979-80 NCEL testing of Navy Stockless, STATO, Moorfast, BRUCE, and STEVIN family anchors (Refs. 7-11 and 7-17); and from some offshore industry tests. The test data and previous interpretations (Refs. 7-1, 7-16, and 7-18) have resulted in the values of m and b shown in Table 7.5-1 for a wide range of drag anchors used in cohesive and cohesionless soils. Table 7.5-1 and Equation 7-2 can be used to predict the holding capacity for large anchors (weight > 200 pounds). The mooring leg, which includes the anchor, however, must be properly designed and the anchors must be properly installed. Figure 7.5-1 and Figure 7.5-2 are a graphical presentation of the data from Table 7.5-1 for cohesive soils (clays and silts) and cohesionless soils (sands), respectively. It is noted that the curves for the Flipper Delta anchors are not the result of field tests but have been estimated.

In Figure 7.5-1, the curves describe the holding capacity of drag anchors in normally consolidated clays and cohesive silts. These curves were developed directly from measured holding capacities. Note the difference in performance for the Stockless anchor with flukes in both fixed and movable positions. The Stockless, with flukes left movable, will develop only half the capacity of the anchor with flukes fixed at 50° in soft soils (Ref. 7-23).

Figure 7.5-2 presents an interpretation of data from field tests in several sand types. Only one curve is used to describe the performance of an anchor type for all types and densities of sands. A more accurate predictive scheme for these variations has not yet been developed. While the holding capacity developed with a given anchor is expected to vary somewhat with sand type and density, the relative capacities should remain about the same. The predictive curves of Figure 7.5-2 are also believed applicable in very stiff to hard clays (i.e., in all hard seafloors). In Table 7.5-1, parameters were assigned to some anchor types where data were not available to develop curves. Based on geometric similarities, the Danforth, G.S., Offdrill II, Flipper Delta, and STEVIN have each been assumed to behave like one of the other anchors which were tested and to have similar values of holding capacity.

The power law method, as used here, includes both buried chain and anchor holding capacity, but makes no separate accounting of anchor and chain contributions. The method assumes that the chain size used is compatible with the holding capacity to be developed. Larger chain sizes may develop somewhat higher holding capacities than those predicted by using Table 7.5-1 (Ref. 7-23). The use of wire rope for the embedded portion of the mooring line, despite the smaller projected area of comparable chain, is not expected to significantly reduce anchor system holding capacity. The smaller diameter wire rope should allow the anchor to penetrate deeper into stronger soils, and an increased anchor resistance at that depth is expected to compensate for any reduction in resistance from the smaller mooring line.

Table 7.5-1. Parameters m and b Used in Equation 7-2: $H_M = m\,(W_a)^b$

Anchor Type[a]	Soft Clays and Mud		Stiff Clays and Sand	
	m	b	m	b
BOSS	24.1	0.94	31.0	0.94
BRUCE Cast	3.9	0.92	39.6	0.80
BRUCE Flat Fluke Twin Shank (FFTS)	30.0	0.92	34.4	0.94
BRUCE FFTS MK4	42.5	0.92	— [c]	— [c]
BRUCE Twin Shank	22.7	0.92	24.1	0.94
Danforth	10.5	0.92	20.0	0.80
Flipper Delta	16.7	0.92	— [c]	— [c]
G.S. (AC-14)	10.5	0.92	20.0	0.80
Hook	22.7	0.92	15.9	0.80
LWT	10.5	0.92	20.0	0.80
Moorfast	10.5	0.92	15.9[d]	0.8
NAVMOOR	24.1	0.94	31.0	0.80
Offdrill II	10.5	0.92	15.9[d]	0.80
STATO	24.1	0.94	28.7[e]	0.94
STEVDIG	16.7	0.92	46.0	0.80
STEVFIX	22.7	0.92	46.0	0.80
STEVIN	16.7	0.92	26.2	0.80
STEVMUD	30.0	0.92	— [f]	— [f]
STEVPRIS MK3 (straight shank)	22.7	0.92	24.1	0.94
STEVPRIS MK5	42.5	0.92	— [c]	— [c]
Stockless (fixed fluke)	5.5	0.92	11.1	0.8
Stockless (movable fluke)	2.9	0.92	11.1[g]	0.8
Stockless (movable fluke)	— [c]	— [c]	7.0[h]	0.8

[a] Fluke angles set for 50-deg in soft soils and per manufacturers' specifications in hard soils, except when otherwise noted.
[b] "b" is an exponent constant in Equation 7-2 and not a footnote.
[c] No data available.
[d] For a 28-deg fluke angle.
[e] For a 30-deg fluke angle.
[f] Anchor not suitable for this seafloor condition.
[g] For a 35-deg fluke angle.
[h] For a 48-deg fluke angle.

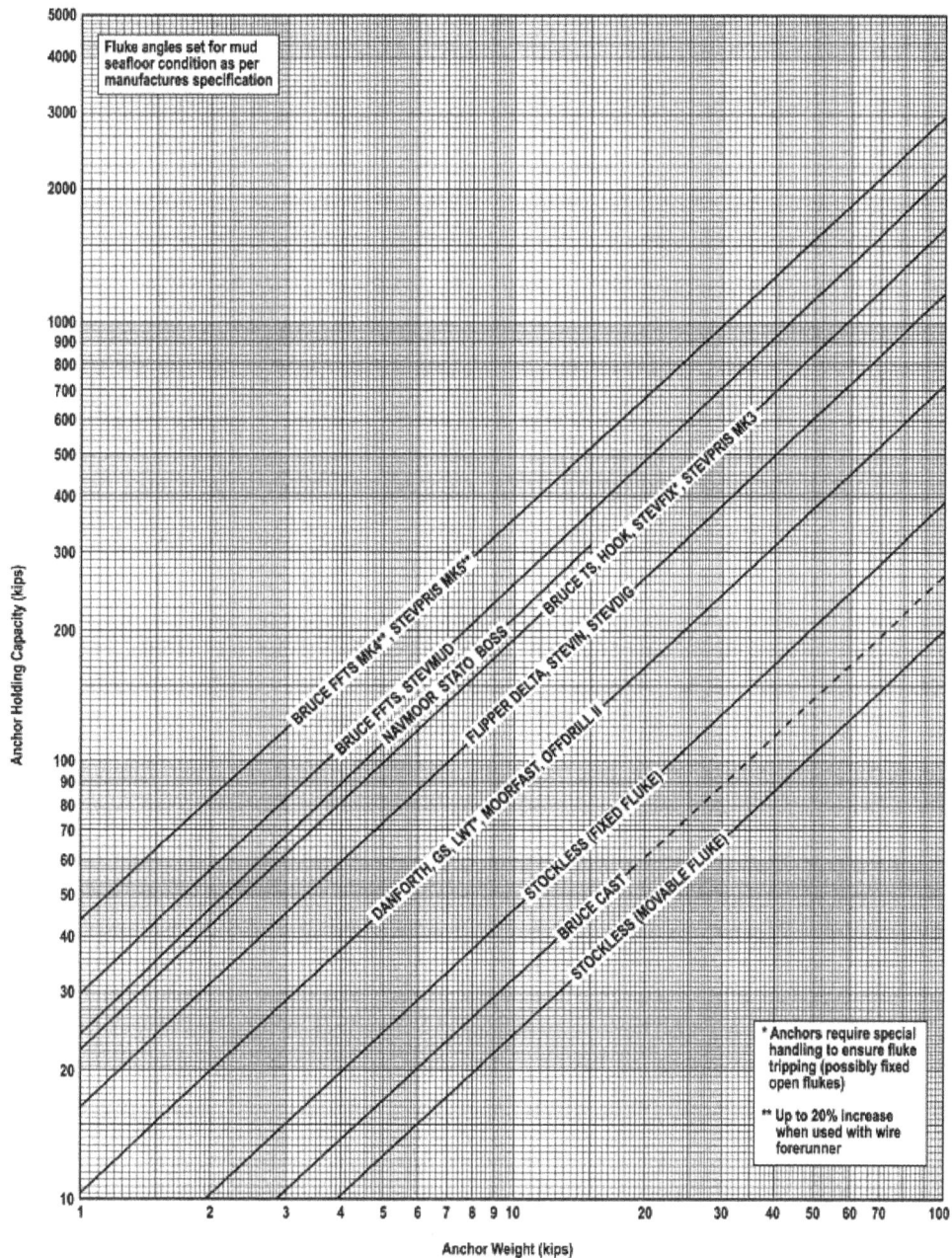

Figure 7.5-1. Anchor chain system holding capacity at the mudline in soft soils.

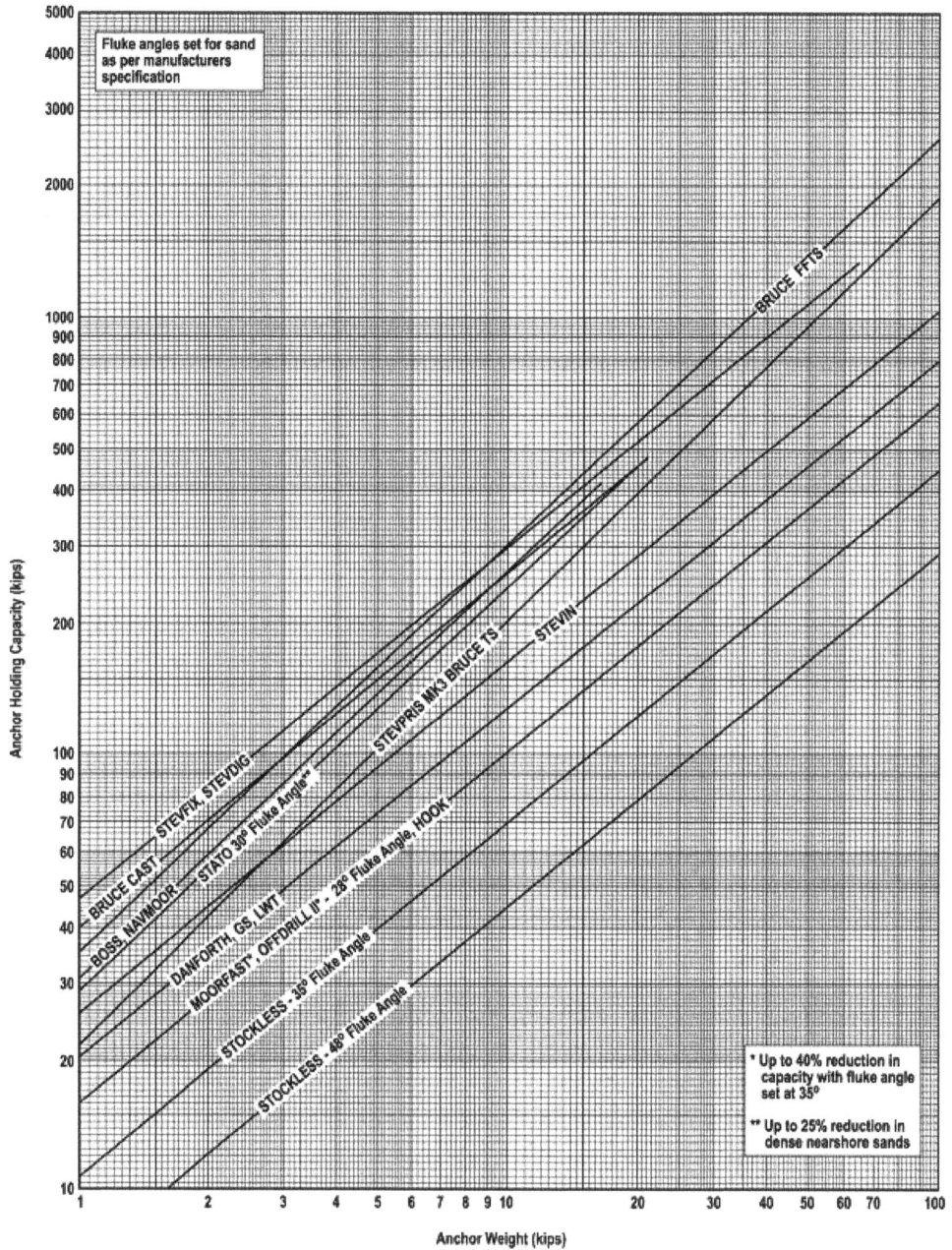

Figure 7.5-2. Anchor chain system holding capacity at the mudline in hard soils.

7-24

7.5.2.1 Holding Capacity for Small Anchors (Weighing Less Than 200-lb)

The holding capacity in kips, H_M, of small anchors (anchor weight < 200-lb) is given by Equation 7-3. In the equation, the anchor efficiency, e, is a dimensionless parameter that varies considerably depending on the soil and anchor type. Table 7.5-2 (Ref. 7-24) lists some values of e for small anchors.

$$H_M = e\,(W_a)$$
(7-3)

where:

e = anchor efficiency, dimensionless

W_a = anchor weight in air (kips) [F]

Table 7.5-2. Average Anchor Efficiency for Small (< 200-lb) Anchors

Anchor Type[a]	Average Anchor Efficiency, e	
	Soft Clays and Mud	Stiff Clays and Sand
BRUCE	6	30
CQR Plow	10	40
Danforth	20-40	50-100
Fortress (Aluminum)	35; 50[b]	100-180
LWT	2-10	40
NAVMOOR 100	25	40-50
STATO 200	25	20; 30[c]
Stockless	2-3	5; 10[d]

[a] Fluke angles set for 50-deg in soft soils and per manufacturers' specifications in hard soils, except when otherwise noted.
[b] For a 45-deg fluke angle.
[c] For a 28-deg fluke angle.
[d] For a 35-deg fluke angle.

7.5.3   Analysis Based on Geotechnical Considerations

A complete method for the prediction of drag anchor holding capacity based on geotechnical considerations is not presently available. The prediction of drag anchor and mooring line penetration is beyond present analytical capability. However, recent work has improved the ability to predict capacity based on geotechnical considerations for anchors embedded in soft soils. The technique for calculating anchor holding capacity is presented here for use with an established anchor, where the depth of penetration is known.

7.5.3.1 Calculations for Soft Cohesive Seafloors

Most soft cohesive seafloors, defined as mud or soft silt to clay size sediments, are normally consolidated to slightly overconsolidated. Drag-embedment anchors will penetrate deeply in these sediments. In this situation, the drag anchors can be expected to behave similar to deeply embedded plate anchors.

Drag anchor flukes differ considerably in their plan shape. In order to adequately describe the projected plan area of the different shapes, a correction factor is applied to the anchor fluke dimensions. The holding capacity of a drag anchor is then expressed as:

$$H_A = N_c(fBL)s_u \qquad\qquad (7\text{-}4)$$

where:

N_c = a constant, depending on the failure mode around the anchor and on the anchor geometry (see Table 7.5-3)

f = correction factor converting the rectangular area B x L to the actual projected fluke area (see Table 7.5-3)

B = width of fluke (see Figure 7.1-5) [L]

L = length of fluke (see Figure 7.1-5) [L]

s_u = undrained shear strength of cohesive soil at the center of area of anchor [F/L^2]

Equation 7-4 does not provide a complete solution to the problem of designing a mooring system because the depth of penetration of the drag anchor/chain system is beyond present predictive capability (save for the estimates shown in Table 7.3-1). For established anchors, where anchor penetration can be measured from a pendant line, Equation 7-4 can be utilized for evaluating the holding capacity. Equation 7-4 predicts only the holding capacity developed at the shank-to-mooring-line connection point. The section of mooring line embedded in the seafloor by the penetrating drag anchor makes a significant contribution to the anchor system's holding capacity. Based on available field data (Refs. 7-11 and 7-17), the mooring line contribution, H_C, in soft muds can be conservatively estimated to be $0.2H_A$ (refer to Section 5.3.3).

7.5.3.2 Considerations for Sands and Stiff Clays

Field measurements (Refs. 7-11, 7-17, and 7-25) show that drag anchors do not penetrate deeply in sand and stiff clays. Often the maximum holding capacity occurs even before the entire fluke is embedded. Presently, a predictive equation which uses geotechnical considerations is not available.

Table 7.5-3. Parameters N_c and f used for Clays and Cohesive Silts in Equation 7-4: $H_A = N_c\,(f\,B\,L)\,s_u$

Anchor Type	N_c	f
Stockless, fixed fluke	13.0	0.54
Danforth	11[a]	0.60
LWT	11[a]	0.60
STATO/NAVMOOR[b]	12	0.95
Moorfast	12	0.95
Offdrill II	12[a]	0.95
STEVFIX	6.4	0.72
STEVMUD	6.8	0.77
Hook	6.2	0.80
BRUCE Cast	4.0	0.36
BRUCE Twin Shank	6.5	0.52

[a] Estimated values.
[b] NAVMOOR has a configuration similar to that of the STATO anchor.

7.5.4 Factor of Safety

A general practice is to size the drag embedment anchor as the "weaker link" of a mooring system. It is preferable to allow the anchor to drag instead of breaking the mooring line. Anchor drag results in redistribution of the overstressed mooring line to its neighboring lines and helps the mooring to survive in storms when environmental loads exceed the design loads. The factors of safety for the mooring line and the drag embedment anchor are selected separately as described below.

7.5.4.1 Mooring Lines

The American Petroleum Institute (Ref. 7-26) and American Bureau of Shipping (Ref. 7-27) recommend limits of line tension and its equivalent factor of safety for mooring lines as listed in

Table 7.5-4. Note that in

Table 7.5-4, the following definitions are applied for the conditions:

- *Intact Condition*: Condition in which all mooring lines are intact.

- *Damaged Condition*: Condition in which the floating platform settles at a new equilibrium position after a mooring line failure.

- *Transient Condition*: Condition in which the floating platform is subjected to transient motions (overshooting) after a mooring line failure before it settles at the new equilibrium position.

Table 7.5-4. Recommended Line Tension Limits and Factor of Safety on Mooring Lines

Condition	Analysis Method	Tension Limit (% of Break Strength)	Equivalent Factor of Safety
Intact	Quasi-static	50	2.0
	Dynamic	60	1.7
Damaged	Quasi-static	70	1.4
	Dynamic	80	1.3
Transient	Quasi-static	85	1.2
	Dynamic	90	1.1

7.5.4.2 Drag Embedment Anchor

Factors of safety for drag embedment anchors are provided in Table 7.5-5 below (Refs. 7-4 and 7-27). Also note that the anchor chain system holding capacity curves shown in Figure 7.5-1 and Figure 7.5-2 do not include a factor of safety.

Table 7.5-5. Recommended Factors of Safety for Drag Embedment Anchors

Condition	Analysis Method	Factor of Safety
(a) Permanent Mooring		
Intact	Quasi-Static	1.8
	Dynamic	1.5
Damaged	Quasi-Static	1.2
	Dynamic	1.0
Transient	Quasi-Static	Not Required
	Dynamic	Not Required
(b) Temporary Mooring		
Intact	Quasi-Static	1.0
	Dynamic	0.8
Damaged	Quasi-Static	Not Required
	Dynamic	Not Required
Transient	Quasi-Static	Not Required
	Dynamic	Not Required

7.6 TROUBLESHOOTING

Section 7.5 offers techniques for predicting the potential holding capacity of drag-embedment anchors, assuming proper penetration and stability of the drag anchor system. Section 7.2 discussed the functioning of drag anchors and identified several penetration and stability problems that could occur in the use of drag-embedment anchors. This section summarizes these problems and potential solutions in a troubleshooting procedure. These are outlined in Table 7.6-1.

Table 7.6-1. Troubleshooting Procedures for Correcting Drag Anchor Performance Problems

Problem	Troubleshooting Procedures[a] for:	
	Cohesive Soils	*Cohesionless Soils*
Failure to Initiate Penetration	1. Proper setting procedure used? 2. Fix fluke in open position 3. Increase stabilizer length	1. Sharpen flukes 2. Reduce fluke angle – to a minimum 27 deg 3. Increase stabilizer length 4. Place barbs on tripping palms 5. Fix fluke open
Failure to Develop Expected Holding Capacity	1. Recover and clean; check for balling up; clean up ball 2. Reset and soak 24 hours 3. Piggyback or change to larger anchor	1. Increase stabilizer length 2. Reduce fluke angle – to a minimum 27 deg 3. Piggyback or change to larger anchor

[a] Troubleshooting procedures listed in order of recommended.

7.6.1   Soft Sediments

Drag anchors in soft seafloors may encounter (1) tripping problems, (2) instability problems, or (3) inability to develop sufficient holding capacity without allowing for sediment strength gain through anchor soaking. Figure 7.6-1 illustrates the effects of tripping and instability problems on the line tension developed while setting an anchor. Failure to trip and initiate penetration is suggested when line tension remains nearly constant at one-half to two times the combined weight of the anchor and the amount of mooring line on the seabed (curve (c) in Figure 7.6-1). Tripping problems are normally avoided by using a proper anchor setting procedure (Figure 7.2-2). When limited support prevents use of this procedure, then correction of a tripping problem requires fixing of the anchor fluke in its open position and, possibly, the lengthening of stabilizers to help right the anchor (Figure 7.2-6b).

Instability problems are suggested by a rise in the line tension followed by a fall to the "failure-to-trip level" (curve (b) in Figure 7.6-1). When the anchor behaves in this manner and fails to develop adequate capacity, the anchor should be recovered and cleaned. In soft soils it is likely that a mud ball has formed on the anchor flukes. In hard soils, the fluke angle should be reduced. The anchor should be reset and dragged until the line tension approaches its prior peak value. The line tension should then be slacked off and the anchor allowed to soak for 24 hours. If this delay is not possible or if, after soaking, the required line tension cannot be developed, then an anchor of higher capacity will be required or the original anchor must be piggybacked (Section 7.7).

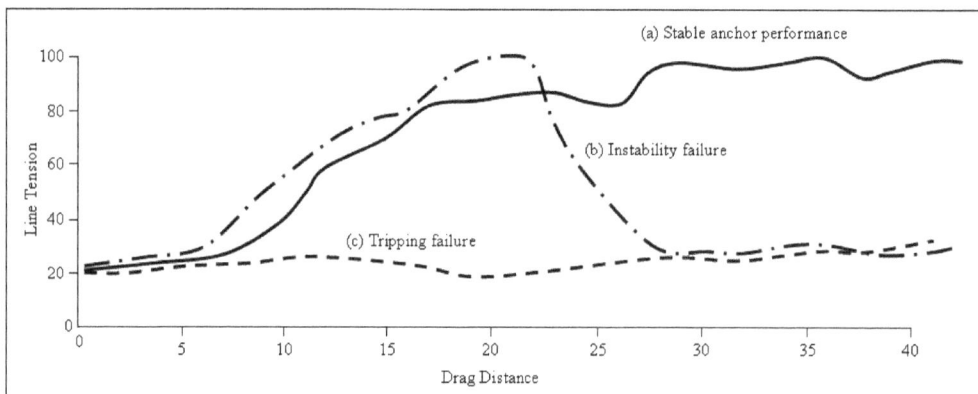

Figure 7.6-1. Typical performance of drag anchors when operating properly and improperly.

7.6.2 Hard Sediments

Tripping, penetration, and instability problems can occur on hard seafloors (including all sands, gravel, and hard clay). Tripping is enhanced by sharpening the fluke tips. Penetration or dig-in is enhanced by reducing the fluke-to-shank angle to as small as 25° to 27° and increasing the stabilizer length to keep the anchor from sliding on its side (Figure 7.2-3). In extreme cases, barbs have been added to the tripping palms to increase the tripping moment, and flukes have been fixed in the open position. On some hard surfaces (e.g., coral or weak rock) these measures cited to initiate penetration will not suffice. Shaped charges have been used on these surfaces to crush the material in front of the anchor and provide an area for fluke embedment.

The primary means for increasing anchor stability in hard seafloors is increasing the stabilizer length. A decrease of the fluke angle to increase penetration has also proven beneficial to stability. Should these steps fail in gaining the required line tension, larger anchors or piggybacked anchors will be required. In extreme cases sufficient anchorage will require shifting to use of a pile.

7.7 PIGGYBACKING

7.7.1 Field Practice

When an anchor will not develop the required capacity on being proof-loaded, it is common practice to install a second anchor on the same mooring leg, in-line, and beyond the first or primary anchor. This is called a piggyback anchor, and sometimes called a tandem or backup anchor. The procedure for installing a piggyback anchor varies.

In the offshore oil industry, many contractors simply detach the existing pendant line surface buoy from the pendant line wire rope of the first anchor (see Figure 7.7-1) and reattach the line to a second anchor. The second anchor is then lowered and set by a second pendant line, while the line to the first anchor is kept tightly stretched. Sometimes it is necessary to retrieve the first anchor, attach a heavier wire rope to connect the two anchors, and set both anchors in sequence. In some cases a chaser system (Figure 7.7-2) is used to install and recover the primary anchors. When using a chaser system, the piggyback anchor is attached directly to the chaser wire. In all of the above techniques, the primary anchor is dragged some short distance to set it before the piggyback anchor is lowered. The Navy, working from available anchor inventory, has found piggybacking necessary for high capacity permanent ship moorings. Techniques for laying the mooring leg in water depths to 100 feet using a crane barge have been demonstrated (Ref. 7-11).

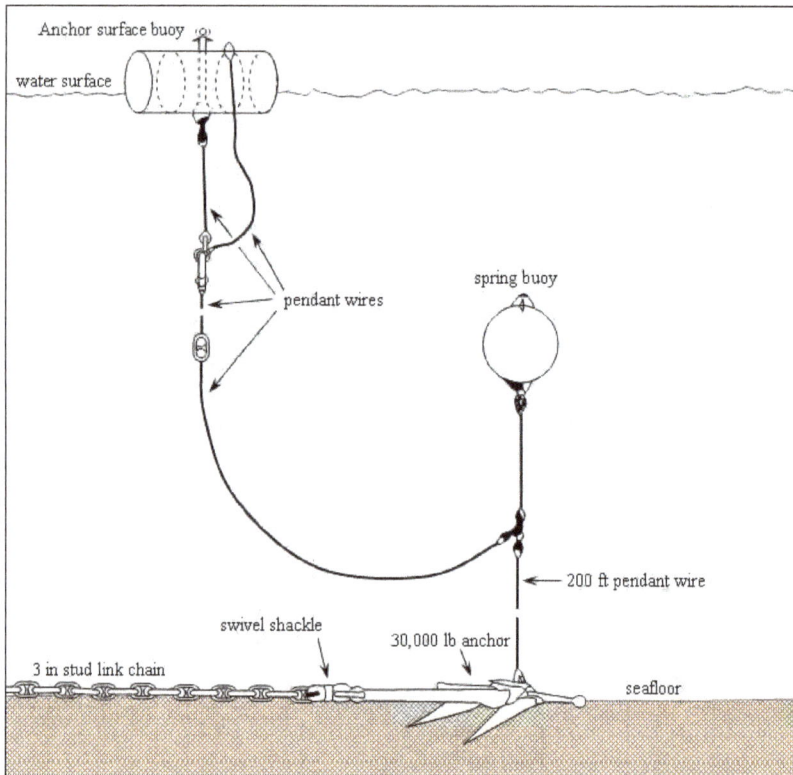

Figure 7.7-1. A pendant line and buoy arrangement for semisubmersibles (Ref. 7-28).

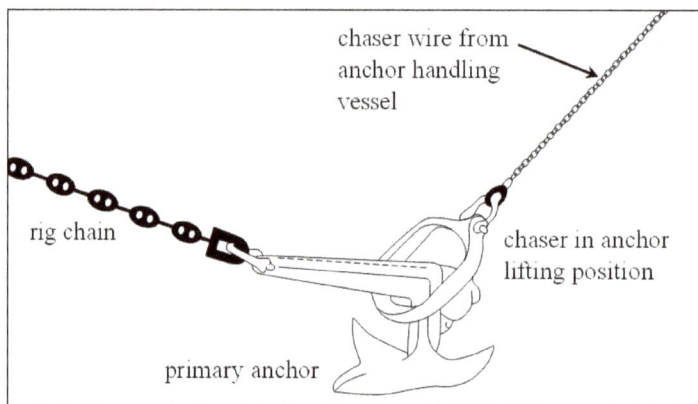

Figure 7.7-2. Chain chaser used to assist anchor deployment and recovery.

7.7.2 Results and Field Problems

The results from piggybacking are strongly dependent on how the primary and secondary anchors are attached to the mooring line. Pendant lines are usually attached to the anchor at the back of the fluke where possible or at the crown end of the shank. This produces mixed results, illustrated by Figure 7.7-3. With many anchors used in this arrangement, the line tension from the piggyback anchor will cause rotation and breakout of the primary anchor (Ref. 7-29). With other anchors, this arrangement works fine and results in a piggyback system capacity equal to or greater than the sum of the holding capacities of the two anchors loaded separately (Ref. 7-30). Once dislodged, however, primary anchors with movable flukes and with load applied directly to the anchor crown will not dig in again, but will slide with the flukes held parallel to the shank by the line tension applied at the pendant padeye (Figure 7.7-3a). Navy tests in sand and mud have shown that the procedure shown in Figure 7.7-3b is suitable provided the inbound anchor is very stable (e.g., BRUCE, Hook, STATO, NAVMOOR) and anchor flukes are prevented from closing up under load. Anchors whose center of fluke area is close to the fluke-to-shank connection point (e.g., Danforth, LWT, STEVIN types, Flipper Delta) are less stable anchors and are not appropriate for shank-to-crown piggyback connections.

For those anchors where the crown attachment (of the piggyback anchor) creates problems, model tests have shown that primary anchor stability will not be significantly affected when the line to the piggyback is attached to the shackle end of the shank of the primary anchor (Figure 7.7-3c). The holding capacity of the anchors attached this way reached the sum of the holding capacities of the individual anchors. This attachment technique should be used only with fixed-fluke anchors to minimize potential for fouling of the second anchor wire or chain with the primary anchor (Ref. 7-29). Parallel tandem anchor arrangements have been suggested (Figure 7.7-4). Full-scale tests show that the anchors tend to come together, and that load equalization between anchors is a problem (Ref. 7-21). Once the anchors do come together they will foul and will not re-embed. If used in this manner, the parallel anchors should be staggered by using different length chain legs to avoid anchor interference. Navy tests (Ref. 7-30) with staggered anchors (separated by at least four fluke lengths) have shown system capacities 15 to 20% greater than the sum of the individual anchor capacities.

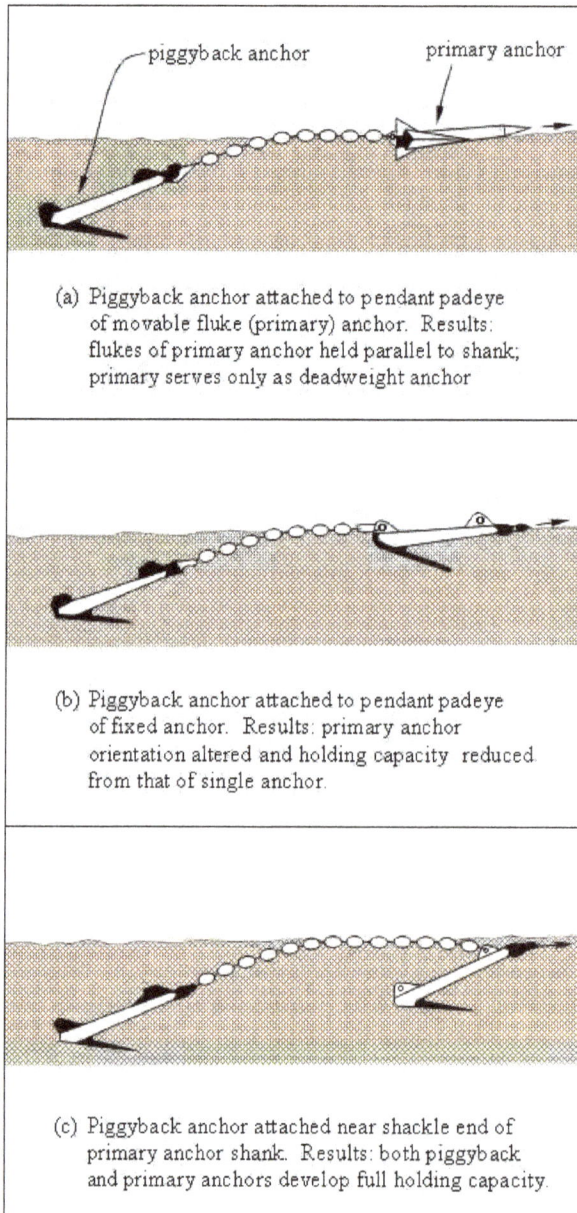

Figure 7.7-3. Tandem/piggyback anchor arrangements (Ref. 7-29).

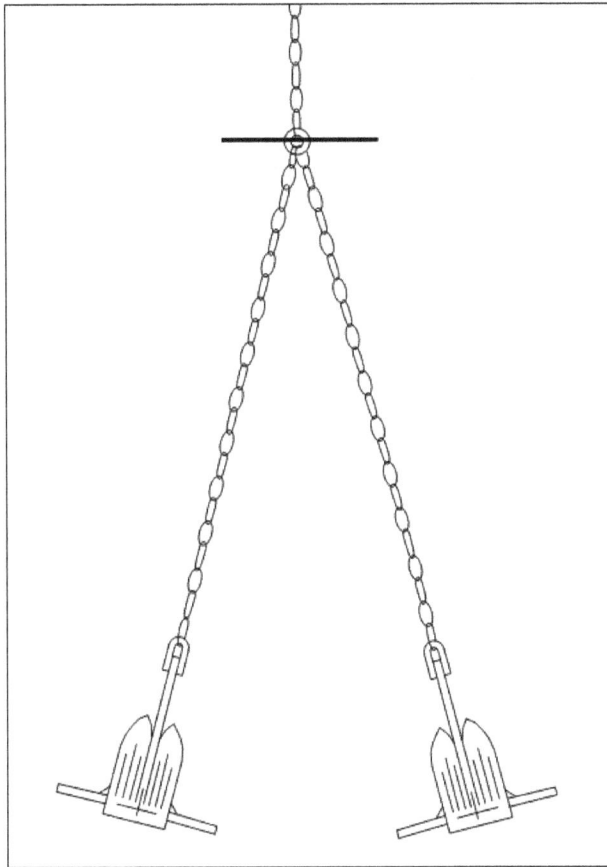

Figure 7.7-4. Parallel anchor arrangements.

7.7.3   Recommended Practice

Careful placement of piggyback anchors is required so that orientation and stability are properly controlled. Attachment of the piggyback anchor at the crown is proper when the primary anchor is of the stabilized Stockless, Hook, STATO, or BRUCE type. Attachment of the piggyback anchor at the shackle end of the primary anchor is proper when the primary anchor is one of the other anchor types. The primary anchors should be set and be well-stabilized before tension from the piggyback anchor is applied.

7.8 REFERENCES

7-1.   Unified Facilities Criteria (UFC): Military Harbors and Coastal Facilities, Naval Facilities Engineering Command (Preparing Activity), UFC 4-150-06. Washington, DC, Dec 2001.

7-2.   "Rules for the Design, Construction, and Inspection of Offshore Structures," Appendix E of Hydrostatic Stability and Anchoring, Det Norske Veritas. Oslo, Norway, 1977.

7-3.   Rules for Building and Classing Mobile Offshore Drilling Units, American Bureau of Shipping, ABS 6-2008. New York, NY, Oct 2008.

7-4.   API Recommended Practice for the Analysis of Spread Mooring Systems for Floating Drilling Units, 2nd Edition, American Petroleum Institute, API RP 2P. Dallas, TX, May 1987.

7-5.   M.A. Childers. "Spread Mooring Systems," in The Technology of Offshore Drilling, Completion and Production, compiled by ETA Offshore Seminars, Inc. Tulsa, OK, Petroleum Publishing Company, 1976, pp. 94-128.

7-6.   "Deep Water Mooring - Parts I, II, III," Petroleum Engineer, Sep-Oct 1974 and May 1975.

7-7.   U.S. Navy Salvage Engineer's Handbook – Appendix G (Anchoring Systems), Naval Sea Systems Command, S0300-A8-HBK-010. Washington, DC, May 1992.

7-8.   Vryhof Anchors. http://www.vryhof.com/. The Netherlands, 2010.

7-9.   The Bruce Anchor Group. http://www.bruceanchor.co.uk/. Isle of Man, England, 2010.

7-10.   Baldt, Inc. http://www.baldt.com/. Baldt, Chester, PA, 2010.

7-11.   R.J. Taylor. Conventional Anchor Test Results at San Diego and Indian Island, Civil Engineering Laboratory, Technical Note N-1581. Port Hueneme, CA, Jul 1980.

7-12.   K.J. Saurwalt. "Movements and Equilibrium of Anchors Holding on an Impervious Sea Bed, Section I," Schip en Werf, Rotterdam, No. 9, 1971, pp. 627-634.

7-13.   R.C. Towne and J.V. Stalcup. New and Modified Anchors for Moorings, Naval Civil Engineering Laboratory, Technical Report R-044. Port Hueneme, CA, Mar 1960.

7-14.   B. Watt Associates. "A Method for Predicting Drag Anchor Holding Capacity," Contractor Report CR 83.036 to Naval Civil Engineering Laboratory, Port Hueneme, CA, Aug 1983.

7-15.   Naval Civil Engineering Laboratory. "The NAVMOOR Anchor" NCEL Technical Data Sheet, (87-05), Port Hueneme, CA, 1987.

7-16. R.W. Beck. "Anchor Performance Tests," in Proceedings of the Offshore Technology Conference, Houston, TX, 1972, pp. 268-276.

7-17. R.J. Taylor. Test Data Summary for Commercially Available Drag Embedment Anchors, Civil Engineering Laboratory. Port Hueneme, CA, Jun 1980.

7-18. M.W. Cole and R.W. Beck. "Small-Anchor Tests to Predict Full Scale Holding Power," Society of Petroleum Engineers, SPE 2637, 1969.

7-19. R.C. Towne and J.V. Stalcup. Tests of BUSHIPS Anchors in Mud and Sand Bottoms. Naval Civil Engineering Laboratory, Technical Note N-195. Port Hueneme, CA, Aug 1954.

7-20. R.C. Towne. Test of Anchors and Ground Tackle Design, Naval Civil Engineering Laboratory, Technical Memorandum M-066. Port Hueneme, CA, Jun 1953.

7-21. R.C. Towne and J.V. Stalcup. Tests of Moorings and Ground Tackle Design in Mud Bottom, Naval Civil Engineering Laboratory, Technical Memorandum M-097. Port Hueneme, CA, Dec 1954.

7-22. R.C. Towne. "Mooring Anchors," paper presented at the Annual Meeting, The Society of Naval Architects and Marine Engineers, New York, NY, 1959.

7-23. R.J. Taylor. "Performance of Conventional Anchors," in Proceedings of the Offshore Technology Conference, Houston, TX, 1981, pp. 363-372.

7-24. Naval Facilities Engineering Command, "Conventional Underwater Construction and Repair Techniques," NAVFAC P-990, Section 2.8.1, Alexandria, VA, May 1995.

7-25. I. Foss, T. Kvalstad, and T. Ridley. "Sea Bed Anchorages for Floating Offshore Structures," FIP Commission on Sea Structures, Working Group on Foundations, Feb 1980.

7-26. API Design and Analysis of Stationkeeping Systems for Floating structures, 3rd Edition, American Petroleum Institute, API RP 2SK. Dallas, TX, May 2008.

7-27. Guide for Building and Classing Floating Production Installations, American Bureau of Shipping, ABS 82-2009. New York, NY, Apr 2010.

7-28. P.G.S. Dove. "Methods in Anchor Handling," Offshore, Mar 1980, pp. 114-128.

7-29. P.J. Klaren. "Anchors in Tandem or the Use of Back-Up Anchors (Piggybacks)," Holland Shipbuilding, Anker Advies Bureau, pp. 230-232.

7-30. R.J. Taylor and G.R. Walker. Model and Small-Scale Tests to Evaluate the Performance of Drag Anchors in Combination, Naval Civil Engineering Laboratory, Technical Note N-1707. Port Hueneme, CA, Oct 1984.

7.9 SYMBOLS

B Width of anchor fluke [L]

b An exponent constant in the holding capacity predictive relationship; value varies with anchor/soil combination (Table 7.5-1)

d_t Depth of penetration of the fluke tip [L]

e Efficiency of drag anchor system, based on anchor air weight

f Correction factor converting the rectangular area BxL to the actual projected fluke area

H_A Drag anchor holding capacity (without chain contribution) [F]

H_C Contribution of embedded chain to holding capacity of drag anchor system [F]

H_M Holding capacity of drag anchor system, anchor plus chain [F]

H_R Holding capacity of 10,000-lb air weight version of the reference anchor, W_A [F]

k A constant in anchor holding capacity predictive equation for sands; value varies with crown embedment, soil friction angle and effective weight, and anchor shape [F/L^3]

L Length of anchor fluke [L]

N_c A constant in anchor holding capacity predictive equation for clays; varies with anchor geometry

s_u Undrained shear strength of soil [F/L^2]

W_A Air weight of anchor (does not include chain weight) [F]

β Angle between fluke and shank of drag anchor [deg]

8   PENETRATION OF OBJECTS INTO THE SEAFLOOR

8.1   INTRODUCTION

8.1.1   Purpose

This chapter presents techniques for predicting the depth of penetration of objects pushed into or impacting on sediments found in the deep ocean. The techniques presented can also be used to predict the force required to push an object to a specified depth within the seafloor and to predict the sediment-related forces acting on a rapidly moving object after it impacts the seafloor surface.

8.1.2   Scope

8.1.2.1 Seafloor Types

Techniques presented here are limited in application to seafloors of unlithified sediments (i.e., the terrigenous and pelagic clays and silts, sands, and deep sea oozes). Special techniques, not covered here, are required to predict penetrations in lithified sediments, coral, basalt, and other rock types. These special techniques are highly empirical and are normally limited in application to a particular projectile (Ref. 8-1). A limited discussion on penetration of plate anchors in coral and rock is given in Chapter 6.

8.1.2.2 Penetrator Types

The prediction techniques used are applicable to objects of all shapes and sizes, ranging from long streamlined objects such as instrumented penetrometers, free-fall corers, and propellant-embedded anchor flukes, to large blunt objects such as gravity anchors, structure bearing pads, and ship hulls.

8.1.2.3 Penetrator Velocities

The techniques presented are applicable for penetration velocities from near zero (normally called "static" penetration) to 400 ft/s. This encompasses the full velocity range expected in Navy deep ocean applications.

8.2 STATIC PENETRATION

8.2.1 Application

The term "static penetration" identifies a category of soil penetration events in which the objects' penetration velocity on impact with the seafloor is less than 3 ft/s. Note that when an object impacts the seafloor at a slow downward velocity, such as 3 ft/s, its velocity will decrease substantially as some of its momentum is transferred to a block of soil beneath it, but the overall effect on penetration will not be significant because the total momentum and frontal geometry are essentially unchanged. For these slow velocities, fluid drag force in the water at the time of impact, soil strain rate, and inertia factors have negligible influence on penetration. This static problem is adequately treated by conventional pier and pile bearing capacity methods. Objects in this category are footings, structure bearing pads, mud mats, raft foundations, gravity anchors, spud cans, jacked piles for offshore platforms, and many bottom-resting devices used to test for properties of seafloor sediments.

8.2.2 Approach

Traditionally, the static penetration problem is subdivided into shallow and deep sediment penetration cases, as defined by the ratio of an object's penetration depth to its diameter (or minimum lateral dimension if not circular), z/B. In shallow penetration, this ratio is less than 2.5. As an expedient, the shallow condition may be treated as a conventional bearing capacity problem with the influence of friction on the sides of the penetrator set equal to zero (Refs. 8-2 and 8-3) (see Section 8.2.3 and Figure 8.2-1 and 8.2-2). This causes the penetration resistance (bearing capacity) to be under-predicted, which is conservative where adequate bearing capacity must be ensured, and non-conservative when adequate penetration depth is the objective.

In deep penetration (i.e., where $z/B \geq 2.5$), the influence of side friction on the penetrator becomes substantial and must be included in computations. More accurate estimates for shallow penetration, as well as deep penetration, are obtained by using the more complicated full universal relationship presented in the next section.

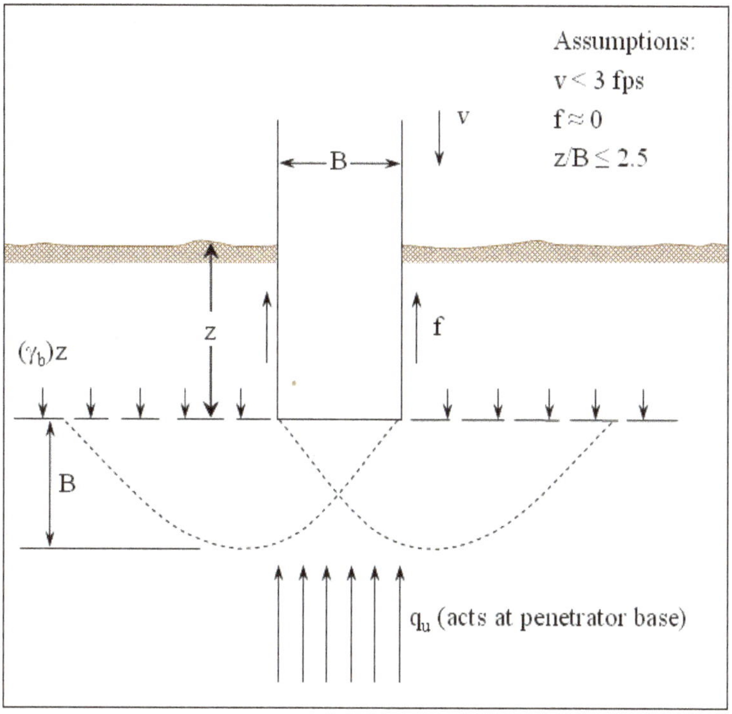

Figure 8.2-1. Shallow static penetration model (Refs. 8-2 and 8-3).

8.2.3 Method for Predicting Static Penetration

8.2.3.1 Descriptive Equations

Static penetration, both shallow and deep, is analyzed using the same fundamental bearing capacity relationships appearing in Chapter 4 (Section 4.3.2). However, for "penetration" a concentric vertical load, a horizontal object base with no skirts, and a flat soil medium are assumed, so the relationships are substantially simplified. Accordingly, the maximum bearing capacity is:

$$Q_u = A_t(q_c + q_q + q_\gamma) + P\,H_s \left(\frac{s_{ua}}{S_t} + \gamma_b\, z_{avg}\, \tan \delta \right) \tag{8-1}$$

where:

A_t = penetrator base area = $B \cdot L$ (rectangular), or $\pi B^2/4$ (circular) $[L^2]$

q_c = bearing capacity stress for cohesion = $s_{uz}\, N_c\, s_c\, d_c$ $[F/L^2]$

q_q = bearing capacity stress for overburden = $\gamma_b\, z\, [1 + (N_q\, s_q\, d_q - 1)\, f_z]$ $[F/L^2]$

q_γ = bearing capacity stress for friction = $\gamma_b\,(B/2)\,N_\gamma s_\gamma\,d_\gamma f_z$ [F/L²]

f_z = depth attenuation factor for the frictional portion of bearing capacity stress, to extend the formulation to any footing depth, as described in Chapter 4, Section 4.3.2.6, Equation 4-24

P = base perimeter = $2B + 2L$ (rectangular) or πB (circular) [L]

H_s = side soil contact height = min (z, H) [L]

s_{ua} = undrained shear strength averaged over the side soil contact zone [F/L²]

S_t = soil sensitivity = ratio of undisturbed to remolded strength

γ_b = buoyant unit weight of soil above the foundation base [F/L³]

z_{avg} = average depth over side soil contact zone = ½ [z + max(0, z −H)] [L]

δ = effective friction angle alongside the footing [deg]

    = $\phi - 5$ deg for rough-sided footings,

    = 0   for smooth-sided footings or where the soil is greatly disturbed

ϕ = soil friction angle ($\phi = \phi_u$ for undrained case; $\phi = \overline{\phi}$ for drained case) [deg]

s_{uz} = undrained shear strength averaged over the base influence zone (normally to a depth 0.7B below the footing base) [F/L²]

z = depth of embedment of foundation [L]

B = base diameter, or minimum of base plan dimensions [L]

L = base diameter, or maximum of base plan dimensions [L]

H = base block height [L]

N_c, N_q, N_γ = bearing capacity factors (Chapter 4, Section 4.3.2.4)

s_c, s_q, s_γ = bearing capacity correction factors for base shape (Chapter 4, Section 4.3.2.5)

d_c, d_q, d_γ = bearing capacity correction factors for base depth (Chapter 4, Section 4.3.2.5)

For cohesive soils (e.g., clays, muds, fine silts), the resistance to penetration in the short term is described by the simplified relationship:

$$Q_u = A_t(s_u N_c{'} + \gamma_b z) + P H_s \left(\frac{s_{ua}}{S_t}\right) \tag{8-2}$$

where:

$$N_c' = N_c s_c d_c = [(2+\pi)]\left[1+\left(\frac{1}{2+\pi}\right)\left(\frac{B}{L}\right)\right]\left[1+\left(\frac{2}{2+\pi}\right)\arctan\left(\frac{D_f}{B}\right)\right]$$

In the long term, Equation 8-1 may be used with long-term cohesive soil properties. However, cohesive soils almost always become stronger under the action of long-term downward loads, so long-term bearing capacity in cohesive soils is almost invariably predicted to be greater than short-term, and long-term penetration in cohesive soils is almost invariably predicted to go no further than short-term.

For cohesionless soils (e.g., coarse silts, sands, and gravels, which are considered to behave as fully drained and therefore the same for short-term and long-term), the resistance to penetration is described by the simplified relationship given in Equation 8-3:

$$Q_u = A_t \gamma_b [z\{1+(N_q s_q d_q - 1)f_z\} + 0.5BN_\gamma s_\gamma d_\gamma f_z] + P\,H_s\left(\gamma_b\, z_{avg} \tan \delta\right) \tag{8-3}$$

The bearing capacity correction factors for depth, d_c and d_q, and the depth attenuation factor, f_z, used in Equations 8-1 through 8-3, are functions of the depth of penetration, z (the factor d_γ is constant, equal to 1). To determine the maximum depth of penetration of an object subject to a given force, the applicable equation must be solved by trial-and-error. An assumed depth is entered into the equation, and it is solved to find the corresponding resisting forces at that penetration. If the resistance forces are smaller than the penetration force, then a deeper penetration is assumed, and the forces are recalculated.

Because the solution is actually trial-and-error, it is often useful to use an iterative approach. This is done by selecting a depth increment and then increasing this increment each time the resistance forces are calculated. In making these calculations, it is helpful to take the assumed penetration depth, z, and corresponding penetration resistance, Q_u, of each calculation set and develop a plot of z versus Q_u. Then, knowing the force available to cause object penetration, the expected depth of penetration can be obtained from the plotted curve. Otherwise, the expected depth can be determined from direct interpolation between the values of Q_u higher and lower than the driving force.

If the ratio z/B is small, the "$P\,H_s$" term in Equations 8-1 through 8-3 may be ignored – the value of S_t may be considered high and the value of δ will be nearly zero because of the high degree of soil disturbance alongside the footing base caused by penetration. However, in the case that the ratio z/B exceeds approximately 2.5, or at shallower depths when environment or installation conditions may have caused the side adhesion or frictional contact to be restored, the soil shear stress on the sides of the object must be considered in the calculation. A flow chart of the calculation process is given in Figure 8.2-2.

IS v < 3 FT/SEC? — NO → USE METHOD FOR DYNAMIC PENETRATION, SECTION 8.3

YES

DETERMINE OBJECT CHARACTERISTICS
L, B, A_t, W_b

OBTAIN SOIL PARAMETERS
(COHESIVE) s_u, γ_b, α
(COHESIONLESS) ϕ, γ_b

SELECT DEPTH INCREMENT Δz

INITIATE $i = 0, z = 0$

$i = i + 1$
$z_i = i (\Delta z)$

IS THIS AN EXPEDIENT ESTIMATE FOR SHALLOW ($z/B < 2.5$) PENETRATION?

YES

DETERMINE VALUES FOR EQUATION 8-1 A_t TERM PARAMETERS (AS NECESSARY)

NO

DETERMINE VALUES FOR FULL EQUATION 8-1 PARAMETERS (AS NECESSARY)

COMPUTE Q_u (AT z_i)

COHESIVE   EQ 8-2
COHESIONLESS   EQ 8-3

COMPUTE Q_u (AT z_i)

PENETRATE ONE MORE DEPTH INCREMENT Δz

IS THE RESISTANCE LARGER THAN THE PENETRATING FORCE?
$Q_u > W_b + F_d$
(AT z_i)

NO

YES

OBTAIN ACTUAL z FROM DATA INTERPOLATION

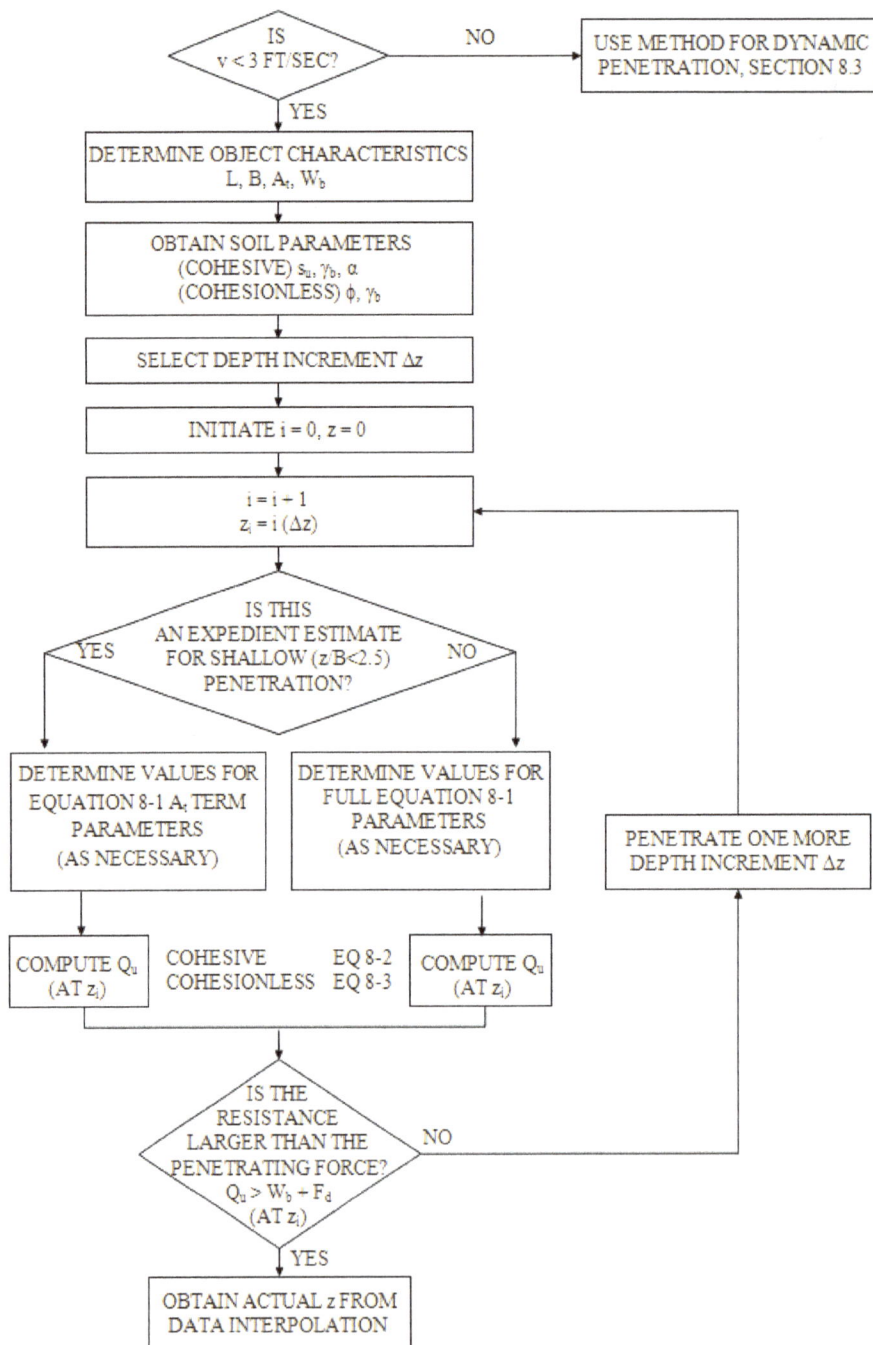

Figure 8.2-2. Flow chart of the calculation procedure for predicting static penetration.

8.2.3.2 Influence of Penetrator Shape

Tapered or rounded penetrators that present an increasing width and area during penetration should have this increasing area considered at each new calculation. The increasing area affects the bearing force, not only directly through the increased bearing area, but also by changing the ratio, z/B, determining the bearing capacity factors.

The penetrator nose shape also influences the shape of the soil failure zone mobilized to resist the penetration. For blunt objects, the soil zone governing the penetration performance reaches a depth of one object width or diameter below the bearing surface. The average of the soil strength values, s_u, over a depth interval of B below z is used in the bearing capacity equation (Figure 8.2-3a). For tapered objects, such as the conical-pointed spud cans on some offshore jack-up platforms, s_u is determined over a depth, B, measured from the bottom of the cylindrical or full section (Ref. 8-3) as shown in Figure 8.2-3b, rather than from below the point of tip penetration.

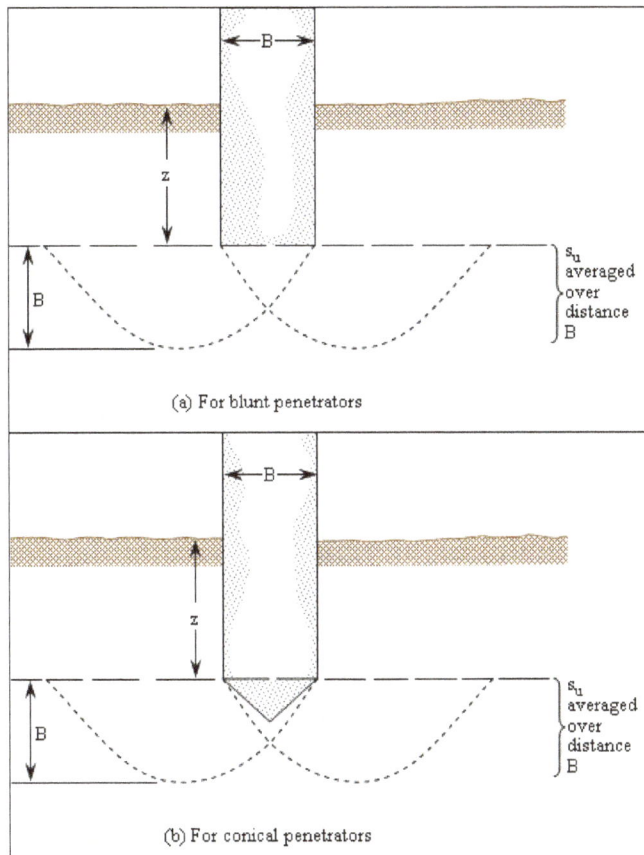

Figure 8.2-3.   Location of the critical shear strength zone B for blunt and conical penetrators (Ref. 8-3).

8.2.3.3 Strength Parameter Selection

Methods for the development of a soil strength profile for undrained shear strength, s_u, and effective stress strength parameters, $\bar{c}$ and $\bar{\phi}$, are given in Chapters 2 and 3. These methods are applicable to the penetration problem when dealing with relatively uniform deposits of clays or sands. However, deep uniform sand deposits are not found often in the deep sea environment. Those sands that do occur are normally the lower portions of turbidite layers and have inter-bedded layers of graded silt to clay sized soil. For such complex sediment profiles, present technology is not sufficiently developed to describe the penetration phenomena. Given the present state-of-the-art, such complex profiles should be treated as cohesive soil when excess penetration is of primary concern and as cohesionless soil when inadequate penetration is the primary concern.

8.2.3.4 Prediction Accuracy

Some field data are available to examine the accuracy of Equation 8-2 in predicting penetration. Equation 8-1 was used to predict the penetration of spud cans for offshore jack-up platforms (Ref. 8-3) in clay. Out of 120 sets of data, 70% of the predictions were found to be within ±25% of the measured penetrations. No prediction was less than 50% of the measured value, and only five predictions exceeded the measured value by more than 50% (Ref. 8-3). Penetration predictions in sand, however, will be considerably less accurate due to difficulties in obtaining accurate in-situ soil strength parameter data in cohesionless soils.

8.2.3.5 Skirt Penetration Prediction

Skirt penetration is treated by the same relationships (Equations 8-1 through 8-3) as object penetration in general. Skirts are generally relatively long, narrow and deep, so it is important to make a full accounting of the effects of shear stresses on the side surfaces. As skirt penetration is a short-term process, in is not necessary to consider long-term soil properties in skirt penetration calculations.

8.3   DYNAMIC PENETRATION

8.3.1   Application

The methods presented in this section are used to predict the penetration of objects entering the seafloor at velocities of 3 ft/s and greater. Examples of objects that undergo dynamic penetration are: objects being lowered rapidly or free falling to the seafloor, such as gravity anchors and ship hulls; other free-falling objects, such as gravity corers and penetrometers; and objects propelled at higher speeds than terminal free-fall velocities, such as propellant-embedded anchor plates. These methods are applicable to cohesive soils, and to granular soils in a more limited sense as discussed in Sections 8.3.3.2 and 8.3.3.7

8.3.2   Approach

The technique presented here predicts both total penetrations and decelerating forces versus depth. Early versions of this technique (Refs. 8-4, 8-5, and 8-6) have been modified (Refs. 8-7 and 8-8) to adapt the technique to velocities up to 400 ft/s.

The approach is basically the same as that used for static penetration prediction, but is extended to account for strain rate effects on the soil shear strength, remolding of the soil on the sides of the penetrator, and transition effects in passing from the overlying fluid drag regime into the soil penetration regime.

An iterative procedure is used for the solution of the dynamic penetration problem because the resisting force terms for nose bearing, side friction, and hydraulic drag, as well as the driving kinetic energy term, are velocity dependent. The procedure is to step the penetrator into the soil in equal finite depth increments or layers (Δz). Resisting soil and hydrodynamic forces are calculated: on the basis of the entry velocity and the soil properties for each step. The energy lost by the penetrator in overcoming the first layer resistance forces is calculated and subtracted from the penetrator kinetic energy on entering the underlying layer. The kinetic energy remaining with the penetrator and its reduced velocity are used for computations on penetrating the second layer. The calculation is repeated for each successive layer until the kinetic energy of the penetrator has been consumed and its velocity reaches zero. The depth at zero velocity is the predicted penetration depth.

The distance penetrated at each step (Δz) must be assigned prior to beginning the calculations. It is suggested that Δz be assigned by dividing an expected (guessed) total penetration into approximately 10 equal increments.

8.3.3   Method for Predicting Dynamic Penetration

8.3.3.1 Forces Acting on the Penetrator

The forces acting on an object penetrating at moderate velocity deep in a soil mass are shown in Figure 8.3-1. The net downward force after full object entry into the soil at distance z_i is given by Equation 8-4:

$$F_i = F_{di} + W_{bi} - Q_{ni} - F_{si} - F_{hi}$$

(8-4)

where:

F_i = net downward force exerted by the penetrator [F]

F_{di} = external driving force, if any (e.g., rocket motor) [F]

W_{bi} = penetrator buoyant weight [F]

Q_{ni} = tip or nose bearing resistance [F] (see Section 8.3.3.2)

F_{si} = side friction or adhesion [F] (see Section 8.3.3.3)

F_{hi} = fluid drag force [F] (see Section 8.3.3.5)

subscript i = ith increment of soil depth

Two of these forces, F_{di} and W_{bi}, are driving the penetrator into the soil mass. The other three, Q_{ni}, and F_{si}, and F_{hi}, are resisting that penetration.

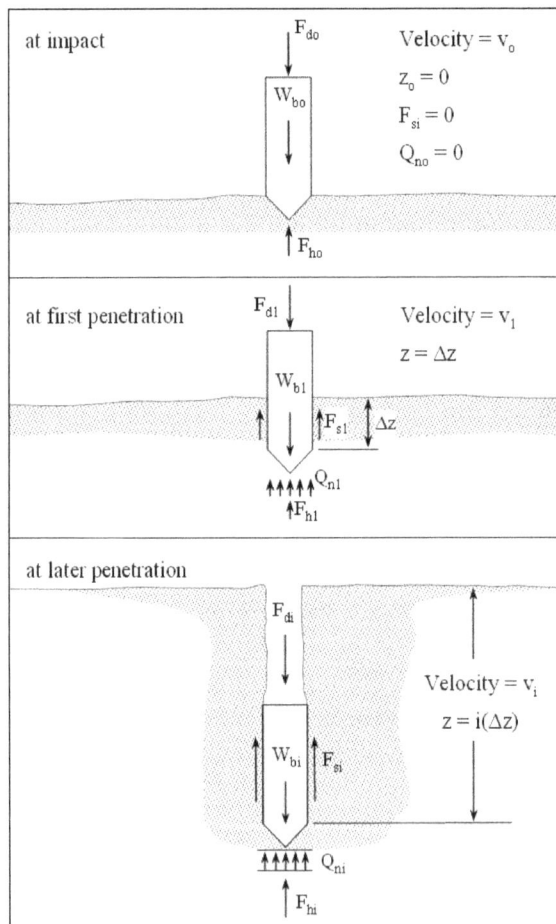

Figure 8.3-1. Forces acting on a penetrator before and after contact with the seafloor.

8.3.3.2 Nose Resistance

The tip or nose bearing resistance force for the *ith* layer, Q_{ni}, is obtained using Equation 8-5:

$$Q_{ni} = s_{ui}(nose)\, S_{\dot{e}i}\, N_{ti}\, A_{t} \tag{8-5}$$

where:

$s_{ui}(nose)$ = soil undrained shear strength at a depth $0.35B$ below z, averaged over *ith* increment of penetration $[F/L^2]$

$S_{\dot{e}i}$ = strain rate factor (see Section 8.3.3.4)

A_t = end area of penetrator (the effective bearing surface) $[L^2]$

N_{ti} = dimensionless nose resistance factor (see Equation 8-6)

In Equation 8-5, the dimensionless nose resistance factor, N_{ti}, is determined by Equation 8-6. Note that the maximum $N_{ti} = 9.9$.

$$N_{ti} = N_c' = N_c s_c d_c = \left[(2+\pi)\right]\left[1+\left(\frac{1}{2+\pi}\right)\left(\frac{B}{L}\right)\right]\left[1+\left(\frac{2}{2+\pi}\right)\arctan\left(\frac{D_f}{B}\right)\right] \tag{8-6}$$

The term s_{ui} in Equation 8-5 represents the undrained shear strength of sands as well as clays. Dynamic penetration in sands is rapid enough to be considered undergoing an undrained type of shear failure. When sands are sheared during a dynamic penetration event, the pore water does not have time to flow, and failure occurs when either the sand grains are crushed or cavitation of the pore water occurs. The forces required to do this are large and hence the undrained shear strength is high (e.g., on an order of magnitude of 20 to 200 psi). A method for obtaining the undrained shear strength of a sand can be found in Reference 8-9 (it is a difficult property to measure). The undrained shear strength is related to the critical confining stress by Equation 8-7:

$$s_{ui} = \left[\frac{\sigma_{cr}(N_\phi - 1)}{2}\right] \tag{8-7}$$

where:

σ_{cr} = critical confining stress $[F/L^2]$ $\approx D_r^{1.7} \cdot 20{,}000$ psf

N_ϕ = $[\tan(\pi/4 + \phi/2)]^2$ and ϕ is in radians

D_r = fractional relative density $\approx (\gamma_b - 56.5 \text{ pcf})/11.5$ pcf

8.3.3.3 Side Friction

The side friction or adhesion force in the ith layer, F_{si}, is obtained from Equation 8-8 (Ref. 8-7). Note that this computation assumes no separation between the soil and the side of the object being penetrated during penetration.

$$F_{si} = \left[\frac{s_{ui}(side)}{S_{ti}} \right] S_{ei} A_{si} \tag{8-8}$$

where:

S_{ti} = soil sensitivity or ratio of undisturbed undrained shear strength to remolded undrained shear strength (obtained normally from soils testing)

A_{si} = side soil contact area of the penetrator when nose is at z_i [L^2]

$s_{ui}(side)$ = soil undrained shear strength averaged over the length of the penetrator in contact with the soil [F/L^2]

8.3.3.4 Strain Rate Factor

Cohesive soil undrained shear strength increases with an increase in the rate of strain. This increase may be as high as a factor of 5.0 when soil shears in response to a rapidly penetrating object (Refs. 8-6 and 8-10). This is more commonly referred to as the strain rate effect on shear strength. A somewhat cumbersome formulation (given in Equation 8-9), based on a best fit to penetration test data (Refs. 8-7 and 8-8), has been developed for calculating the strain rate factors $S_{\dot{e}i}$:

$$S_{\dot{e}i} = \frac{S_{\dot{e}}^{*}}{1 + [C_{\dot{e}} v_i /(s_{ui} D_e) + C_o]^{-0.5}} \tag{8-9}$$

Note that the strain rate factor has a minimum value of $S_{\dot{e}i} = 1$. Also, in Equation 8-9,

$S_{\dot{e}}^*$ = maximum strain rate factor, from Table 8.3-1

$C_{\dot{e}}$ = empirical strain rate coefficient, from Table 8.3-1 [F·T/L^2]

v_i = velocity at depth z_i [F/L]

C_o = empirical strain rate constant, from Table 8.3-1

s_{ui} = soil undrained shear strength equal to $s_{ui}(nose)$ or $s_{ui}(side)$, depending on which of these that $S_{\dot{e}i}$ is modifying [F/L^2]

D_e = equivalent diameter of penetrator as determined by Equation 8-10 [L]

8-12

$$D_e = (4A_t / \pi)^{0.5}$$ (8-10)

The appropriate values of S_e^*, C_e, and C_o to use in Equation 8-9 are specified in Table 8.3-1, as are the conditions of their use. In Table 8.3-1, long cylindrical penetrators are categorized separately from "all other object shapes." This simple geometric shape has a large side surface area in comparison to its frontal area and lends itself better to this predictive technique (see Section 8.3.3.7).

Table 8.3-1. Values of Constants Used in Equation 8-9

Condition for Use in Rapid Penetration Problems	Parameter Value		
	S_e^*	C_e (lb-sec/ft^2)	C_o
Problems with long, cylindrical penetrators (Ref. 8-8)	4	4	0.11
All other object shapes where inadequate penetration is of concern (Ref. 8-11)	3	10	0.25
All other object shapes where excess penetration is of primary concern (Ref. 8-11)	2	40	1.0

8.3.3.5 Fluid Drag Force

The fluid drag force acting on a penetrator while moving through water is assumed to continue to exist as it moves through the soil. The fluid drag force is calculated as:

$$F_{hi} = (0.5)C_D \rho A_t (v_i)^2$$ (8-11)

where:

C_D = dimensionless fluid drag coefficient (the same as that in seawater)

ρ = mass density of the soil, the "fluid" being accelerated [FT^2/L^4]

v_i = penetrator velocity after penetrating the *ith* layer [F/L]

Values for C_D values are best obtained, where possible, by back figuring from measured terminal velocities in water. In absence of measurements, use C_D values from a hydrodynamics reference (e.g., Ref. 8-12).

8.3.3.6 Method of Solution

An iterative process is used for the solution of penetration problems because the major resisting force terms (Q_{ni}, F_{si}, and F_{hi}) are velocity or depth dependent, requiring the input of new values with penetrator travel. The net downward force is an inertial force related to the deceleration of the penetrator and can be obtained from the following modification of Newton's second law (modified to eliminate the parameter time):

$$F_i = M v_i \left(dv / dz \right) \tag{8-12}$$

where:

M = penetrator mass [FT^2/L]

dv/dz = instantaneous change in velocity [L/T]

For making incremental calculations, dv/dz is replaced with $(2\Delta v)/(2\Delta z)$. The double increments are used to minimize deviations in the prediction caused by minor errors in the assumed penetration velocity (Ref. 8-7). Then, after reorganization of terms:

$$2\Delta v = \frac{2\,\Delta z}{M} \left(\frac{F_i}{v_i} \right) \tag{8-13}$$

The new velocity for the $(i + 1)th$ increment is given by Equation 8-14:

$$v_{i+1} = v_{i-1} + 2\Delta v_i \tag{8-14}$$

To begin the incremental calculations, the velocity, v_1, at the end of the first increment of seabed penetration, z_1, must be generated. An approximation for v_1 can be obtained from:

$$v_1 = v_0 + (1/v_0)[(\Delta z / M)(F_{d5} + W_{b5} - Q_{n5} - F_{s5} - F_{h5})] \tag{8-15}$$

where:

v_0 = initial penetrator velocity on entering the seafloor [F/L]

$F_{d.5}, W_{b5}, Q_{n5}, F_{s.5}, F_{h5}$ = initial estimates of the respective force values based on conditions at mid-depth in the first layer of penetration [F]

This approximation for v_1 is then used to calculate F_{d1}, W_{b1}, Q_{n1}, F_{s1} and F_{h1}. At this point, all the forces necessary to calculate the net downward force have been determined. The first iteration for values of F_i, $2\Delta z$, and v_{i+1} can then be completed (Equations 8-11, 8-13, and 8-14). Note that for these calculations, the shear strength at $z = z_i$ is applied to obtain changes from $z = z_{i-1}$ to $z = z_{i+1}$, so it is not necessary to use average shear strength over the $2\Delta z$ depth change. Subsequent iterations are made by recalculating F_{di}, W_{di}, Q_{ni}, F_{si}, and F_{hi} for the next increment of penetration. When the computation for v_{i+1} produces a negative velocity, the iterative procedure is completed. A flow chart for this procedure is shown in Figure 8.3-2. The maximum penetration of the object is then obtained by interpolating between the last two velocity values as follows:

$$z = z_i + \Delta z \left(\frac{v_i}{v_i - v_{i+1}} \right) \qquad \text{(8-16)}$$

8.3.3.7 Prediction Accuracy

The dynamic penetration prediction technique provides reasonable penetration estimates for long slender penetrators in cohesive sediments, such as hemipelagic and pelagic clays and in fine silts. This is true because the predictive technique uses empirical data ($S_{\dot{e}}^*$, $C_{\dot{e}}$, and C_o) from field tests with penetrators of this shape in soils of these types. For differently shaped penetrators and for penetration in sands and other noncohesive materials, the accuracy will not be as good. Little field data exist for object penetration into these granular sediments and for their in-situ properties. In addition, the undrained shear strength developed during the rapid penetration is a very difficult parameter to obtain. Prediction of penetration in oozes is particularly difficult. In foraminiferal oozes, limited data suggest this method will underpredict penetration – possibly by as much as a factor of two (Ref. 8-13).

Penetration of the Navy's Doppler penetrometer, a long slender penetrometer shape, in various seafloors is compared with in-situ conditions in Reference 8-13.

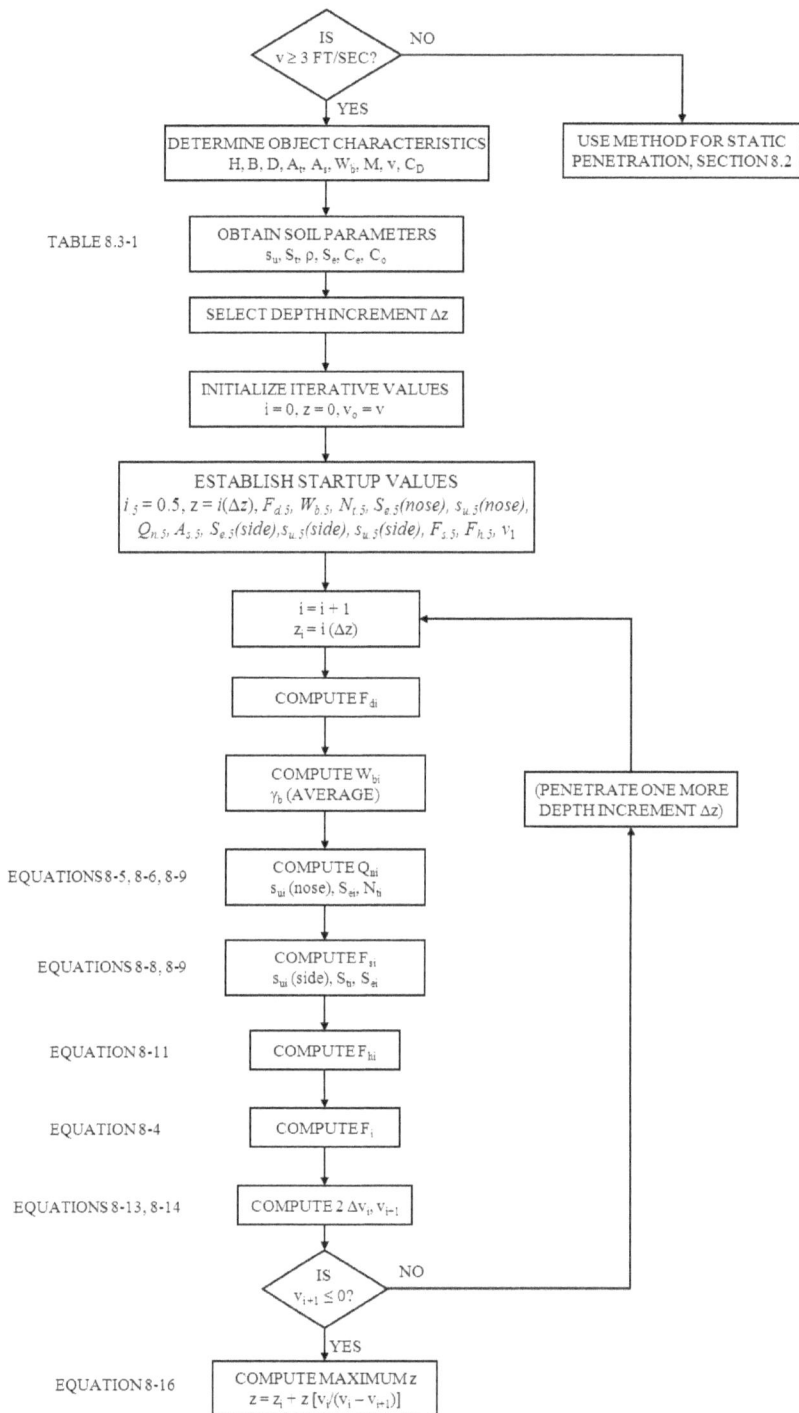

Figure 8.3-2. Flow chart of the calculation procedure for predicting dynamic penetration.

8.4   EXAMPLE PROBLEMS

8.4.1   Problem 1 – Slow Penetration of a Long Cylinder

8.4.1.1 Problem Statement

Determine how deep into the seafloor a cylinder will penetrate after being slowly lowered to the bottom.

Data: An electric power source (EPS) is to be lowered to a specific seafloor location at a speed of 2 ft/s. An estimation of the depth of static penetration z is required to verify that cooling water intakes remain above the mudline and to provide input data for calculating lateral stability and breakout load. The EPS container is 4 feet in diameter by 12 feet high and weighs 20 kips in seawater. The sediment at the site is a pelagic clay. Soil properties at the site have been determined from a 40-foot piston core. A schematic diagram for this problem and the results of laboratory testing for soil shear strength and density are shown in Figure 8.4-1.

8.4.1.2 Problem Solution

The analytical procedures and computations used to solve this problem are shown below. They follow the procedure outlined in Section 8.2 and summarized by the flowchart in Figure 8.2-3. Because the installation is immediate (but still static), the expedient (conservative when excessive penetration is a concern) form of Equation 8-2 is used for "shallow" trial depths less than 2.5 times the diameter. <u>Note</u>: It is helpful to keep track of calculated values in tabular form. This is shown for Problem 1 in Table 8.4-1.

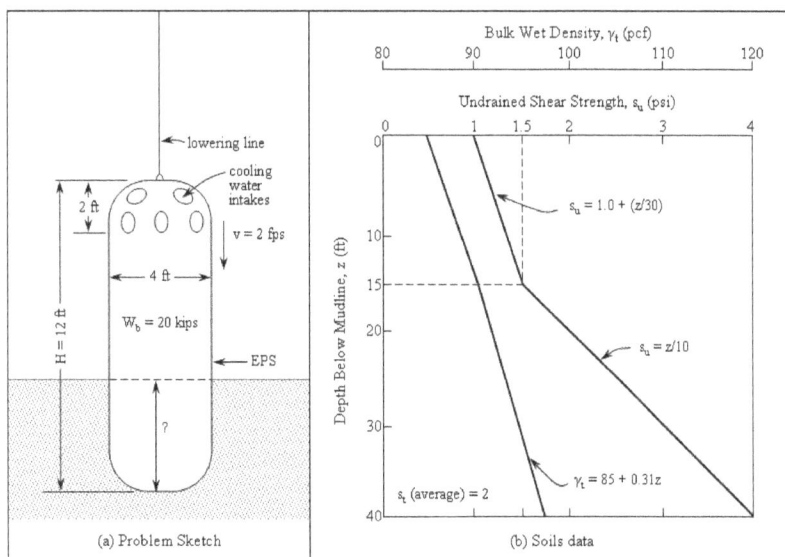

Figure 8.4-1. Problem sketch and soils data for example Problem 1.

8-17

Problem 8.4-1

ANALYTICAL PROCEDURES	COMPUTATIONS
1. Is this a case of "static" penetration? Is $v_o < 3$ ft/sec?	YES, $v_o = 2$ ft/sec < 3 ft/sec
2. Determine object characteristics. H (known) B (known) $A_t = (\pi B^2)/4$ W_b (known) F_d (known)	$H = 12$ ft $B = 4$ ft $A_t = (3.14)(4$ ft$)^2/4 = 12.6$ ft^2 $W_b = 20,000$ lb $F_d = 0$
3. Obtain soil profiles for s_u and γ_b.	Profiles are shown in Figure 8.4-1(b) for s_u and γ_b. γ_b is equal to $(\gamma_t - 64$ pcf$)$.
4. Select depth increment Δz.	$\Delta z = 2.0$ ft (a round number for convenience)
5. Calculate embedment depth: $z_1 = (1)\Delta z$	$z_1 = (1)(2.0$ ft$) = 2.0$ ft
6. Is this a case of shallow penetration? Is $z/B < 2.5$?	YES, $z_1/B = 2$ ft $/ 4.0$ ft $= 0.5$

7. Determine values for:
 	
s_{uz} averaged over z_1 to $0.7B$ below z_1
    (same as at depth 0.35 B below z_1)
    from Figure 8.4-1(b)
 	
γ_{ba} averaged over 0 to z_1 (from
    Figure 8.4-1(b))
 	
N_c' at z_1 (from Equation 8-2) (N_c' must not
    exceed 9.9)
	@ z_1, $z + 0.35B = 2$ ft $+ 0.35(4$ ft$) = 3.4$ ft
	
	s_{uz} @ 3.4 ft $= (1 + 3.4/30)$ psi $(144$ psf/psi$)$
	              $= 160.3$ psf
	
	γ_b @ 0 ft $= [85 + 0.31(0)]$ pcf $- 64$ pcf
	           $= 21.0$ pcf
	
	$\gamma_{ba} = 21.0$ pcf $+ (0.31$ pcf/ft$)(2$ ft/2$) = 21.3$ pcf
	
	$N_c' = (2+\pi)[1 + (4$ ft/4 ft$) / (2+\pi)]$
	        $[1 + 2 \arctan(2$ ft $/ 4$ ft$) / (2+\pi)]$
	    $= 7.25$

Problem 8.4-1

ANALYTICAL PROCEDURES	COMPUTATIONS
8. Calculate Q_u, the resistance to penetration at $z = 2.0\,ft$ (from Equation 8-2 with no side shear stress). $Q_u = A_t(s_{uz} N_c' + \gamma_b z)$	$Q_u = 12.6\,ft^2\,[(160.3\,psf)(7.25)$ $\quad + (21.3\,pcf)(2.0\,ft)] = 15{,}140\,lb$
9. Does Q_u exceed the driving forces? Is $Q_u > W_b + F_d$?	NO. $W_b + F_d = 20{,}000\,lb + 0 = 20{,}000\,lb > 15{,}140\,lb$
10. Continue iterative process. Increase the assumed penetration by Δz.	$z_1 = 2\Delta z = (2)(2.0\,ft) = 4.0\,ft$
11. Is this still shallow penetration? Repeat steps 6 through 10 for $z_2 = 4\,ft$ Calculate s_u at depth $B/2$ below z averaged over 0 to z_2 Calculate γ_b averaged over 0 to z_2. Calculate N_c' at z_2. Calculate Q_u at z_2. Is $Q_u > W_b + F_d$?	YES, $z/B = 4.0\,ft\,/\,4.0\,ft = 1.0\ < 2.5$ $z_2 + 0.35B = 4\,ft + 0.35\,(4\,ft) = 5.4\,ft$ s_u @ 5.4 ft $= (1 + 5.4/30)\,psi\,(144\,psf/psi)$ $\qquad = 169.9\,psf$ $\gamma_{ba} = 21.0\,pcf + (0.31\,pcf/ft)(4\,ft\,/\,2) = 21.6\,pcf$ $N_c' = (2+\pi)\,[1 + (4\,ft\,/\,4\,ft)/(2+\pi)]$ $\qquad [1 + 2\arctan(4\,ft/4\,ft)\,/\,(2+\pi)] = 8.0$ $Q_u = (12.6\,ft^2)[(163.2\,psf)(8.0)$ $\qquad + (21.6\,pcf)(4.0\,ft)] = 18{,}210\,lb$ NO, $W_b + F_d > Q_u$ $(W_b + F_d$ remains 20,000 lb)
12. Repeat steps 6 through 10 for $z_3 = 6\,ft$. Is $Q_u > W_b + F_d$?	$z_3 + 0.35B = 6\,ft + 0.35\,(4\,ft) = 7.4\,ft$ s_u @ 7.4 ft $= 179.5\,psf$ $\gamma_{ba} = 21.0\,pcf + (0.31\,pcf/ft)(6\,ft\,/\,2) = 21.9\,pcf$ $N_c' = 8.5$ $Q_u = 20{,}810\,lb$ YES, $Q_u > W_b + F_d$. Therefore, the predicted penetration falls between 4 and 6 ft.

Problem 8.4-1

ANALYTICAL PROCEDURES	COMPUTATIONS
13. Actual penetration can be obtained by direct interpolation of the last two iterations or, graphically, from a plot of all the developed Q_u values (see Figure 8.4-2).	The predicted penetration is equal to 5.4 ft (from Figure 8.4-2).

SUMMARY

The cooling water intakes are 10 ft above the EPS base. Therefore, they should be 4.6 ft above the seafloor (10 ft – 5.4 ft = 4.6 ft). Because the EPS is considered a very important installation, great care should be taken that the lowering speed at impact is as slow as possible. A more rapid lowering could add a dynamic force component and increase penetration.

Table 8.4-1. Summary of Calculations for Problem 1

z (ft)	$z + 0.35B$ (ft)	Effective s_{uz} (psf)	Average γ_b (pcf)	N_c'	Q_u (lb)
2	3.4	160.3	21.3	7.2	15,140
4	5.4	169.9	21.6	8.0	18,210
6	7.4	179.5	21.9	8.5	20,810

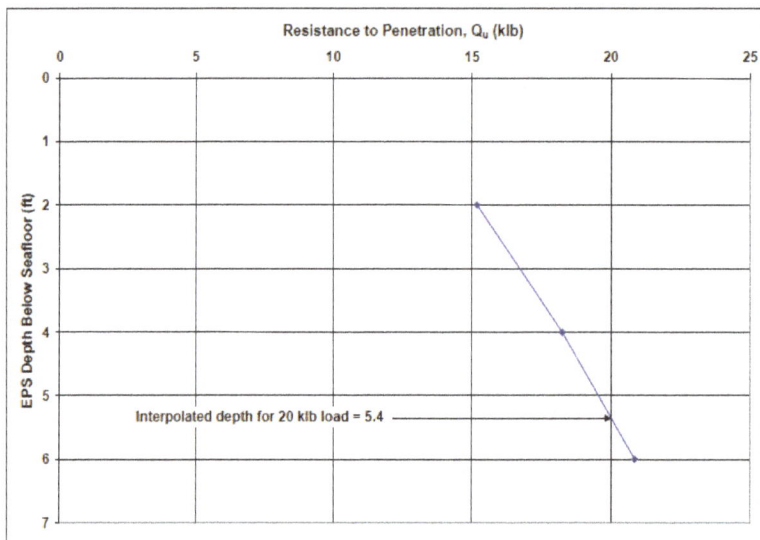

Figure 8.4-2. Plot of predicted soil resistance to EPS penetration.

8-20

8.4.2 Problem 2 – Rapid Penetration of a Long Cylinder

8.4.2.1 Problem Statement

Determine how deep into the seafloor a large long cylinder will penetrate if it falls to the seafloor at terminal velocity.

Data: The electric power source (EPS) in Problem 1 was being lowered to the seafloor when the lowering line was severed. The EPS could not be located after falling to the seafloor. In order to decide whether to attempt a recovery or to abandon or destroy the device, the depth of its penetration into the seafloor needs to be estimated. The EPS is suspected to have hit the seafloor in the same orientation in which it was being lowered due to a concentrated mass at its lower end and drag from the severed lowering line. The terminal velocity and drag coefficient for this orientation have been calculated as 40 ft/s and 1.0, respectively. The soil properties are the same as those shown in Figure 8.4-1b. A schematic diagram for this problem is shown in Figure 8.4-3.

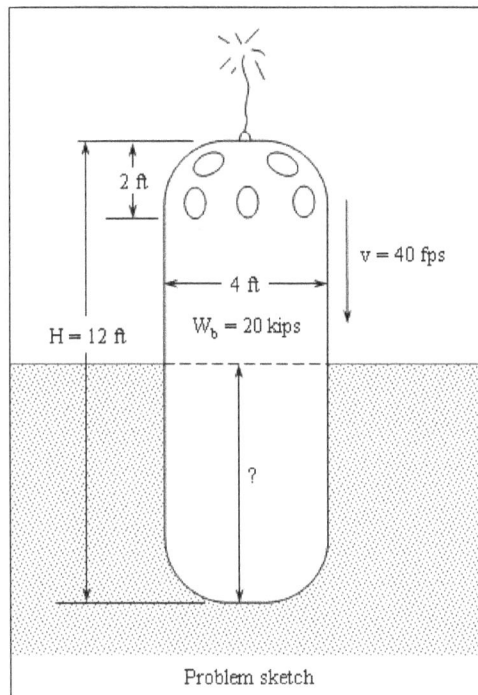

Figure 8.4-3. Sketch for example Problem 2.

8.4.2.2 Problem Solution

The analytical procedures and computations used to solve this problem are shown below. They follow the procedures outlined in Section 8.3 and summarized by the flowchart in Figure 8.3-2.

Problem 8.4-2

ANALYTICAL PROCEDURES	COMPUTATIONS
1. Is this a case of dynamic penetration? Is $v_o > 3\ ft/sec$?	YES, $v_o = 40\ ft/sec$
2. Determine object characteristics. H (known) B (known) $A_t = (\pi B^2)/4$ $A_s = \pi B z$ W_b (known) $M = $ (dry weight)$/32.2\ fps^2$ V (known) C_D (known)	$H = 12\ ft$ $B = 4\ ft\quad D = 4.0\ ft$ $A_t = (3.14)(4\ ft)^2/4 = 12.6\ ft^2$ $A_s = (3.14)(4\ ft)\ z = 12.6\ (z)\ ft^2$ $W_b = 20{,}000\ lb$ (Assume dry weight $= W_b +$ weight of water displaced by the near-cylindrical EPS) EPS Volume $\cong A_t\ (12\ ft) = 151\ ft^3$ Water Weight $\cong (64\ pcf)(151\ ft^3) = 9600\ lb$ Dry Weight $\cong W_b + 9600\ lb = 29{,}660\ lb$ $M = (29{,}660\ lb) / (32.2\ fps^2) = 921\ slugs$ $v = 40\ fps$ $C_D = 1.0$

Problem 8.4-2

ANALYTICAL PROCEDURES	COMPUTATIONS

3. Obtain soil parameters:
 	
   s_u (known)
 	
   S_t (known)
 	
   $\rho = (\gamma_t) / (32.2\ fps^2)$
 	
   $S_e{}^*, C_e, C_o$ (from Table 8.3-1)
	s_u from Figure 8.4-1b
	
	$S_t = 2$ (from Figure 8.4-1b)
	
	(It was seen from Problem 1 that γ_t varied little with depth. Therefore, say $\gamma_t = \gamma_t$ at 5 ft)
	
	$\rho = [85 + 0.31(5)]\ pcf\ /\ (32.2\ fps^2)$
	   $= 2.69\ slugs/ft^3$
	
	$S_e{}^* = 4.0$
	
	$C_e = 4\ lb\text{-}sec/ft^2$ (Constants for a long cylindrical penetrator)
	
	$C_o = 0.11$

4. Select depth increment, Δz (a round number to limit computations to about 10 iterations). Δz = (assumed penetration)/(10)	Assume a maximum penetration depth (20 ft) $\Delta z = (20\ ft)/(10) = 2\ ft$
5. Initialize values: $i = 0$, $z_0 = 0$, $v_0 = v$	$v_0 = 40\ fps$

6. Begin computations to estimate v_1 based on conditions at the mid-depth of the first increment:
 	
   Increase i:
 	
   $i = 0.5$
 	
   $z_i = i\ (\Delta z)$
	$i = 0.5$
	
	$z_{0.5} = 0.5\ (2\ ft) = 1\ ft$

7. Compute F_{d5}	$F_{d5} = 0$ (no external driving force)

Problem 8.4-2

ANALYTICAL PROCEDURES	COMPUTATIONS
8. Compute W_{b5} @ $z_1/2$ (from Figure 8.4-1b) $W_{b5} = W_b - $ (displaced soil weight) $= W_b - \gamma_{b5}$(avg @ z_{05})(z_{05})$(\pi)(B^2)/4$	γ_b @ 0 ft = $\gamma_t - 64\ pcf$ = 85 pcf − 64 pcf $\qquad = 21.0\ pcf$ γ_{b5}(avg @ z_{05}) = γ_b @ ($z_{05}/2$) = γ_b @ (1 ft /2) $\qquad\qquad = 21.0\ pcf + (0.31\ pcf/ft)(1\ ft\ /\ 2)$ $\qquad\qquad = 21.2\ pcf$ W_{b5} = 20,000 lb − (21.2 pcf)(1 ft)$(\pi)(4\ ft)^2/4$ $\qquad = 19,730\ lb$
9. Compute Q_{n5} (Equation 8-5). $Q_{n5} = s_{u5}(nose) S_{e0} \, N_{t5} A_t$ $s_{u5}(nose)$ (using Figure 8.4-1b) S_{e5} (Equation 8-9 using v_0) N_{t5} (Equation 8-6)	$s_u(nose)$ @ $z_1/2$ = s_u @ (1 ft+0.35 B) $\qquad\qquad = s_u$ @ 2.4 ft s_u @ 2.4 ft = (1 + 2.4/30) psi = 155.5 psf $S_{e5} = \dfrac{S_e^{\,*}}{1 + [C_e\, v_0/(s_{u5}B) + C_o]^{-0.5}}$ $S_{e5} = \dfrac{4}{1 + \left(\dfrac{(4lb*s/ft^2)(40\,fps)}{(155.5\,psf)(4\,ft)} + 0.11 \right)^{-0.5}} = 1.51$ N_{t5} = (2+π) [1 + (B/L) / (2+π)] $\qquad\qquad$ [1 + 2 arctan($z_{05}/2/B$) / (2+π)] $\qquad$ = (2+π) [1 + (4 ft /4 ft) / (2+π)] $\qquad\qquad\qquad$ [1 + 2 arctan(1 ft /4 ft) / (2+π)] $\qquad$ = 6.73 Q_{n5} = (155.5 psf)(1.51)(6.73)(12.6 ft^2) $\qquad$ = 19,840 lb

Problem 8.4-2

ANALYTICAL PROCEDURES	COMPUTATIONS
10. Compute $F_{s\,5}$ (Equation 8-8). $$F_{s\,5} = [s_{u\,5}(side)/S_{t\,5}]S_{e\,5}\,A_{si}$$ $s_{u\,5}(side)$ (using Figure 8.4-1b @ $z_{0\,5}/2$) $S_{e\,5}$ (Equation 8-9. Uses the same values as in Step 8 except where $s_{u\,5}(side)$ is substituted for $s_{u\,5}(nose)$). $A_{s\,5} = 12.6(z_{0\,5})\,ft^2$	$s_{u\,5}(side) = s_u\,(avg\ @\ z_{0\,5}) = s_u\ @\ (z_{0\,5}/2)$ $\qquad = [1 + (1\,ft\,/2)/30\,ft]\,psi = 146.4\,psf$ $$S_{e\,5} = \frac{4}{1 + \left(\dfrac{(4lb*s/\,ft^2)(40\,fps)}{(146.4\,psf)(4\,ft)} + 0.11\right)^{-0.5}} = 1.53$$ $A_{s\,5} = (12.6\,ft)(1.0\,ft) = 12.6\,ft^2$ $F_{s\,5} = (146.4\,psf\,/2)(1.53)(12.6\,ft^2) = 1{,}410\,lb$
11. Estimate $F_{h\,5}$ using v_0 (Equation 8-11). $$F_{h\,5} = (0.5)C_D\,\rho\,A_t\,(v_0)^2$$	$F_{h\,5} \cong 0.5(1.0)(2.69\ slugs/ft^3)(12.6\ ft^2)(40\ ft/s)^2$ $\cong 27{,}040\,lb$
12. Estimate the velocity v_1 (Equation 8-15). $$v_1 = v_0 + (1/v_0)[(\Delta z/M)(F_{d\,5} + W_{b\,5} - Q_{n\,5}$$ $$- F_{s\,5} - F_{h\,5})]$$	$v_1 = 40\,ft/s + (1/40\,ft/s)[(2\,ft/921\,slugs)$ $\cdot (0 + 19{,}730\,lb - 19{,}840\,lb - 1{,}410\,lb$ $- 27{,}040\,lb)] = 40\,ft/s - 1.55\,ft/s$ $= 38.45\,ft/s$
13. Begin computations for the double velocity increment $2\Delta v_1$ based on conditions at z_1, the mid-depth of the first double depth increment, and the value of v_1 from step 12. Continue repeating from Step 21 back to this step, using the succeeding values of v_i from step 21, until $v_{i+1} < 0$: Increase i: $i = 1$ $z_i = i\,(\Delta z)$	$i = 1$ $z_1 = 1\,(2\,ft) = 2\,ft$
14. Compute F_{d1}.	$F_{d1} = 0$ (no external driving force)

Problem 8.4-2

ANALYTICAL PROCEDURES	COMPUTATIONS
15. Compute W_{b1} @ z_1 from Figure 8.4-1b. $W_{b1} = W_b -$ (displaced soil weight) $\quad = W_b - \gamma_{b1}$(avg @ z_1)$(\pi)(B^2)/4$	γ_{b1}(avg @ z_1) = γ_b @ $(z_1/2)$ $\quad = 21.0\ pcf + (0.31\ pcf/ft)(2\ ft\ /\ 2)$ $\quad = 21.3\ pcf$ W_{b1} = 20,000 lb − (21.3 pcf)(2 ft)(π)(4 ft)^2/4 $\quad = 19,460\ lb$
16. Compute Q_{n1} (Equation 8-5). $Q_{n1} = s_{u1}(nose)S_{e1}\ N_{t1}A_t$ $s_{u1}(nose)$ (using Figure 8.4-1b) S_{e1} (Equation 8-9, using v_1 estimated in step 12) N_{t1} (Equation 8-6)	$s_u(nose)$ @ $z_1 = s_u$ @ $(2\ ft + 0.35\ B) = s_u$ @ 3.4 ft s_u @ 3.4 ft = (1 + 3.4/30) psi = 160.3 psf $S_{e1} = \dfrac{4}{1+\left(\dfrac{(4lb*s/ft^2)(38.45\,ft/s)}{(160.3\,psf)(4\,ft)}+0.11\right)^{-0.5}} = 1.49$ N_{t1} = $(2+\pi)\ [1 + (B/L)\ /\ (2+\pi)]$ $\qquad [1 + 2\ \arctan(z/B)\ /\ (2+\pi)]$ $\quad = (2+\pi)\ [1 + (4\ ft\ /4\ ft)\ /\ (2+\pi)]$ $\qquad [1 + 2\ \arctan(2\ ft\ /4\ ft)\ /\ (2+\pi)]$ $\quad = 7.25$ Q_{n1} = (160.3 psf)(1.49)(7.25)(12.6 ft^2) $\quad = 21,710\ lb$
17. Compute F_{s1} (Equation 8-8). $F_{s1} = [s_{u1}(side)/S_{t1}]S_{e1}\ A_{s1}$ $s_{u1}(side)$ (using Figure 8.4-1b) S_{e1} (Equation 8-9. Uses the same values as in step 8 except where $s_{u1}(side)$ is substituted for $s_{u1}(nose)$) $A_s = 12.6(z)\ ft^2$	$s_{u1}(side) = s_u$ (avg @ z_1) = s_u @ $(z_1/2)$ $\quad = [1 + (2\ ft\ /\ 2)\ /30\ ft]\ psi$ = 148.8 psf $S_{e1} = \dfrac{4}{1+\left(\dfrac{(4lb*s/ft^2)(38.45\,ft/s)}{(148.8\,psf)(4\,ft)}+0.11\right)^{-0.5}} = 1.51$ A_s = (12.6 ft)(2.0 ft) = 25.1 ft^2 F_{s1} = (148.8 psf / 2)(1.51)(25.1 ft^2) = 2,830 lb
18. Compute F_{h1} (Equation 8-11). $F_{h1} = (0.5)C_D\ \rho\ A_t\ (v_1)^2$	F_{h1} = 0.5(1.0)(2.69 $slugs/ft^3$)(12.6 ft^2) $\qquad$ (38.45 ft/s)^2 $\quad = 24,990\ lb$

Problem 8.4-2

ANALYTICAL PROCEDURES	COMPUTATIONS
19. Compute F_1 (Equation 8-4). $$F_1 = F_{d1} + W_{b1} - Q_{n1} - F_{s1} - F_{h1}$$	F_1 = 0 + 19,460 lb – 21,710 lb – 2,830 lb – 24,990 lb = - 30,060 lb
20. Compute the double velocity increment $2\Delta v_1$ (Equation 8-13). $$2\Delta v_1 = (2\Delta z/M)(F_1/v_1)$$	$2\Delta v_1$ = [(2)(2.0 ft)/921 $slugs$] · (-30,140 lb / 38.45 ft/s) = -3.40 ft/s
21. Compute v_2, the velocity for the next iteration (Equation 8-14). $$v_{i+1} = v_{i-1} + 2\Delta v_1$$	$v_2 = v_0 + 2\Delta v_1$ = 40 ft/s + (-3.40 ft/s) = 36.60 ft/s
22. This is the end of the first iteration. If a table is being kept of the computed values (such as Table 8.4-2), the values should be entered. For the second iteration, steps 13 through 21 are repeated using v_2 and applicable values for a nose penetration of 4.0 ft. These steps summarized below.	

Problem 8.4-2

ANALYTICAL PROCEDURES	COMPUTATIONS
(Repeat steps 13 through 21) $i = i + 1 = (1 + 1) = 2$ $z_i = i\,(\Delta z)$ F_{d2} $W_{b2} = W_b - \gamma_{b2}(\text{avg @ } z_2)(z)(\pi)(D^2)/4$ $Q_{n2} = s_{u2}(nose)\,S_{\cdot e2}\,N_{t2}\,A_t$ $S_{\cdot e2} = \dfrac{S_{\cdot e}^{*}}{1 + [C_{\cdot e}\,v_1/(s_{u2}D) + C_o]^{-0.5}}$ $N_{t2} = (2 + \pi)\left[1 + \left(\dfrac{1}{2+\pi}\right)\!\left(\dfrac{B}{L}\right)\right]\left[1 + \left(\dfrac{2}{2+\pi}\right)\arctan\!\left(\dfrac{D_f}{B}\right)\right]$ $F_{s2} = [s_{u2}(side)/S_{t2}]\,S_{\cdot e2}\,A_s$ $A_{s2} = 12.6(z)\,ft^2$ $F_{h2} = (0.5)C_D\,\rho\,A_t\,(v_2)^2$ $F_2 = F_{d2} + W_{b2} - Q_{n2} - F_{s2} - F_{h2}$ $2\Delta v_2 = (2\Delta z/M)(F_2/v_2)$ $v_3 = v_1 + 2\Delta v_2$	$z = 2(2\,ft) = 4\,ft$ $F_{d2} = 0$ $\gamma_{b2}(\text{avg @ } z_2) = \gamma_b$ @ $(z_2/2) = 21.0\,pcf +$ $\qquad (0.31\,pcf/ft)(4\,ft/2) = 21.6\,pcf$ $W_{b2} = 20{,}000\,lb - (21.6\,pcf)(4\,ft)(\pi)(4\,ft)^2/4$ $\qquad = 18{,}910\,lb$ $s_{u2}(nose) = s_u$ @ $5.4\,ft = (1 + 5.4/30)\,psi$ $\qquad = 169.9\,psf$ $S_{\cdot e2} = \dfrac{4}{1 + \left(\dfrac{(4lb*s/ft^2)(36.60\,ft/s)}{(169.9\,psf)(4\,ft)} + 0.11\right)^{-0.5}} = 1.45$ $N_{t2} = (2+\pi)\,[1 + 0.2(4\,ft/4\,ft)/(2+\pi)]$ $\qquad [1 + 2\arctan(4\,ft/4\,ft)/(2+\pi)] = 8.0$ $Q_{n2} = 169.9\,psf\,(1.45)(8.0)(12.6\,ft^2) = 24{,}880\,lb$ $s_{u2}(side) = s_u$ (avg @ z_2) = s_u @ $(z_2/2)$ $\qquad = [1 + (4\,ft/2)/30\,ft]\,psi = 153.6\,psf$ $S_{t2} = 2$ $S_{\cdot e2} = \dfrac{4}{1 + \left(\dfrac{(4lb*s/ft^2)(36.60\,ft/s)}{(153.6\,psf)(4\,ft)} + 0.11\right)^{-0.5}} = 1.48$ $A_{s2} = (12.6\,ft)(4.0\,ft) = 50.3\,ft^2$ $F_{s2} = (153.6\,psf/2)(1.48)(50.3\,ft^2) = 5{,}730\,lb$ $F_{h2} = (0.5)(1.0)(2.69\,slugs)(12.6\,ft^2)(36.60\,fps)^2$ $\qquad = 22{,}650\,lb$ $F_2 = 0 + 18{,}910\,lb - 24{,}880\,lb - 5{,}730\,lb$ $\qquad - 22{,}650\,lb = -34{,}340\,lb$ $2\Delta v_2 = [2(2.0\,ft)/921\,slugs]$ $\qquad \cdot (-34{,}340\,lb/36.30\,fps) = -4.07\,fps$ $v_3 = 38.45\,fps + (-4.07\,fps) = 34.38\,fps$

Problem 8.4-2

ANALYTICAL PROCEDURES	COMPUTATIONS

SUMMARY

This is the end of the second iteration. Repeating steps 13 through 21 is continued until the value of v_{i+1} obtained in step 21 becomes negative. Calculations made during subsequent iterations are not shown. The values obtained are shown in Table 8.4-2.

On the tenth iteraton, the EPS has a small negative velocity. The calculation for $2\Delta v_i$ on the eleventh iteration produces a large positive number because that calculation blows up at a velocity near zero. (Note that in other cases for which the next-to-last iteration produces a small positive velocity, the calculation for $2\Delta v_i$ produces a large negative number.) Interpolation between velocity values for the ninth and tenth iterations using Equation 8-16 shows the maximum value of penetration to be 19.7 ft. (It is only coincidence that this is nearly the same as the assumed penetration used in Step 4).

CONCLUSION: It is likely the EPS has completely penetrated into the seafloor and lies buried under about 8 ft of soil.

Table 8.4-2. Summary of Calculations for Problem 2

i	z	v	γ_{bavg}	W_b	s_{unose}	$S_{e.nose}$	N'	Q_n	s_{uside}	$S_{e.side}$	A_s	F_s	F_h	F	$2\Delta v$
(-)	(ft)	(fps)	(pcf)	(lb)	(psf)	(-)	(-)	(lb)	(psf)	(-)	(ft2)	(lb)	(lb)	(lb)	(fps)
0	1	40.00	21.2	19,734	155.5	1.51	6.73	19,842	146.4	1.53	12.6	1,407	27,043	--	--
1	2	38.45	21.3	19,464	160.3	1.49	7.25	21,711	148.8	1.51	25.1	2,825	24,987	-30,059	-3.40
2	4	36.60	21.6	18,913	169.9	1.45	8.02	24,876	153.6	1.48	50.3	5,731	22,647	-34,340	-4.07
3	6	34.38	21.9	18,347	179.5	1.42	8.49	27,153	158.4	1.46	75.4	8,690	19,972	-37,469	-4.73
4	8	31.87	22.2	17,764	189.1	1.38	8.79	28,854	163.2	1.42	100.5	11,678	17,168	-39,935	-5.44
5	10	28.93	22.6	17,166	198.7	1.34	8.99	30,138	168	1.39	125.7	14,649	14,149	-41,770	-6.27
6	12	25.60	22.9	16,553	208.3	1.30	9.13	31,104	172.8	1.35	150.8	17,558	11,077	-43,187	-7.33
7	14	21.61	23.2	15,924	217.9	1.26	9.23	31,726	177.6	1.30	175.9	20,304	7,891	-43,997	-8.84
8	16	16.76	23.5	15,279	227.5	1.20	9.31	31,937	182.4	1.24	201.1	22,739	4,746	-44,142	-11.44
9	18	10.17	23.8	14,619	237.1	1.12	9.37	31,398	187.2	1.15	226.2	24,426	1,747	-42,952	-18.35
10	20	-1.59	24.1	13,943	246.7	0.97	9.42	28,448	192	0.97	251.3	23,336	43	-37,883	103.27
interp	19.73														

8.5 REFERENCES

8-1.   J.F. Wadsworth III and R.M. Beard. Propellant-Embedded Anchors: Prediction of Holding Capacity in Coral and Rock Seafloors, Civil Engineering Laboratory, Technical Note N-1595. Port Hueneme, CA, Nov 1980.

8-2.   K. Terzaghi. Theoretical Soil Mechanics. New York, NY, John Wiley and Sons, 1943.

8-3.   J.P. Gemenhardt and J.A. Focht Jr. "Theoretical and Observed Performance of Mobile Rig Footings on Clay," in Proceedings of the Offshore Technology Conference, Houston, TX, 1970. (Paper 1201).

8-4.   W.E. Schmid. "The Penetration of Objects into the Ocean Bottom," in Civil Engineering in the Oceans II. New York, NY, American Society of Civil Engineers, 1970, pp. 167-208.

8-5.   R.J. Smith. Techniques for Predicting Sea Floor Penetration, U.S. Naval Post Graduate School. Monterey, CA, Jun 1969.

8-6.   H.J. Migliore and H.J. Lee. Seafloor Penetration Tests: Presentation and Analysis of Results, Naval Civil Engineering Laboratory, Technical Note N-1178. Port Hueneme, CA, Aug 1971, p. 59.

8-7.   D.G. True. Undrained Vertical Penetration into Ocean Bottom Soils, Ph.D. Thesis, University of California, Berkeley. Berkeley, CA, 1976.

8-8.   D.G. True. Penetration of Projectiles into Seafloor Soils, Civil Engineering Laboratory, Technical Report R-822. Port Hueneme, CA, May 1975, p. 45.

8-9.   H.B. Seed and K. L. Lee. "Undrained Strength Characteristics of Cohesionless Soils," Journal of the Soil Mechanics and Foundations Division, American Society of Civil Engineers, Vol. 93, No. SM6, Nov 1967.

8-10.   U. Dayal, J.H. Allen, and J.M. Jones. "Use of the Impact Penetrometer for the Evaluation of the In Situ Strength of Marine Sediments," Marine Geotechnology, Vol. 1, No. 2, 1975, pp. 73-89.

8-11.   D.G. True. Personal communication, Dec 1982.

8-12.   S.F. Hoerner. Fluid Dynamics Drag. Brecktown, NJ, Hoerner, 1965.

8-13.   R.M. Beard. Expendable Doppler Penetrometer for Deep Ocean Sediment Strength Measurements, Naval Civil Engineering Laboratory, Technical Report R-905. Port Hueneme, CA, Feb 1984.

8.6 SYMBOLS

A_s Side area of penetrator in contact with soil [L^2]

A_t End area of penetrator (the effective bearing surface) [L^2]

B Base diameter, or minimum of base plan dimensions [L]

C_D Fluid drag coefficient

$C_{\dot{e}}$ Empirical strain rate coefficient [FT/L^2]

C_o Empirical strain rate constant

$\bar{c}$ Effective soil cohesion [F/L^2]

D_e Equivalent diameter of penetrator [L]

D_r Fractional relative density

d_c, d_q, d_γ Correction factor for depth of base embedment

F Net downward force on penetrator [F]

F_d External driving force on penetrator [F]

F_h Fluid drag force on penetrator [F]

F_i Net downward force exerted by the penetrator [F]

F_s Side friction or adhesion force on the penetrator [F]

f_z Attenuation factor for bearing capacity stress due to friction at depth

H Base block height [L]

H_s side soil contact height = min (z, H) [L]

i ith increment of soil depth

L Base diameter, or maximum of base plan dimensions [L]

M Penetrator mass [FT^2/L]

$N_c, N_q, N_\gamma, N_\phi$ Bearing capacity factors

N_c' Bearing capacity factor for static penetration in cohesive soils

N_t Nose resistance factor

P Base perimeter = $2B + 2L$ [L]

Q_n Nose resistance during dynamic penetration [F]

Q_u Net resistance to (static) penetration [F]

q_c Standard cone penetration resistance [F/L^2]

q_q Bearing capacity stress for overburden [F/L^2]

q_γ Bearing capacity stress for friction [F/L^2]

$S_{\dot{e}}$ Strain rate factor

$S_{\dot{e}}*$ Maximum strain rate factor

S_t	Soil shear strength sensitivity
s_c, s_φ, s_γ	Correction factors for shape of base
s_u	Soil undrained shear strength [F/L^2]
s_{ua}	Undrained shear strength averaged over the side soil contact zone [F/L^2]
s_{uz}	Undrained shear strength averaged over the base influence zone [F/L^2]
v	Penetrator velocity [L/T]
v_0	Penetrator velocity at soil contact [L/T]
W_b	Buoyant unit weight of penetrator [F]
z	Depth of embedment of foundation [L]
z_{avg}	Average depth over side soil contact zone = ½ [z + max(0, $z - H$)] [L]
Δz	Incremental change in penetration [L]
γ_b	Soil buoyant unit weight [F/L^2]
δ	Effective friction angle alongside the footing [deg]
ρ	Mass density of the material being accelerated [FT^2/L^4]
σ_{cr}	Critical confining pressure [F/L^2]
ϕ	Soil friction angle [deg]
$\bar{\phi}$	Effective or drained friction angle [deg]
ϕ_u	Undrained friction angle of cohesionless soil [deg]

9 BREAKOUT OF OBJECTS FROM THE SEAFLOOR

9.1 INTRODUCTION

Planning for the removal of objects embedded in or resting on the seafloor requires determination of the breakout force. If immediate breakout is needed, the immediate breakout force can be estimated. Alternatively, if the immediate breakout force is higher than the available lifting capacity, the time required to achieve breakout can be estimated for a specified force. Also, the force required to achieve breakout can be estimated for a specified duration. The force-time prediction is highly inaccurate due to the complex nature of the breakout process and uncertainties resulting from soil embedment, and object variations. Immediate breakout (the initial condition, which causes the largest force) is addressed in this chapter as well as extended-term breakout (which occurs at a reduced force).

The total force required to achieve breakout of an object from the seafloor includes the buoyant weight of the object and the soil-generated breakout resistance force. The buoyant weight is the submerged weight of the object, which equals the weight in air less the weight of water displaced by the object. <u>Note</u>: All forces discussed in this chapter are static forces; dynamics due to sea state or other causes are not considered.

9.1.1 Applications

The determination of breakout force is key in two general situations: (1) where retrieval is desired, and (2) where it is desired that the object remain in place. The more common of the two cases is the retrieval of objects embedded in the seafloor. These objects may include sunken ships, submarines, airplanes, weapons, previously placed instrumentation packages, or foundations for seafloor structures. The goal in planning the retrieval operation is to predict and provide sufficient lifting force for immediate breakout, or to determine an adequate amount of time for the application of a lower force to cause breakout. A conservative estimate for this operation is the upper limit on the force or time required for breakout to occur. This upper limit is based on multiplying the best estimated breakout force by a factor of two, and is described in Section 9.4.

The second type of breakout situation involves objects that are expected to remain in place and not become dislodged from the seafloor. An example would be a foundation which could experience an uplift force in excess of its underwater weight. For this application, a conservative estimate for breakout is the lower force limit. That is, the object is expected to remain in place if no uplift forces are exerted in excess of this value. The lower force limit is based on dividing the best estimated breakout force by a factor of two; however this application is not described further in this chapter. The procedures presented for evaluating the problem and determining values for breakout force or time for breakout are outlined in Figure 9.1-1.

DETERMINE OBJECT AND EMBEDMENT CHARACTERISTICS V_c, A, B, D, L, W_s

IS D/B < 2.5? — NO → THIS IS NOT TREATED AS A BREAKOUT PROBLEM. SEE CHAPTER 5 OR 6.

YES

IS Y_{tops} < 3 FPT? — NO

YES

IS D/B < 1? — NO → USE DYNAMIC PENETRATION AND BURIED OBJECTS (SECTION 9.4.3)

YES

WHAT IS THE SOIL TYPE? — COHESIONLESS → IMMEDIATE BREAKOUT FORCE APPROXIMATELY EQUALS OBJECT UNDERWATER WEIGHT. $F_b = \gamma V_s$

COHESIVE

DETERMINE SOIL CHARACTERISTICS

SOLVE FOR F_q EQUATION 9-1

CHECK RESIDENCE TIME — WEEKS TO YEARS

HOURS TO DAYS

SHORT RESIDENCE TIME (SECTION 9.4.1)

LONG RESIDENCE TIME (SECTION 9.4.2)

WHAT IS THE SOIL TYPE? — COHESIONLESS

COHESIVE

SOLVE SIDE ADHESION FORCE, F_s EQUATION 9-9

SOLVE SIDE ADHESION FORCE, F_s EQUATION 9-10

SOLVE BASE SUCTION FORCE, F_{bs} EQUATION 9-13

BASE SUCTION FORCE EQUALS ZERO $F_{bs} = 0$

SOLVE SOIL WEIGHT ABOVE OBJECT, F_s EQUATION 9-14

SOLVE RECOVERY LINE FORCE FOR IMMEDIATE BREAKOUT, F_b

END OF PROBLEM

IS D/B < 0.25? — NO → SOLVE F_{ts} EQUATION 9-3

YES

SOLVE F_{ts} EQUATION 9-2

SOLVE F_{bs} EQUATION 9-4

IS D/B < 0.25? — NO → SOLVE F_{sLR} EQUATION 9-7

YES

SOLVE F_{sLR} EQUATION 9-5

SOLVE F_{bsLR} EQUATION 9-6

IS THIS LOAD ACHIEVEABLE? — YES → APPLY THIS LOAD AND IMMEDIATE BREAKOUT SHOULD OCCUR

NO

END OF PROBLEM

BREAKOUT TIME USING LESS THAN IMMEDIATE BREAKOUT FORCES (SECTION 9.5)

BREAKOUT TIME USING LESS THAN IMMEDIATE BREAKOUT FORCES (SECTION 9.5)

DETERMINE F_{Lb}

SOLVE FOR t_{bo} EQUATION 9-17?

DETERMINE REQUIRED LEVEL OF CONFIDENCE AND APPLY CONFIDENCE FACTOR EQUATION 9-18

IS THIS t_b TOO LONG? — YES → USE BREAKOUT AIDS TO LOWER t_b OR F_{Lb} (NOTE: IT MAY BE POSSIBLE TO BREAK OBJECT OUT WITH LIMITED F_{Lb} OR t_b)

NO

APPLY LINE LOAD F_{Lb} FOR TIME t_b AND BREAKOUT SHOULD OCCUR

END OF PROBLEM

END OF PROBLEM

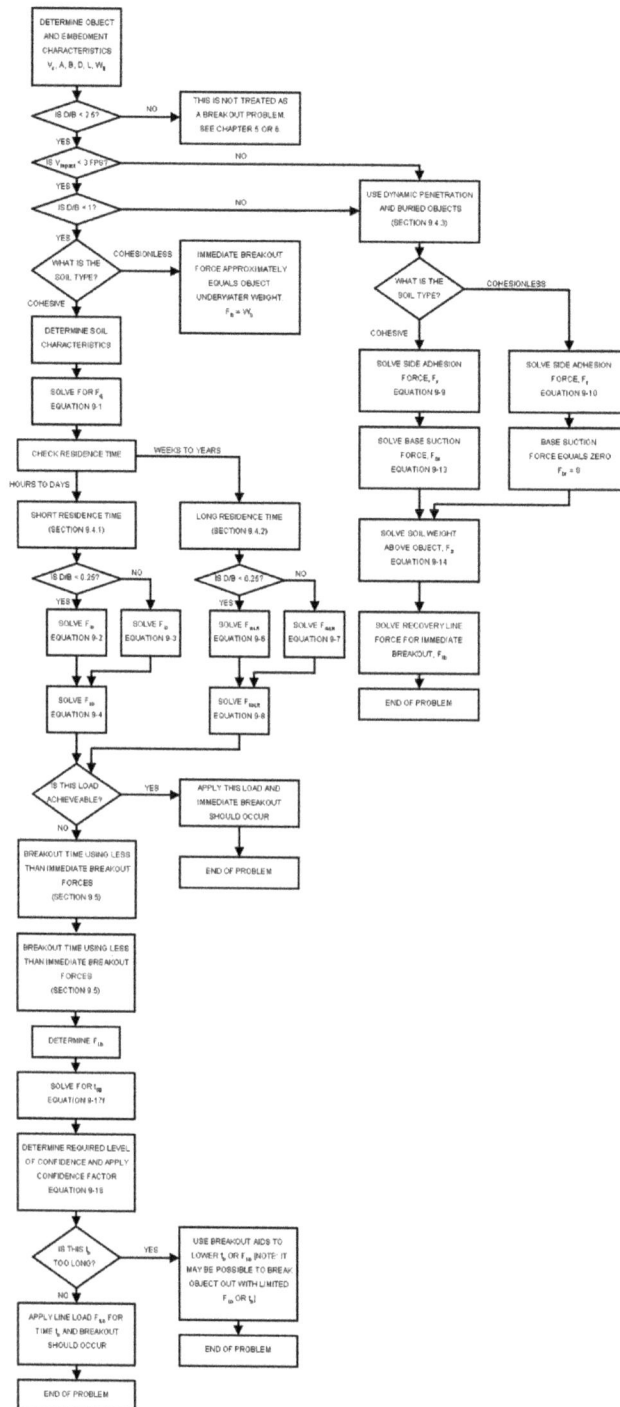

Figure 9.1-1. Breakout analysis flowchart.

9.1.2 Definitions

The following definitions are used throughout this chapter. Figure 9.1-2 presents a graphical representation of some of the terms.

9.1.2.1 Breakout Force

The soil breakout resistance force, herein referred to as breakout force, is the resistance of the soil to removal of an object from the seafloor. The breakout force includes a base suction force, side adhesive or frictional force, and the buoyant weight of any soil adhering to or lifted with the object. Of these, the base suction force normally provides the largest component of the breakout force.

9.1.2.2 Recovery Line Force

The recovery line force is the sum of the breakout force and the object's buoyant weight, less the displaced soil buoyant weight. The recovery line force is defined at the lift point on the object.

9.1.2.3 Immediate Breakout

Immediate breakout is achieved within seconds or minutes after application of the recovery line force, without time for any significant drainage of pore water or associated relief of the base suction force (not to be confused with the length of time after object placement or installation).

Note: "Suction" here means that the force is applied downward to the base of the object as it is lifted. More precisely, when breakout occurs, it does so by soil shear failure and/or water flow, but not by overcoming the (almost always much greater) limiting suction which is defined by the ambient hydrostatic pressure.

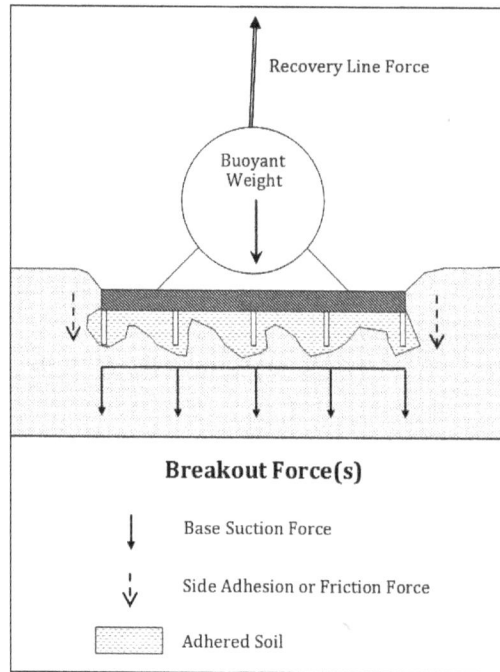

Figure 9.1-2. Illustration of breakout forces.

9.2 GENERAL CONCEPTS

An object cannot be lifted from the seafloor unless either water or soil moves into the space occupied by the object. Whether breakout occurs as a result of soil or water flow depends on time, soil permeability (the rate at which water flows through the soil), soil strength, and object to soil contact. An uplift force in excess of object weight will induce pore water pressure changes in the soil around the object. These will diminish with time as water flows to dissipate the negative pore pressures.

In high permeability cohesionless soils (sand and gravel), water flow is rapid; thus, the suction force dissipates rapidly and is commonly ignored. In low permeability cohesive soils (silt and clay), water flow will be slow. Because of the slow water flow, waiting for the suction force to dissipate is unreasonable for an expeditious recovery operation. For immediate breakout in cohesive soil, soil will be pulled into the space vacated by the object. This soil flow is akin to a bearing capacity type failure in reverse.

Typically, breakout occurs via some combination of water flow and soil flow. Once the object starts to move and soil flow initiates, flow channels can form. Flow channels may be preexisting when an object rests on a more competent seafloor, where contact between the base of the object and the (not perfectly flat) seafloor is not established over the entire base surface. These channels serve to rapidly reduce the suction force, thereby accelerating breakout. As reducing the suction force at the object base is the key to achieving breakout, several of the breakout force mitigation techniques suggested in Section 9.6 of this chapter are designed to speed water flow to the base of the object.

In Section 9.4, procedures are described for determining the maximum suction force and the soil strength upon which it directly depends. These procedures apply to cohesive soils only, as suction forces disappear almost immediately because of rapid pore pressure dissipation in the more permeable cohesionless soils (sandy soils).

In Section 9.5, a procedure is described for determining the time to breakout if a force less than the immediate breakout force is applied. This procedure only applies to objects that are embedded up to one times their width. The force-time prediction is highly inaccurate due to the complex nature of the breakout process and uncertainties resulting from soil embedment, and object variations.

9.3   SETTLEMENT AND BREAKOUT FORCE PREDICTIONS

9.3.1   Settlement Relationships

Immediate settlement of any object on soil is comprised of elastic and plastic settlement components. Elastic settlement occurs prior to soil failure and is recoverable upon load removal. Plastic settlement occurs when the pressure of the object is greater than the bearing capacity of the soil causing the object to penetrate until the bearing capacity of the soil is equal to the weight of the object. The summation of these two types of settlement equals the total immediate settlement. Elastic settlement is a minor part of total settlement, and thus is not usually considered. Secondary compression or long-term creep in cohesive soils will result in minor additional object settlement and an associated increase in breakout force. It is not practical to obtain the soil information necessary to estimate secondary compression. Its influence should be less than 10%, thus it is not considered quantitatively in this chapter.

Because breakout force can be a function of embedment depth, calculated or observed settlement is important. Refer to Chapter 4 for settlement calculations[3].

9.4   BREAKOUT RELATIONSHIPS

9.4.1   Short Term Residence (Hours to Days) Immediate Breakout for Shallow Foundations (D/B <1)

The recovery line force is the total retrieval line force required to extract an object from the seafloor. For cohesive soils, it is equal to the breakout force plus the object's submerged

---

[3] Chapter 4 does not directly address solving for the embedment depth when plastic failure occurs. The assumption is that the object will embed into the soil until the bearing capacity equals the bearing load. If the object embeds deeper than its width (D/B > 1), use Chapter 5 for calculating the bearing capacity.

weight minus the buoyant weight of displaced sediment. Although the buoyant weight of the displaced sediment is included, it is often minor and may be excluded for simplicity in most cases. However, as can be is significant in weak soils, it is included here.

The immediate breakout force required to remove an object from the seafloor soil, F_{Ib} has been shown to be a function of the object's maximum historical net downward force, F_q, (Equation 9-1) carried by the soil before breakout is attempted.

There is some experimental evidence (Ref. 9-1) that suggests that this historical force should be reduced when the embedment of the object (relative to its width) is low ($D/B < 0.25$). However, these results were based primarily on laboratory test data obtained with the object at neutral buoyancy during setup, such that base contact likely was not well established over time. This assumption led to non-conservative (low) predictions breakout forces at low values of D/B. So alternatively, two different equations are now recommended when estimating F_{Ib} (Equations 9-2 and 9-3).

$$F_q = W_b - W_s \tag{9-1}$$

For $0 \leq D/B \leq 0.25$, use Equation 9-2 to estimate the immediate breakout force, F_{Ib}. For $0.26 \leq D/B \leq 1$, use Equation 9-3 to estimate F_{Ib}:

$$F_{Ib} = \frac{F_q}{2} \tag{9-2}$$

$$F_{Ib} = F_q \left(1 - e^{-2.75(D/B)} \right) \tag{9-3}$$

where:

F_q = maximum historical net downward force [F]

F_{Ib} = immediate breakout force of the soil [F]

W_b = buoyant weight of the foundation or anchor [F]

W_s = buoyant (wet) weight of the displaced soil = displaced soil volume $\cdot \gamma_b$ [F]

γ_b = buoyant (wet) unit weight of soil; if unknown see Chapter 2 for guidance [F/L^3]

As breakout forces can vary greatly due to permeability, it is recommended that a factor of safety of two be applied to Equations 9-2 or 9-3 when estimating the needed recovery line force to ensure immediate breakout. This factor of safety is included in Equation 9-4 for the recovery line force for immediate breakout, F_{IIb}:

$$F_{IIb} = 2F_{Ib} + W_b + W_c - W_s \tag{9-4}$$

where:

W_b = buoyant weight of the object [F]

W_c = buoyant weight of any contained or adhering soil [F]

W_s = buoyant weight of displaced soil [F]

Note: If the soil bearing capacity is large enough (> W_b) to prevent bearing capacity failure leading to significant penetration of the object when it is placed on the seafloor ($D/B <$ 0.25), then no soil is displaced or contained and Equation 9-4 simplifies to:

$$F_{llb\,(surface)} = 2.0W_b \qquad (9\text{-}5)$$

Using this relationship to estimate the recovery line force for immediate breakout has been referred to as the "rule of thumb."

As stated at the outset, these relationships are valid only for cohesive soil. On or in cohesionless soils suction will dissipate rapidly, so the recovery line force will simply be equal to W_b. Also, on very strong cohesive soils, even though the theoretical immediate value of $2W_b$ may exist momentarily, the observed breakout force will likely be lower (closer to W_b) because the contact area will be so small that progressive release of adhesion and inflow of water will occur almost as quickly as the recovery line force can be applied.

9.4.2 Long Term Residence (Months to Years) Immediate Breakout for Shallow Foundations (D/B <1)

Special consideration must be given when objects are retrieved after extended time periods. For such objects, the equations stated in Section 9.4.1 are not conservative enough, because over longer periods of time, immediate breakout forces tend to go up for one of the following two reasons.

Condition 1. When the initial bearing capacity is about equal to the bearing load, the shear strength can decrease due to creep. Therefore, settlement and the effective contact area will increase, increasing the immediate breakout force. For normally consolidated marine clays, the strength can _decrease_ by a maximum of 15% for each order of magnitude decrease in the load time. Therefore, the shear strength after, for example, two years of residence is weaker than the shear strength after 10 minutes (five orders of magnitude longer) by a factor of 1.15 raised to the fifth power, which is approximately equal to two. Thus, the shear strength would be reduced by half, increasing the settlement and effective contact area, resulting in a long residence immediate breakout force of approximately twice the short term immediate breakout.

Under Condition 1, use Equation 9-6 to estimate the long-term residence immediate breakout force (F_{IbLR}) when $0 \leq D/B \leq 0.25$. Alternately, Equation 9-7 is used to estimate F_{IbLR} when $0.26 \leq D/B \leq 1$.

$$F_{IbLR} = F_q \qquad\qquad (9\text{-}6)$$

$$F_{IbLR} = 2 \, F_q \left(1 - e^{-2.75(D/B)}\right) \qquad\qquad (9\text{-}7)$$

Condition 2. When the initial bearing capacity is greater than the bearing load, the shear strength can increase over time due to consolidation. Therefore, increasing the bearing capacity has been shown to increase the immediate breakout force. For situations where the cohesive soil may be strengthened over time from normal object-induced consolidation, there is no simple quantitative means readily available for estimating the amount of strengthening. Therefore, use of Equations 9-6 and 9-7 is recommended as a reasonable upper bound on the immediate breakout force.

As breakout forces can vary greatly due to permeability, it is recommended that a factor of safety of two be applied to Equations 9-6 or 9-7 when estimating the needed recovery line force for long term residence to ensure immediate breakout for either Condition 1 or 2 (Equation 9-8).

$$F_{IIbLR} = 2 \, F_{IbLR} + W_b + W_c - W_s \qquad\qquad (9\text{-}8)$$

9.4.3   Dynamic Penetration and Buried Objects Immediate Breakout ($1 < D/B < 2.5$)

Special consideration must be given to recovery line forces when objects are retrieved after dynamic penetration (i.e., when the impact velocity > 3ft/s) (Ref. 9-2) or when the object is completely buried.

The total recovery line force for a buried object, F_{IIb}, is the sum of the side adhesion force, the base suction force, the soil weight above the object, the soil weight trapped within the object, and object buoyant (wet) weight (Equation 9-9).

$$F_{IIb} = F_s + F_{bs} + F_a + W_b \qquad\qquad (9\text{-}9)$$

where:

F_s = side adhesion force; Equation 9-10 for cohesive soils or Equation 9-11 for cohesionless soils [F]

F_{bs} = base suction force; F_{bs} = 0 for cohesionless soils or Equation 9-12 for cohesive soils [F]

F_a = soil weight above object (plus any soil weight trapped within the object); Equation 9-13 [F]

W_b = object buoyant (wet) weight [F]

For objects penetrating the seafloor at velocities greater than 3 ft/s or buried by means other than settlement, side adhesion can play a large role in breakout force. For cohesive soils, the side adhesive force, F_s, is given by:

$$F_s = s_{u\,avg} D P_s \qquad (9\text{-}10)$$

where:

$s_{u\,avg}$ = average shear strength over the burial depth, D [F/L^2]

D = depth of burial to object base (Note: if D is greater than the object height, use H instead of D in this equation) [L]

P_s = side perimeter of object (e.g., $P_s = 2B + 2L$ for a rectangular prism object) [L]

B = width of the object [L]

L = length of the object [L]

For cohesionless soils (sand), the side adhesive force, F_s, is given by Equation 9-11:

$$F_s = \gamma_b \left(\frac{D}{2} \right) D P_s (\tan \phi) \qquad (9\text{-}11)$$

where:

ϕ = soil friction angle

D = depth of burial to the object base [L]

P_s = side perimeter of object (e.g., $P_s = 2B + 2L$) [L]

γ_b = buoyant (wet) unit weight of soil [F/L^3]

For cohesive soils, suction force plays a factor due to low permeability. Whereas bearing capacity of the soil holds the object up, the suction between the soil and object holds the object down. The base suction force, F_{bs}, is obtained using Equation 9-12.

For cohesionless soils, suction force is dissipated quickly due to high permeability. Therefore, $F_{bs} = 0$ for cohesionless soils.

$$F_{bs} = 5.14 A_b \left[s_{uo} + g\left(\frac{2D+B}{2} \right) \right] \cdot \left[1 + 0.2\left(\frac{D}{B} \right) \right] \cdot \left[1 + 0.2\left(\frac{B}{L} \right) \right] - \gamma_b A_b D \qquad (9\text{-}12)$$

where:

A_b = horizontal area of object base touching the soil [L²]

s_{uo} = shear strength of the soil at the soil surface [F/L²]

g = shear strength gradient of the cohesive soil [(F/L²)/L]

D = depth of burial to object base [L]

B = width of base bearing area [L]

L = length of base bearing area [L]

γ_b = buoyant (wet) unit weight of soil [F/L³]

If there is soil above the object due to burial or trapped within the object, it will also need to be lifted to raise the object and will increase the recovery line force. For both cohesive and cohesionless soils, this additional force attributed to the soil weight above object (plus any soil weight trapped within the object) is given by:

$$F_a = \gamma_b (D - H) A_p + W_c \tag{9-13}$$

where:

γ_b = buoyant (wet) unit weight of soil [F/L³]

D = depth of burial to object base [L]

H = height of object [L]

A_p = effective plan cross-sectional area of object [L²]

W_c = buoyant weight of contained or adhering soil [F]

9.5 BREAKOUT TIME USING LESS-THAN-IMMEDIATE BREAKOUT FORCES FOR SHALLOW FOUNDATIONS (D/B < 1)

9.5.1 Introduction

Often the force required for immediate breakout exceeds the maximum available lift force. Several breakout force mitigation techniques are available in these situations, including eccentric lifting, jetting, and the use of water flow paths and/or pumping to relieve the base suction. However, in some cases simply waiting for a short time may reduce the required force to match the available lift.

The procedure discussed in Section 9.5.2 evolved from one previously available in Reference 9-2. The previous procedure is a tool for predicting the time needed to achieve breakout with a known force that less than the immediate breakout force. No specific indication of confidence is included but the usefulness of the time estimate is limited by the caveat that it is highly imprecise and the underlying data in Reference 9-1 varies by as much as two orders of magnitude from the fitted relationship that is the basis of the procedure. The procedure presented in Section 9.5.2 improves upon the previous one in three ways:

1) The fitted relationship that is its basis is better aligned with the data so deviations of measurements from back-predictions are minimized.

2) The form of the relationship is simpler facilitating calculations.

3) The new procedure provides estimates of breakout time at specified levels of confidence that the time estimate will not be exceeded whereas the previous procedure simply provides a most-likely (50% confidence) time estimate.

Note: The equations presented in the following sections (Section 9.5.2 through 9.5.3) were developed from empirical data and curve fitting techniques. Therefore, it is imperative that the units specified for each equation are the ones used in calculations.

9.5.2 Basis of Development of New Procedure

Reference 9-1 documents breakout data for partially embedded objects tested under laboratory conditions and in the field within the San Francisco Bay and the Gulf of Mexico. These tests provide the basis for the extended breakout relationship previously recommended in Reference 9-2, which is given by Equation 9-14a using the specific units indicated.

$$\log_{10}\left(F_{Lb} / F_{Ib}\right) = -0.193 \cdot \left(\log_{10} T - 3.84\right) \tag{9-14a}$$

Equation 9-14a can be rewritten in a more convenient form as:

$$\left(F_{Lb} / F_{Ib}\right)^{pf} = 10^{pt} \Big/ T \tag{9-14b}$$

where:

F_{Lb} = long-term lift breakout applied force (lb)
   = *Available Long-Term Recovery Force* $- W_b + W_s$

F_{Ib} = immediate breakout force; see Equation 9-3[4] (lb)

pf = power of force ratio (dimensionless) = 5.18 per References 9-1 and 9-2

---

[4] For loads less than the immediate breakout force, Equation 9-3 is used for F_{Ib} regardless of burial depth.

pt = power of ten (dimensionless)

   = 3.84 for Reference 9-1 field data fit and Reference 9-2, or

   = 4.24 for Reference 9-1 lab data fit

T = time parameter computed using Equation 9-15 (lb·min/ft^4)

The time parameter, T, used in Equations 9-14a and 9-14b, is computed using the following relationship (Equation 9-15). Again, the equation only applies using the units specified, and yields T in units of *lb·min/ft^4*.

$$T = \left(pt_b / D^2\right)\left(B/D\right)^2 = \left(pt_b / B^2\right)/(D/B)^4 \tag{9-15}$$

where:

p = suction in pore fluid beneath base $\approx F_{Lb}/A$ (psf)

t_b = time to breakout (min)

B = width of object base (ft)

D = depth of embedment (ft)

A = area of object base (ft^2)

Because the data are considerably scattered, breakout times predicted using this relationship carry a large range of uncertainty. Estimates using this relationship should be used only to determine whether or not other means to reduce breakout force should be employed. If the embedment depth is known, the methods of Section 9.5.3 should be used to estimate time required for breakout to occur.

9.5.3   Development of New Relationship Involving Known Embedment Depth

As a first step in the development of an improved procedure towards estimating the time to breakout, Equations 9-3, 9-14b, and 9-15 were combined and rearranged as shown in Equation 9-16. In the equation, F_q is the maximum historical net downward force.

$$\left(F_{Lb} / F_q\right)^{pf+1} = 10^{pt}\left[1.0 - 0.97e^{(-2.75D/B)}\right]^{pf}(D/B)^4 /\left[\left(F_q t_b\right)/\left(AB^2\right)\right] \tag{9-16}$$

Equation 9-16 shows the full structure of the previous relationship, i.e., that the normalized force raised to a power is equal to a constant multiplied by a depth-dependent term divided by a relatively simple time-dependent term.

Simplification of Equation 9-16 is achieved by replacing the numerator on the right-hand side by a constant multiplied by a complex depth-dependant term and divided by a relatively simple power of D/B. This simplified format is:

$$R_f{}^{pf'} = C'/T'$$
(9-17)

where:

R_f = force ratio (dimensionless) = F_{Lb}/F_q

pf' = power of force ratio (dimensionless)

C' = constant (lb·min/ft^4)

T' = time parameter computed using Equation 9-18 (lb·min/ft^4)

The time parameter, T', used in Equation 9-17, is computed using the following relationship (Equation 9-18). The equation only applies using the units specified, and yields T' in units of *lb·min/ft^4*.

$$T' = \frac{F_q\, t_b\, /(AB^2)}{(D/B)^{pd'}}$$
(9-18)

where:

F_q = maximum historical net downward force (lb)

t_b = time to breakout (min)

A = area of object base (ft^2)

B = width of object base (ft)

D = depth of embedment (ft)

pd' = power of depth-to-width ratio (dimensionless)

Next, the original data as tabulated in Reference 9-1 were replotted on axes of R_f vs. T' and parameter adjustments in the fitting process were made such as to minimize the overall scatter and maximize the overall alignment for all of the data at once (Ref. 9-2). For a fit of 50% level of confidence[5] ($LC = 50$), the following values of the parameters apply:

---

[5] A 50% level of confidence indicates there is a 50% chance the object will break out under the specified loading conditions. Similarly, a 95% level of confidence indicates there is a 95% chance the object will break out under the specified conditions.

$C' = 7{,}700 \text{ lb·min/ft}^4$

$pd' = 4.54$

$pf' = 4.80$

By combining Equations 9-17 and 9-18, and using the values of the parameters C', pd' and pf' associated with a 50% level of confidence, the following equation is formed:

$$\left(F_{Lb}/F_q\right)^{4.8} = \frac{7{,}700(D/B)^{4.54}}{[F_q\, t_b\, /(AB^2)]} \tag{9-19}$$

Equation 9-19 can be rewritten in a more convenient form as shown in Equation 9-20. Note that in this equation, t_b is replaced by t_{50}, indicating the breakout time associated with a 50% level of confidence.

$$t_{50}' = \frac{7{,}700}{\left(F_{Lb}/F_q\right)^{4.8}}(D/B)^{4.54}\, AB^2\, /\, F_q \tag{9-20}$$

In order to calculate breakout time using a selected level of confidence above 50%, a factor f_{tLC} is obtained from Table 3.2-1 and applied as shown.

$$t_{LC} = t_{50} \cdot f_{tLC} \tag{9-21}$$

Table 9.5-1. Factors for Determining Long-Term-Lift Breakout Times at Selected Levels of Confidence

Level of Confidence, LC (%):	50	75	90	95	99
Level of Confidence Time Factor, f_{tLC} (used in Equation 9-21)	1	2.25	6.09	14.9	117

9.6 BREAKOUT AIDS

Breakout aids are any operations that reduce the force required for immediate breakout or the time required for long-term breakout. The following discussion provides suggestions for achieving breakout when the immediate breakout force is too high to be applied or the long-term breakout time for the force that can be applied is too long. Although most discussion is on reducing breakout force, most breakout aids also reduce breakout time at a particular force level. It should also be noted that reduction in breakout force, by one-half, for example, is not the same as a lowering of the retrieval line load by one-half. Line load is equal to breakout force plus object buoyant weight minus displaced soil weight, as defined in Equation 9-4 or 9-8.

9.6.1 Jetting and Drainage Tubes

An effective way to reduce the force required for breakout is to improve the flow of water into the soil around the object. This can be done either passively with drainage tubes or actively with water jets (Figure 9.6-1). Both consist of tubes that are forced beneath the object, or may be a part of the object if a breakout problem is anticipated. Preferably, the tubes should have openings along the side to improve water flow all along the tube.

With water jetting, a pump is connected to the end of the tube and water is forced through the tube into the soil. Jetting is preferable because the positive pressure is more effective, and the turbulence of water flow may cause some soil erosion and reduce the object-to-soil contact area.

With drainage tubes no pumps are used. In this technique, the act of pulling up on the object draws water through the tubes into the spaces below the object. It is essential for this technique that the free end of the tube does not clog.

Field tests (Ref. 9-4) have demonstrated that a 50 to 77% reduction of the breakout force is possible with both water jets and drainage tubes. The reduction for rounded objects was greater than for square blocks. It should be expected that the effectiveness of these aids will vary significantly, depending on specific soil and embedment conditions.

No guidelines exist for the design of either water jets or drainage tubes. Where used, however, the spacing between openings (both along the tubes and between the tubes) should be minimized and kept relatively constant over the entire, contact area of the object.

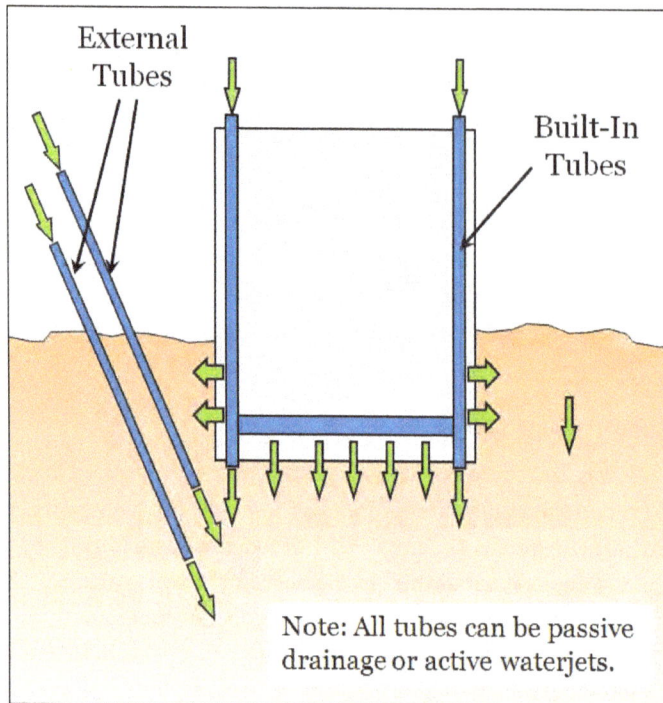

External
Tubes

Built-In
Tubes

Note: All tubes can be passive
drainage or active waterjets.

Figure 9.6-1. Water flow techniques.

9.6.2   Eccentric Loading

A reduction in the required breakout force will result if the object is lifted from one end rather than through its-center. In Chapter 4, the concept of effective area is introduced with respect to bearing capacity prediction for eccentrically loaded foundations. This concept also applies to breakout prediction. The uplift force is essentially applied to a reduced area, raising the applied stress in that area. After that section is broken free, the force is transferred to the remaining area and breaks it free. A rough estimate of the effect of this procedure is that breakout force may be reduced by up to 50%. Long, narrow objects should see the most effect. In many field cases, this may be art easy breakout aid to use; however structural limitations on the object could prevent eccentric attachment of the applied load (Figure 9.6-2).

9.6.3   Cyclic Loading

The strength of some cohesive sediments can be reduced through cyclic application of uplift force and resulting development of positive sediment pore water pressures. For example, based on earthquake stability analysis (Ref. 9-5), breakout force may be reduced by 30 to 40% with as few as ten load applications for clayey silt soils. This procedure will be less effective with soils which are more fine-grained and more plastic. During the time between load applications, an uplift force equal to the buoyant weight of the object should be maintained, and the loads should be applied every 10 minutes or less. Benefits from this effect may occur where wave actions affect the recovery vessel and cause unsteady recovery line loads to be applied (Figure 9.6-2).

9.6.4 Rocking or Rolling

Prior to lifting, if a significant lateral force can be applied to an object embedded in relatively stiff soil ($s_u > 3$ psi), openings may develop along the object's side. These openings will facilitate water flow and reduce the object-soil contact area. A lateral force greater than the passive resistance of the sediment (estimated as $2DLs_u$) is likely necessary to achieve adequate movement. If possible, the object should be loaded alternately in opposite directions to increase the development of openings along the sides. This process of rocking or rolling may also result in a reduction of object embedment depth and, therefore, an additional reduction in necessary breakout force (Figure 9.6-2).

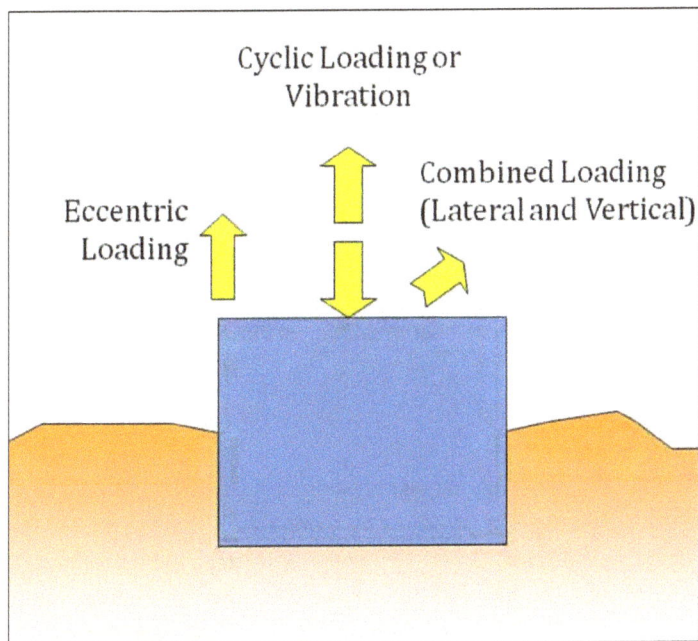

Figure 9.6-2. Soil strength reduction techniques.

9.6.5 Breakaway Parts

Objects that are meant to be placed on the seafloor can be designed with "breakaway parts." That is, the portion of the object in contact with the sediment is designed to separate from the remainder of the object upon application of uplift loading. The immediate breakout force is effectively reduced to the level required to break any connectors to the parts that remain behind. This situation can be achieved with weak links or connectors that corrode in seawater and leave the object parts to be recovered unrestrained for vertical motion. This procedure can be useful for instrumentation packages that are on the seafloor for long periods of time before retrieval (Figure 9.6-3).

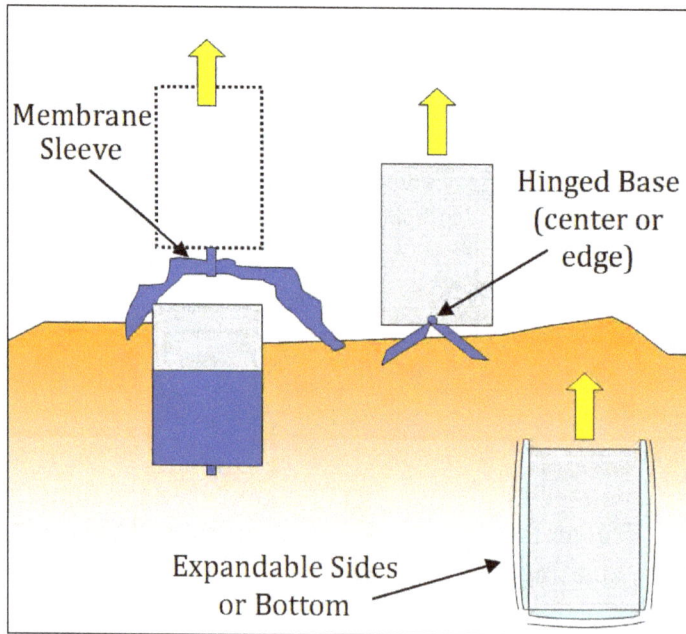

Figure 9.6-3. Breakaway techniques.

9.6.6   Altering Buoyant Weight

The retrieval line force may be significantly reduced if it is possible to decrease the object's underwater weight. This can be done by pumping air into enclosed spaces, attachment of lift bags, or removal of heavy parts of the object prior to attempting breakout. These methods do not alter the breakout force required but may be useful where the breakout force is not the largest component in the retrieval line load. Air- or lift-bag-assisted recovery may introduce potentially dangerous situations where retrieval line control can be lost. For example, trapped air will expand in volume as the object rises from the seafloor or can be lost if the object's orientation changes. A positively buoyant object will become dangerous to anything above it as it moves toward the water surface (Figure 9.6-4).

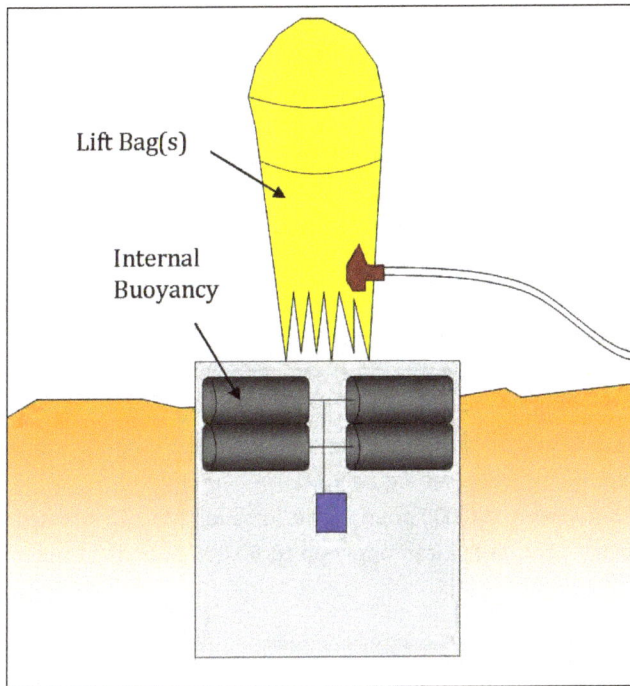

Figure 9.6-4. Buoyancy techniques.

9.7 OTHER FACTORS

9.7.1 Irregular Shape or Non-Uniform Embedment Depth

Laboratory and field tests used to define the empirical equations in this chapter have not been conducted with objects that are irregular in shape or that have a non-uniform embedment depth. The equations presented, and the procedures of Section 9.4 for calculating breakout forces can still be used for these unusual cases. However, this introduces a higher level of uncertainty and results in the computational procedures becoming less accurate.

9.7.2 Foundation Skirts

If a foundation has skirts extending below its base (useful to provide an increased resistance to lateral loads), the skirts will raise the required immediate breakout force. The skirts force the soil failure surfaces to a greater subbottom depth. The sediment contained within the skirts must be considered as part of the object. The procedures outlined in this chapter apply as if the foundation plus trapped soil is a solid object and embedment is considered to be at the depth of skirt penetration. The skirted foundation may also introduce problems during recovery operations. Soil trapped within the skirt system may drop away at any time following breakout. (It usually ends up on the deck of the recovery vessel.)

9.8 EXAMPLE PROBLEMS

9.8.1 Problem 1 – Recovery of a Large Cylinder (Short Residence)

9.8.1.1 Problem Statement

Determine if it is possible to quickly lift an embedded, cylindrically shaped object within the capacity of an available recovery vessel and if this is not possible, to estimate how long it will take for the object to break free of the seafloor at a lower force.

Data: A large, solid cylinder – 10 feet in diameter, 20 feet long, with an underwater weight of 46,000 pounds was slowly laid on the seafloor for 20 hours.

The soil is a very soft clay with a shear strength at the mudline of 0 psf that increases with depth into the seafloor at 10 psf/ft and a soil unit weight of 20 pcf at the mudline that increases with depth at 0.2 pcf/ft. During the first 10 minutes on the seafloor, the cylinder embedded 2 feet into the seafloor (see Figure 9.8-1).

The vessel can apply a 100,000 pound uplift force for 10 to 15 minutes without seriously damaging the recovery gear. But, it can apply up to 60,000 pounds of uplift force for several days.

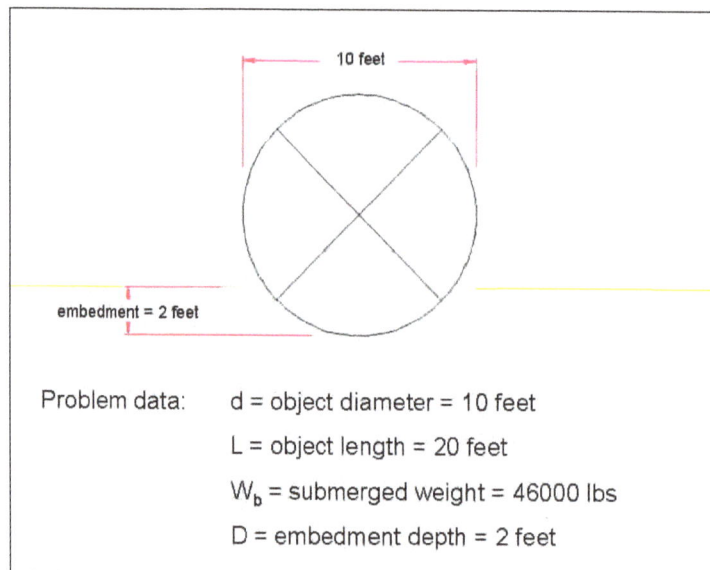

Problem data: d = object diameter = 10 feet

L = object length = 20 feet

W_b = submerged weight = 46000 lbs

D = embedment depth = 2 feet

Figure 9.8-1. Problem sketch and data for example Problem 1.

9.8.1.2 Problem Solution

The analytical and computational procedures used to solve this problem are shown below. They follow the procedures outlined by the flow chart in Figure 9.1-1.

9-20

Problem 9.8-1

ANALYTICAL PROCEDURES	COMPUTATIONS
1. Determine the object and embedment characteristics: The object is circular and therefore equivalent dimensions of a rectangular object of same volume and weight must be solved for. V_s = displaced volume of soil L = equivalent length B = width D = embedment depth W_b = submerged weight The first dimension required is B, the width at the seafloor surface which is a function of embedment depth and is found using the Pythagorean Theorem. d = diameter of object 	$D = 2\,ft$ $$\left(\frac{B}{2}\right)^2 = \left(\frac{d}{2}\right)^2 - \left(\frac{d}{2} - D\right)^2$$ $$\left(\frac{B}{2}\right)^2 = \left(\frac{10}{2}\right)^2 - \left(\frac{10}{2} - 2\right)^2$$ $B = 8\,ft$

ANALYTICAL PROCEDURES	COMPUTATIONS
2. An equal rectangular foundation which displaces the same amount of soil is required. Find the volume of the circular object below the soil surface (V_s).	$\sin\left(\dfrac{\beta}{2}\right) = \dfrac{4}{5}$ β = 1.85 radians Area below surface: $= \pi \dfrac{d^2}{4} \times \dfrac{\beta}{2\pi} - \dfrac{B}{2}\left(\dfrac{d}{2} - D\right)$ $= \pi \dfrac{10^2}{4} \times \dfrac{1.85}{2\pi} - \dfrac{8}{2}\left(\dfrac{10}{2} - 2\right)$ $= 11.13\ ft^2$ Effective depth: $D' = \dfrac{Area}{B} = \dfrac{11.13}{8} = 1.4\ ft$ L = 20 ft $V_s = B \times L \times D' = 8 \times 20 \times 1.4 = 224\ ft^3$ W_b = 46,000 lbs
3. Check if this is a breakout problem. Is D/B <2.5?	Using the new D and B of a rectangular object: D'/B = (1.4 ft)/(8 ft) = 0.175 YES, D/B < 2.5
4. Check if this is a problem of dynamic penetration. Is V_{impact} > 3 ft/s?	Object was laid slowly. Assume V_{impact} < 3 ft/s NO, this is not a case of dynamic penetration.
5. Check if this is a problem of shallow embedment. Is D/B <1?	D'/B = (1.4 ft)/(8 ft) = 0.175 YES, D/B < 1
6. Determine soil type.	Soil is cohesive.

Problem 9.8-1

ANALYTICAL PROCEDURES	COMPUTATIONS
7. Determine the soil characteristics. $\gamma_b = 20 + 0.2(z)$ *pcf* γ_b averaged from 0 to D'	$\gamma_b = (0.2z + 20)$ *lb/ft³* γ_b @ 0 ft = [(0.2)(0) +20] *lb/ft³* = 20 *lb/ft³* γ_b @ 1.4 ft = [(0.2)(1.40)+20]*lb/ft³* = 20.3 *lb/ft³* γ_b (avg) = (20 + 20.3 *lb/ft³*)/2 = 20.2 *lb/ft³*
8. Determine weight of displaced soil (W_s). $W_s = V_s \times \gamma_b$	$W_s = (224ft³)(20.2$ *lb/ft³*$) \cong 4{,}500$ *lb*
9. Determine the immediate breakout force (F_{Ib}) for short residence (Equations 9-1 and 9-2). $F_q = W_b - W_s$ (9-1) $F_{Ib} = F_q/2$ (9-2) *for D/B < 0.25*	$F_q = 46{,}000 - 4{,}500 = 41{,}500$ *lb* $F_{Ib} = 41{,}500$ *lb/2 = 20,750 lb*
10. Determine the short residence recovery line force for immediate breakout, F_{IIb} (Equation 9-4). $F_{IIb} = 2F_{Ib} + W_b + W_c - W_s$	Assuming there is no soil adhesion, $W_c = 0$ *lbs* $F_{IIb} = 2(20{,}750) + 46{,}000 + 0 - 4{,}500$ $= 83{,}000$ *lbs*

SUMMARY

The 100,000 pound line load can be applied by the vessel to immediately breakout the object. Application of a lower force for a longer period is not required.

9.8.2 Problem 2 – Recovery of a Large Cylinder (Long Residence)

9.8.2.1 Problem Statement

Determine if it is possible to quickly lift an embedded, cylindrically shaped object within the capacity of an available recovery vessel and if this is not possible, to estimate how long it will take for the object to break free of the seafloor at a lower force.

Data: The same object as described in Example 9-1 was slowly laid on the seafloor. Over a few years it embedded 2 feet in the seafloor (see Figure 9.8-2).

The soil is a very soft clay with a shear strength at the mudline of 0 psf that increases with depth into the seafloor at 10 psf/ft and a soil unit weight of 20 pcf at the mudline that increases with depth at 0.2 pcf/ft. During the first 10 minutes on the seafloor, the cylinder embedded 2 feet into the seafloor (see Figure 9.8-2).

The vessel can apply a 100,000 pound uplift force for 10 to 15 minutes without seriously damaging the recovery gear. But, it can apply up to 60,000 pounds of uplift force for several days.

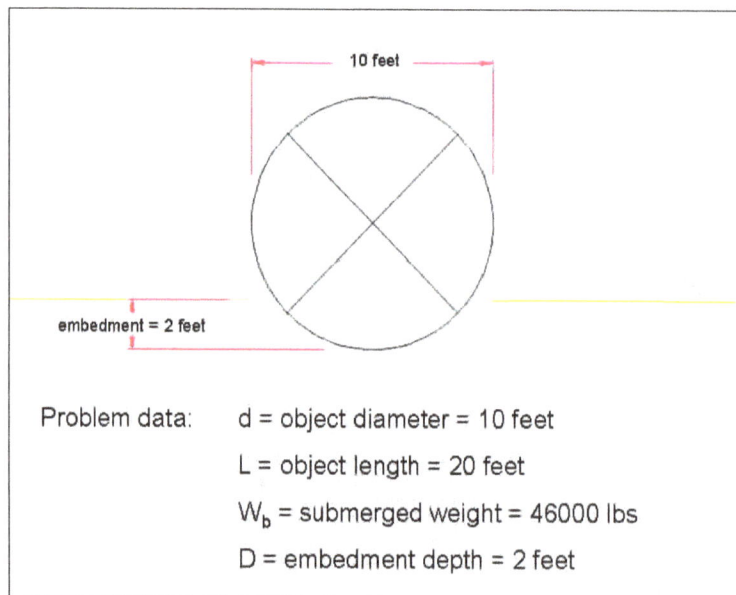

Problem data:
d = object diameter = 10 feet
L = object length = 20 feet
W_b = submerged weight = 46000 lbs
D = embedment depth = 2 feet

Figure 9.8-2. Problem sketch and data for example Problem 2.

9.8.2.2 Problem Solution

The analytical and computational procedures used to solve this problem are shown below. They follow the procedures outlined by the flow chart in Figure 9.1-1 and discussed in this chapter.

Problem 9.8-2

ANALYTICAL PROCEDURES	COMPUTATIONS
1. Determine the object and embedment characteristics: From Problem 9.8-1, all steps are the same except calculation of immediate breakout for long residence, F_{IbLR}, (Equation 9-6 or 9-7) and recovery line force for immediate breakout, F_{IIbLR} (Equation 9-8). Given from Problem 9.8-1: W_s = 4,500 lb W_b = 46,000 lb W_c = 0 D/B < 0.25; So Equation 9-6 is used to compute F_{IbLR}. $F_{IbLR} = F_q$ (9-6) $F_{IIbLR} = 2F_{IbLR} + W_b + W_c - W_s$ (9-8)	In order to compute F_{IbLR}, the bearing capacity, F_q, must first be determined (Equation 9-1). $F_q = W_b - W_s = 46{,}000 - 4{,}500 = 41{,}500$ lb F_{IbLR} = 41,500 lb F_{IIbLR} = 2(41,500) + 46,000 + 0 - 4,500 = 124,500 lb
INITIAL CONCLUSION – The required line recovery force of 124,500 lbs is greater than the available line force. The recovery vessel is unlikely to be able to quickly lift the object from the seafloor by applying a 100,000 pound line load. Therefore, the length of time for breakout using the constant 60,000 lb force must be considered	
2. Determine time to breakout at constant load (Equation 9-20). $$t_{50} = \frac{7{,}700}{\left(F_{Lb}/F_q\right)^{4.8}}(D/B)^{4.54} AB^2 / F_q$$ From the problem statement constant load can be held at 60,000 lb.	F_{Lb} = 60,000 lb – 46,000 lb + 4,500 lb = 18,500 lbs $$t_{50} = \frac{7{,}700}{\left(18{,}500\big/41{,}500\right)^{4.8}}\left(1.4\big/8\right)^{4.54}(20\cdot8)(8^2)/41{,}500$$ $t_{50} = 33.6\,min$

Problem 9.8-2

ANALYTICAL PROCEDURES	COMPUTATIONS

3. Determine time to breakout at 90% confidence and 95% confidence (Equation 9-21).
 	
$$t_{LC} = t_{50} \cdot f_{tLC}$$
 	
The level of confidence time factor, f_{tLC}, is given by Table 3.2-1.
	From Table 9-5-1,
	    f_{tLC} = 6.09 for 90% confidence
	    f_{tLC} = 14.7 for 95% confidence
	
	$t_{90} = (33.6\,\text{min})(6.09) = 205\,\text{min}$
	
	$t_{95} = (33.6\,\text{min})(14.9) = 500\,\text{min}\ or\ 8.3\ hours$

SUMMARY

If it is reasonable under other (unspecified) conditions that exist, the 100,000 lb line load should be applied to the object for as long as it can safely be sustained. The object has a chance of breaking free within minutes. This chance can be improved by applying the load eccentric to the object's center of gravity or by using other breakout aids discussed in Section 9.6 (For example, repetitive short-duration applications of the 100,000-lb force).

If it does not break free at this load, there is an even better chance for breakout within 34 minutes to 8 hrs at a line load of 60,000 lbs. To be reasonably sure of breakout at this lower line load, the load may have to be exerted for up to 8 hours. Breakout aids, if they can be applied, should speed up the recovery. Note: If the object does not break free under the 100,000 lb load applied for 10 to 15 minutes, this effort is not "wasted." This load application acts as a type of breakout aid which lowers the amount of time the 60,000 lb load will have to be applied before breakout occurs.

9.8.3 Problem 3 – Recovery of a Buried Foundation

9.8.3.1 Problem Statement

Determine if it is reasonable to expect quick recovery of a small foundation from the seafloor with the limited capacity of a specific workboat.

Data: The foundation is a heavy, small circular footing that will be buried for a 3-day experiment and then recovered.

The soil is a very soft clay with a shear strength at the mudline of 0 psf that increases with depth into the seafloor at 10 psf/ft and a soil unit weight of 20 pcf at the mudline that increases with depth at 0.2 pcf/ft.

The object will be embedded 6 feet below the surface. The footing is 4 feet in diameter, 4 feet tall, and has an underwater weight of 3,000 pounds (see Figure 9.8-3). The vessel that will install and remove the foundation is capable of lifting 15,000 pounds off the seafloor and maintaining that force level indefinitely.

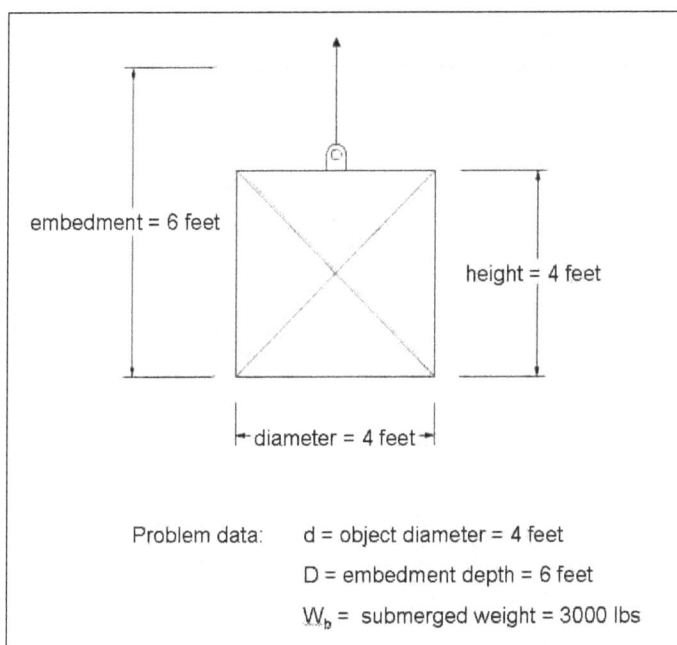

Problem data: d = object diameter = 4 feet

D = embedment depth = 6 feet

W_b = submerged weight = 3000 lbs

Figure 9.8-3. Problem sketch and data for example Problem 3.

9.8.3.2 Problem Solution

The analytical and computational procedures used to evaluate this problem are shown below. They follow the procedures outlined by the flow chart in Figure 9.1-1.

Problem 9.7-3

ANALYTICAL PROCEDURES	COMPUTATIONS
1. Determine the object and embedment characteristics (V_s, A, L, B, D, d, and W_b). V_s = displaced volume of soil A = horizontal area L = equivalent length B = equivalent width D = equivalent depth below seafloor H = height of object W_b = submerged weight d = diameter of object	$V_s = \dfrac{\pi(4\,ft)^2}{4}(4\,ft) = 50.3\,ft^3$ $A = \dfrac{\pi(4\,ft)^2}{4} = 12.6\,ft^2$ For a circular foundation, $B = L = (A)^{\frac{1}{2}} = (12.6\,ft^2)^{\frac{1}{2}} = 3.55\,ft$ D = 6 ft H = 4 ft W_b = 3,000 lb d = 4 ft
2. Check if this is a breakout problem. Is D/B <2.5?	$D/B = (6\,ft)/(3.55\,ft) = 1.69$ YES, $D/B < 2.5$
3. Check if this is a problem of dynamic penetration. Is V_{impact} > 3 ft/s?	Object was laid slowly. Assume $V_{impact} < 3\,ft/s$ NO, this is not a case of dynamic penetration.
4. Check if this is a problem of burial. Is D/B <1?	$D/B = (6\,ft)/(3.55\,ft) = 1.69$ NO, $D/B > 1$
5. Is the soil cohesive?	YES, soil is cohesive; use Equation 9-10 for side friction.

Problem 9.7-3

ANALYTICAL PROCEDURES	COMPUTATIONS
6. Determine side friction, F_s, for cohesive soils (Equation 9-10). $F_s = s_{u\,(avg)} D P_s$ <u>Note</u>: If D > object height, H, use H instead of D in Equation 9-10. $P_s = \pi d$ The soil type is given as: $s_u = 0 + 10(z)$ psf	$(D - H) = 6\,ft - 4\,ft = 2\,ft$ s_u @ 2 ft = 0 + 10(2) psf = 20 psf s_u @ 6 ft = 0 + 10(6) psf = 60 psf $s_{u(avg)}$ = (20 + 60 psf)/2 = 40 psf $P_s = \pi(4) = 12.6\,ft$ $D > H$, so use H is Equation 9-10 F_s = (40 psf)(4 ft)(12.6 ft) ≈ 2010 lb
7. Determine base suction force, F_{bs}, from Equation 9-12. $F_{bs} = 5.14 A_b \left[s_{uo} + g\left(\frac{2D+B}{2}\right) \right] \cdot \left[1 + 0.2\left(D/B\right)\right] \cdot \left[1 + 0.2\left(B/L\right)\right] - \gamma_b A_b D$ $s_u = 0 + 10(z)$ psf $\gamma_b = 20 + 0.2(z)$ lb/ft³	$(2D + B)/2 = [2(6\,ft) + 3.55\,ft]/2 = 7.78\,ft$ s_u @ 7.78 ft = 0 + 10(7.78) psf = 77.8 psf $s_{uo} = s_u$ @ 0 ft = 0 psf γ_b @ D = [20+.2(6)] lb/ft³ = 21.2 lb/ft³ γ_b @ (D-H) = [20+.2(6-4)] lb/ft³ = 20.4 lb/ft³ γ_b (avg) = (21.2 + 20.4 lb/ft³)/2 = 20.8 lb/ft³ $F_{bs} = 5.14(12.6\,ft^2)[0 + 10\,psf\,/\,ft(7.78\,ft)] \cdot$ $\left[1 + 0.2\left(\dfrac{6\,ft}{3.55\,ft}\right)\right]\left[1 + 0.2\left(\dfrac{3.55\,ft}{3.55\,ft}\right)\right]$ $- \left[(20.8 lb\,/\,ft^3)(12.6\,ft^2)(6\,ft)\right]$ F_{bs} = 6,530 lb

8. Determine the force from the soil weight above the object, F_a, (Equation 9-13).
 	
$F_a = \gamma_b (D - H) A_p + W_c$
 	
$\gamma_b = 20 + 0.2(z)$ pcf
 	
γ_b averaged from 0 to (D-H)
	From step 7,
	    γ_b (avg) = 20.2 lb/ft³
	
	F_a = (20.2 lb/ft³)(6 − 4 ft)(12.6 ft²) + 0 lb
	    ≈ 510 lb

Problem 9.7-3

ANALYTICAL PROCEDURES	COMPUTATIONS
9. Determine the recovery line force for immediate breakout, F_{llb}, from Equation 9-9). $F_{llb} = F_s + F_{bs} + F_a + W_b$	$F_{llb} =$ 2,010 *lb* + 6,530 *lb* + 510 *lb* + 3,000 *lb* $=$ 12,050 *lb*

SUMMARY

The immediate breakout force of the buried object is estimated to be 12,050 lb, which is less than the lifting capacity of the vessel, therefore a quick recovery is reasonable. However, for planning purposes a factor of safety of 2 should be applied, such that the lifting capacity of the equipment should be 24,000 lbs.

9.9 REFERENCES

9-1.   H. J. Lee. Unaided Breakout of Partially Embedded Objects from Cohesive Seafloor Soils, Naval Civil Engineering Laboratory, Technical Report R-755. Port Hueneme, CA, Feb 1972.

9-2.   K. Rocker, Jr. Handbook for Marine Geotechnical Engineering, Deep Ocean Technology, Naval Civil Engineering Laboratory. Port Hueneme, CA, Mar 1985.

9-3.   D.G. True. "Appendix C – Time for Breakout Using Less-Than-Immediate Breakout Forces," Naval Facilities Engineering Service Center, Deep Ocean Technology Handbook. Port Hueneme, CA, Apr 2004.

9-4.   K.D. Vaudrey. "Evaluation of Bottom Breakout Reduction Methods", Naval Civil Engineering Laboratory, Technical Note N-1227. Port Hueneme, CA, Apr 1972.

9-5.   H.J. Lee, B.D. Edwards, and H.E. Field. "Geotechnical Analysis of a Submarine Slump, Eureka, California," in Proceedings of the Offshore Technology Conference, Houston, TX, 1981 (OTC 4121).

9.10 SYMBOLS

A	Horizontal area touching soil [L^2]
A_b	Horizontal area touching soil [L^2]
A_p	Effective plan cross-sectional area of object [L^2]
B	Width of object [L]
C'	Constant [FT/L^4]
D	Depth of burial to object base [L]
F_a	Soil weight above object (plus any soil weight trapped within the object) [F]
F_{bs}	Base suction force [F]
F_{Ib}	Immediate breakout force of the soil [F]
F_{IbLR}	Long term residence immediate breakout force of the soil [F]
F_q	Maximum historical net downward force [F]
F_{Lb}	Long-term-lift breakout applied force [F]
F_{IIb}	Recovery line force for immediate breakout [F]
F_{IIbLR}	Recovery line force for long term residence immediate breakout force of the soil [F]
F_s	Side adhesive force [F]
f_{tLC}	Level of confidence time factor
g	Shear strength gradient of the cohesive soil [(F/L^2)/L]
H	Height of object [L]
L	Length of object [L]
N'_c	Dimensionless bearing capacity factor
P	Suction in pore fluid beneath base [F/L^2]
pf	Power of force ration
P_s	Side perimeter of object [L]
pt	Power of ten
R_f	Force ratio
pd'	Power of depth-to-width ratio
Q_u	Bearing capacity [F]
s_u	Shear strength [F/L^2]
$s_{u\ avg}$	Average shear strength over depth burial D [F/L^2]
s_{u0}	Shear strength of the soil at the soil surface [F/L^2]
T, T'	Time Parameters [FT/L^4]
t_b	Time to breakout [T]

t_{50}	Breakout time at a level of confidence of 50% [T]
t_{LC}	Breakout time at a level of confidence [T]
W_b	Buoyant weight of the foundation or anchor [F]
W_c	Buoyant weight of any contained or adhering soil [F]
W_s	Buoyant weight of the displaced soil [F]
γ_b	Buoyant (wet) unit weight of soil [F]
ϕ	Soil friction angle [deg]

[This page intentionally left blank]

10 SCOUR

10.1 INTRODUCTION

10.1.1 Background

Scour is the process of erosion in soil. A general lowering of the seafloor naturally occurs due to erosion of the surface soil as the ocean water flows across it. This type of scour is called general scour. When a structure, such as a pier or pipeline, is placed on or in the seafloor, an obstacle is created. The water will then have to change its pattern to move around the obstacle. This interaction can lead to local scour of the supporting soil around the obstacle. If the scour becomes too great, structural instability and failure can occur. To reduce structural failures, scour must be accounted for in the design of any marine structure.

The depth of scour can be estimated so that a structure can be designed to withstand the maximum possible scour. The methods used to estimate scour are based on equations and computer programs. Besides designing the structure to withstand maximum scour, countermeasures can be used to mitigate scour problems. The most common countermeasure consists of placing a filter and then an armor layer often made of rocks or riprap to protect the soil from eroding. The countermeasure chosen depends on the soil type, the water conditions, and the obstacle. Further discussion on scour estimates and countermeasure selection is presented later in this chapter.

10.1.2 Scope

The objectives of this chapter are to: (1) describe the fundamentals of scour and erosion, (2) provide the most common techniques to estimate scour, (3) present countermeasure methods for minimizing scour, and (4) give several examples of scour calculations.

10.2 FUNDAMENTALS OF SCOUR AND EROSION

There are three components involved in any scour problem: the soil, the water, and the obstacle. Each component plays a role in scour development. The fundamental concepts associated with each component are discussed next.

10.2.1 The Soil

10.2.1.1 Critical Shear Stress and Critical Velocity

The threshold of erosion is considered one of the most important soil parameters in studying erosion. The threshold is described as either the critical shear stress (τ_c), or as the critical velocity (V_c). Below the threshold, erosion will not occur. Above the threshold, erosion will occur.

Hjulström's diagram summarizes the relationship between grain size and water velocity (Figure 10.2-1). It specifies whether the sediment, based on the flow velocity and the diameter of the grain, will be eroded, transported, or deposited (Ref. 10-1). The top curve in Figure 10.2-1 represents the threshold for erosion.

The threshold depends largely on soil properties. For example, Figure 10.2-2 and Figure 10.2-3 from Reference 10-2 show the relationship between mean grain size (D_{50}) for a soil particle and the critical velocity and critical shear stress, respectively. For coarse-grained soils (i.e., sands), there is a fairly linear relationship with the erosion threshold. For fine-grained soils (i.e., clays), however, there is more scatter in the data. This suggests that the threshold values depend on more than the grain size.

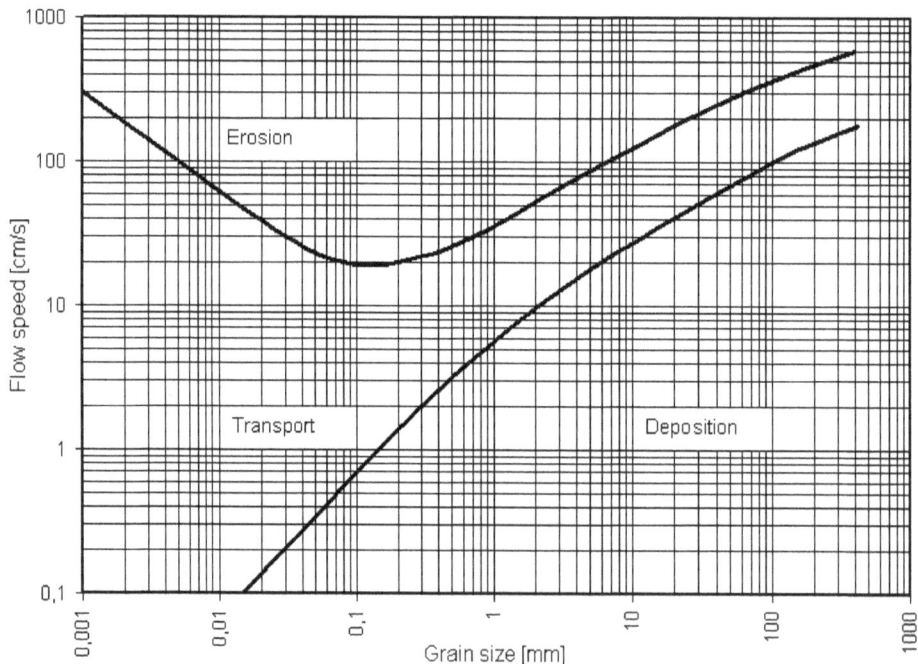

Figure 10.2-1. Average velocity as a function of mean grain size (Hjulström's diagram, Ref. 10-1).

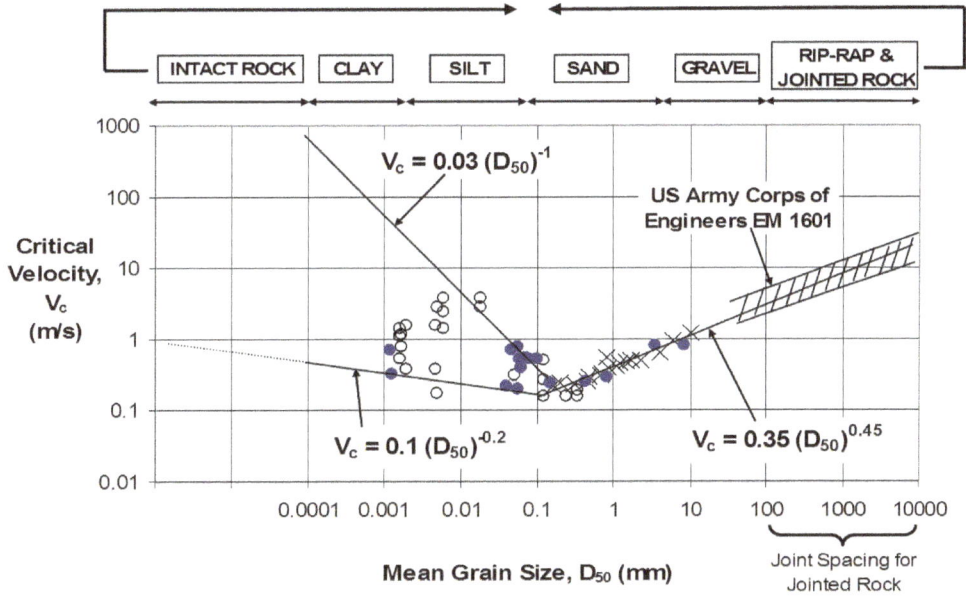

Figure 10.2-2. Critical velocity as a function of mean grain size (Ref. 10-2).

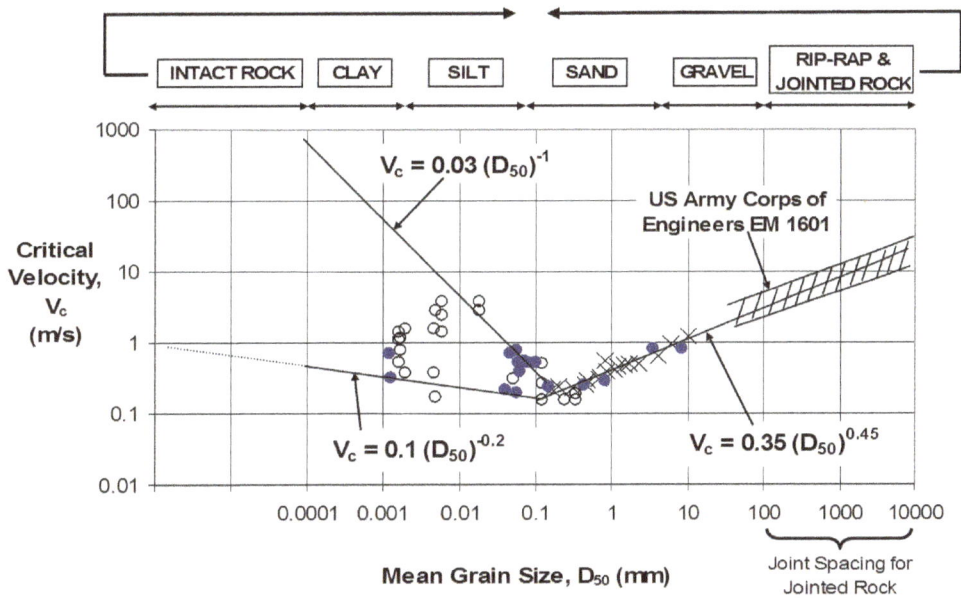

Figure 10.2-3. Critical shear stress as a function of mean grain size (Ref. 10-2).

Both the water and the soil can impact the critical shear stress and the erosion rate. For example, an increase in the salt concentration may increase the critical shear stress and decrease the erosion rate (Ref. 10-3). Correlations to soil properties other than grain size, such as undrained shear strength (s_u), plasticity index (PI), water content (w), and percent passing sieve No. 200 (#200), have been attempted yet have failed (Ref. 10-3). This is because erodibility depends on multiple soil parameters all involved in the resistance to erosion. A general consensus on the impact of certain parameters, however, is shown in Table 5.2-1. It is often preferable to measure the erosion function directly in an apparatus such as the Erosion Function Apparatus (EFA) (Ref. 10-2).

Table 10.2-1. Factors Influencing Erodibility (Ref. 10-2)

When This Parameter Increases	Erodibility
Soil unit weight	Decreases
Soil plasticity index	Decreases
Soil void ratio	Increases
Soil swell[6]	Increases
Soil percent passing sieve #200	Decreases
Soil dispersion ratio	Increases
Soil sodium absorption ratio	Increases
Soil temperature	Increases
Water temperature	Increases

10.2.1.2 Erosion Categories

Erosion categories, based on 15 years of erosion testing experience, have been proposed to serve as a classification system for the erodibility of soils (Ref. 10-2). This classification system can be presented in terms of velocity (Figure 10.2-4) or shear stress (Figure 10.2-5).

These categories are largely based on erodibility testing in the laboratory. The Erosion Function Apparatus, or EFA, was developed in the early 1990s to measure the erosion function of soil (Figure 10.2-6). Once samples from a site have been collected, the sampling tubes are brought back to the laboratory. One sampling tube is then placed in the EFA, and the sample is pushed out of the sampling tube only as fast as it is eroded by the water flowing over it. The water velocity is controlled by the user. An erosion rate is then measured for each velocity and

---

[6] Soil swell is the increase in volume of soil, usually caused by disturbance of the soil. Soil swell creates more air pockets and results in an effective increase in the soil's void volume. An increase in volume also results in a decrease in soil density.

the soil-water interface shear stress is calculated. This is repeated using different water velocities. Point by point, the erosion function is obtained (Figure 10.2-7). In the absence of lab testing, use the proposed erosion category charts (Figure 10.2-4 and Figure 10.2-5).

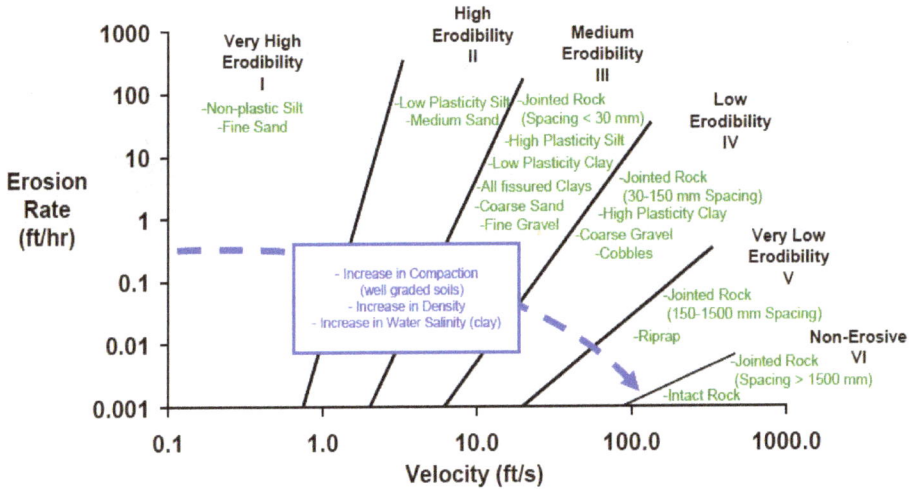

Figure 10.2-4. Proposed erosion categories for soils and rocks based on velocity (Ref. 10-2).

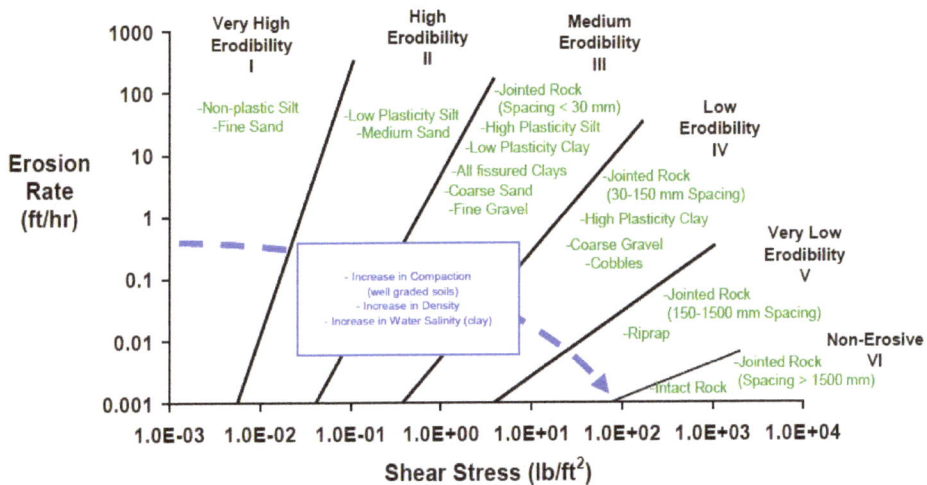

Figure 10.2-5. Proposed erosion categories for soils and rocks based on shear stress (Ref. 10-2).

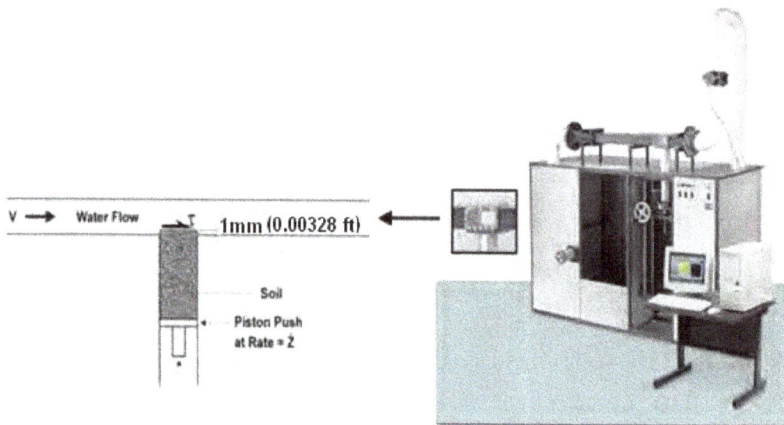

Figure 10.2-6. Erosion Function Apparatus (Ref. 10-2).

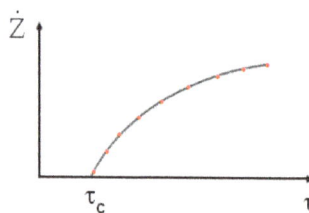

Figure 10.2-7. Erosion function as measured in the EFA.

10.2.2 The Water

10.2.2.1 Hydraulic Shear Stress

The horizontal velocity of water (V_x) is largest near the top of the water column and zero at the bottom (Figure 10.2-8). Conversely, the shear stress is largest near the bottom of the water column and zero at the top. Looking at an element of water (Figure 10.2-8), the shear stress (τ) causes a shear strain (γ).

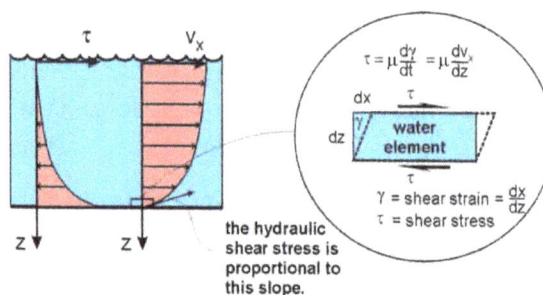

Figure 10.2-8. Velocity and shear stress profile versus flow depth (Ref. 10-2).

The shear strain represents the ratio of the change in horizontal displacement (dx) between two points to the vertical distance (dz) separating them as the element is sheared ($\gamma = dx/dz$). Since the water is flowing, the shear strain changes with time. In water, the shear stress is proportional to the rate of shear strain (Equation 10-1).

$$\tau = \mu \left(\frac{d\gamma}{dt} \right) \qquad (10\text{-}1)$$

where μ is the dynamic viscosity of water [FT/L^2], and $d\gamma/dt$ is the rate of shear strain [1/T].

The rate of shear strain can also be expressed in terms of horizontal velocity ($V_x = dx/dt$). The shear stress is thus proportional to the gradient of the velocity profile with flow depth (Equation 10-2).

$$\tau = \mu \left(\frac{dV_x}{dz} \right) \qquad (10\text{-}2)$$

The magnitude of these erosive shear stresses is fairly small as compared to those seen in other areas of geotechnical engineering (Figure 10.2-9). This difference in shear stress magnitudes is because, in erosion studies, the resistance of a single particle or small cluster of particles is examined, whereas in the other areas, the resistance of a much larger soil mass is examined.

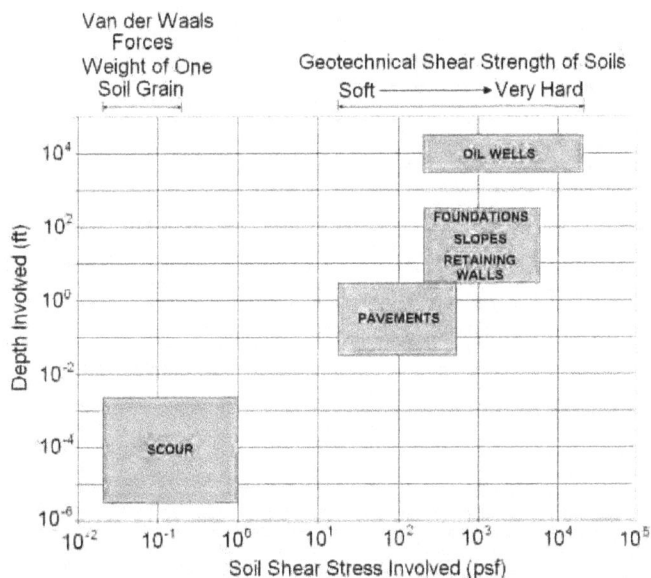

Figure 10.2-9. Range of shear stresses encountered in geotechnical engineering (Ref. 10-2).

10.2.2.2 Waves

Wave action impacts scour. Waves can be a single event such as in a flood or storm surge, or they can be the periodic gravity waves seen in open waters. Scour depth at a structure will change if waves are present as compared to scour under a current only. Wave characteristics such as the wave height (H), wave length (L), wave period (T), phase angle (θ_w), orbital velocity (V_{orb}), and semi-orbital length (a) are therefore important in wave induced scour analysis (Figure 10.2-10, Figure 10.2-11, and Figure 10.2-12). The depth of water (d) and the size or diameter of the structure the waves are impacting are also significant parameters for wave scour.

Under a current, the velocity of the water particles is basically horizontal. The water particles within a wave, however, experience an orbital velocity (V_{orb}). The amount of motion, or the size of the orbit, decreases with water depth. In shallow waters, the orbital motion is elliptical and transitions to circular in deep waters (Figure 10.2-11). Below a depth of half of the wave length ($L/2$), there is no noticeable orbital motion (Ref. 10-4).

According to Reference 10-5, the maximum local horizontal orbital velocity (V_{orb}) occurs when the wave is at its crest, or the phase angle (θ_w) is equal to 0, 2π, etc. (Figure 10.2-12). Conversely, the maximum local vertical orbital velocity ($V_{orb,v}$) occurs when the phase angle is equal to $\pi/2$, $3\pi/2$, etc. (Figure 10.2-12).

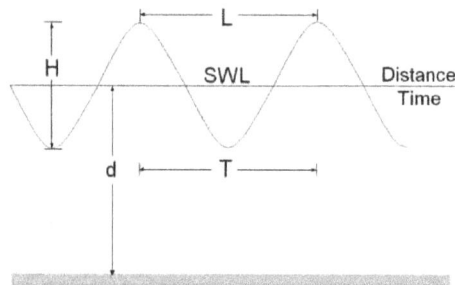

Figure 10.2-10. Wave Parameters.

Shallow-water waves occur when the water depth (d) is less than 0.05 times the wave length (L), or d < $L/20$ (Ref. 10-5). Deep-water waves occur when the water depth is greater than 0.5 times the wave length, or $d > L/2$. In between these two ($L/20 < d < L/2$), transitional waves occur. It is important to remember that deep and shallow-water waves are not defined by the absolute depth of water, but by the ratio of water depth to wave length (d/L). Oftentimes, the deepwater wave length (L_o) is known, but the wave length at a specific depth is unknown. The ratio of the water depth to the wave length d/L can easily be estimated using Table 10.2-2.

For each case (shallow, deep, or transitional), some water characteristics will change (Table 10.2-3). The variable z in this table represents the depth of interest (Figure 10.2-11). For example, when looking at the maximum horizontal orbital velocity at the bed ($V_{orb,bm}$), z in the "Wave Particle Horizontal Orbital Velocity" equation given by Table 10.2-3 equals the negative total water depth ($z = -d$), and the phase angle (θ_w) is equal to 0, 2π, etc. (Figure 10.2-12).

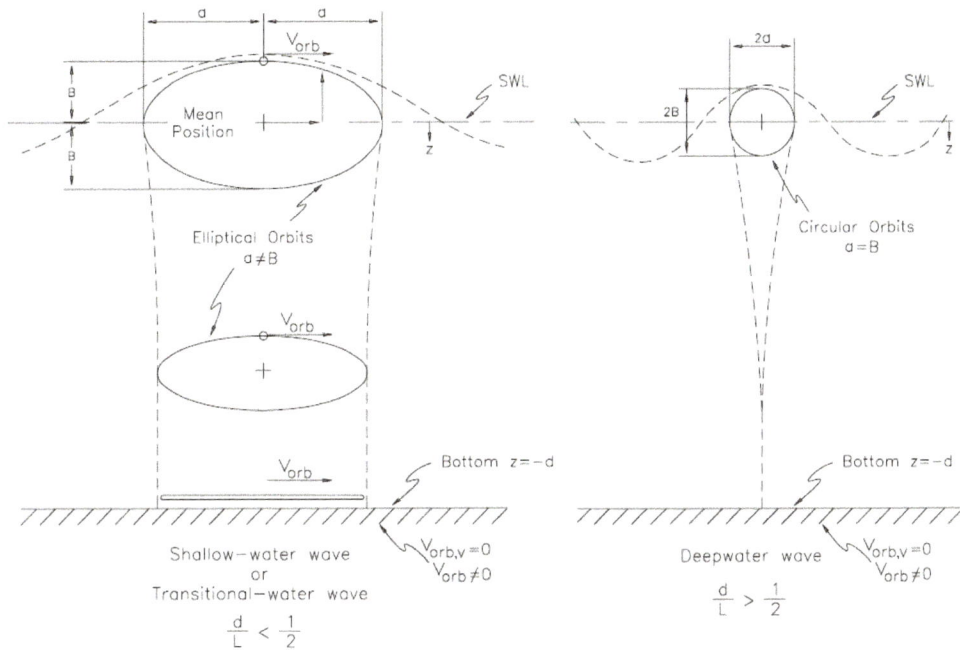

Figure 10.2-11. Wave orbital velocity description (Ref. 10-5).

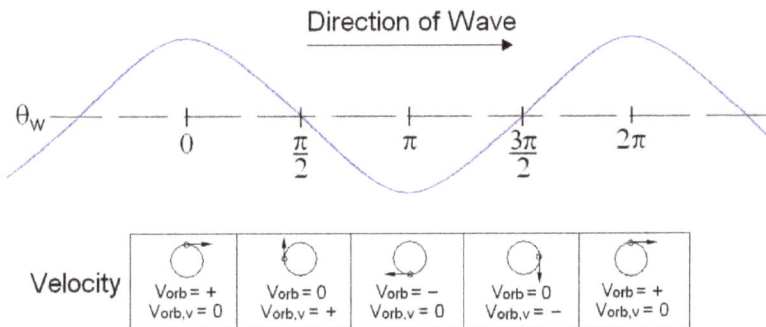

Figure 10.2-12. Local Wave Velocities (Ref. 10-5).

The equations presented in Table 10.2-3 assume a linear wave theory to predict and describe wave behavior. Linear theory is valid when the wave height is small relative to the wave length. Higher-order wave theories are available (Ref. 10-5) and should be used when the problem requires more accuracy.

The Keulegan-Carpenter number (KC) is an important parameter in marine scour. The KC number (Equation 10-3) governs the wake pattern produced by oscillatory flow, such as seen in waves (Ref. 10-6).

$$KC = \frac{V_m T}{D} \qquad\qquad (10\text{-}3)$$

where:

V_m = maximum velocity of interest (e.g., the maximum orbital velocity at the bed, $V_{orb,bm}$) [L/T]

T = wave period [T]

D = diameter of the structure [L]

Small KC numbers ($KC < 6$) indicate that the orbital motion is small in comparison to the size of the structure. Large KC numbers ($KC \geq 6$) indicate that the orbital motion is large in comparison to the size of the structure (Ref. 10-7). When this occurs, vortices will form around the structure leading to erosion (Ref. 10-6).

Table 10.2-2. Relationship Between d/L_0 and d/L

d/L_0	d/L	d/L_0	d/L	d/L_0	d/L	d/L_0	d/L	d/L_0	d/L	d/L_0	d/L	d/L_0	d/L	d/L_0	d/L	d/L_0	d/L	d/L_0	d/L
0.0000	0.0000	0.0079	0.0358	0.0680	0.1120	0.1470	0.1808	0.2260	0.2472	0.3050	0.3166	0.3840	0.3898	0.4630	0.4657	0.5420	0.5432	0.6210	0.6215
0.0001	0.0040	0.0080	0.0360	0.0690	0.1130	0.1480	0.1816	0.2270	0.2480	0.3060	0.3175	0.3850	0.3907	0.4640	0.4666	0.5430	0.5442	0.6220	0.6225
0.0002	0.0056	0.0081	0.0362	0.0700	0.1139	0.1490	0.1825	0.2280	0.2489	0.3070	0.3184	0.3860	0.3917	0.4650	0.4676	0.5440	0.5452	0.6230	0.6235
0.0003	0.0069	0.0082	0.0364	0.0710	0.1149	0.1500	0.1833	0.2290	0.2498	0.3080	0.3193	0.3870	0.3926	0.4660	0.4686	0.5450	0.5461	0.6240	0.6245
0.0004	0.0080	0.0083	0.0367	0.0720	0.1158	0.1510	0.1841	0.2300	0.2506	0.3090	0.3202	0.3880	0.3936	0.4670	0.4696	0.5460	0.5471	0.6250	0.6255
0.0005	0.0089	0.0084	0.0369	0.0730	0.1168	0.1520	0.1850	0.2310	0.2515	0.3100	0.3212	0.3890	0.3945	0.4680	0.4705	0.5470	0.5481	0.6260	0.6265
0.0006	0.0098	0.0085	0.0371	0.0740	0.1177	0.1530	0.1858	0.2320	0.2523	0.3110	0.3221	0.3900	0.3955	0.4690	0.4715	0.5480	0.5491	0.6270	0.6275
0.0007	0.0106	0.0086	0.0373	0.0750	0.1186	0.1540	0.1866	0.2330	0.2532	0.3120	0.3230	0.3910	0.3964	0.4700	0.4725	0.5490	0.5501	0.6280	0.6285
0.0008	0.0113	0.0087	0.0376	0.0760	0.1195	0.1550	0.1875	0.2340	0.2540	0.3130	0.3239	0.3920	0.3974	0.4710	0.4735	0.5500	0.5511	0.6290	0.6295
0.0009	0.0120	0.0088	0.0378	0.0770	0.1205	0.1560	0.1883	0.2350	0.2549	0.3140	0.3248	0.3930	0.3983	0.4720	0.4744	0.5510	0.5521	0.6300	0.6305
0.0010	0.0126	0.0089	0.0380	0.0780	0.1214	0.1570	0.1891	0.2360	0.2558	0.3150	0.3257	0.3940	0.3993	0.4730	0.4754	0.5520	0.5531	0.6310	0.6315
0.0011	0.0132	0.0090	0.0382	0.0790	0.1223	0.1580	0.1900	0.2370	0.2566	0.3160	0.3266	0.3950	0.4002	0.4740	0.4764	0.5530	0.5540	0.6320	0.6324
0.0012	0.0138	0.0091	0.0384	0.0800	0.1232	0.1590	0.1908	0.2380	0.2575	0.3170	0.3275	0.3960	0.4012	0.4750	0.4774	0.5540	0.5550	0.6330	0.6334
0.0013	0.0144	0.0092	0.0386	0.0810	0.1241	0.1600	0.1916	0.2390	0.2584	0.3180	0.3284	0.3970	0.4021	0.4760	0.4783	0.5550	0.5560	0.6340	0.6344
0.0014	0.0149	0.0093	0.0389	0.0820	0.1250	0.1610	0.1925	0.2400	0.2592	0.3190	0.3293	0.3980	0.4031	0.4770	0.4793	0.5560	0.5570	0.6350	0.6354
0.0015	0.0155	0.0094	0.0391	0.0830	0.1259	0.1620	0.1933	0.2410	0.2601	0.3200	0.3303	0.3990	0.4040	0.4780	0.4803	0.5570	0.5580	0.6360	0.6364
0.0016	0.0160	0.0095	0.0393	0.0840	0.1268	0.1630	0.1941	0.2420	0.2609	0.3210	0.3312	0.4000	0.4050	0.4790	0.4813	0.5580	0.5590	0.6370	0.6374
0.0017	0.0165	0.0096	0.0395	0.0850	0.1277	0.1640	0.1950	0.2430	0.2618	0.3220	0.3321	0.4010	0.4059	0.4800	0.4822	0.5590	0.5600	0.6380	0.6384
0.0018	0.0170	0.0097	0.0397	0.0860	0.1286	0.1650	0.1958	0.2440	0.2627	0.3230	0.3330	0.4020	0.4069	0.4810	0.4832	0.5600	0.5610	0.6390	0.6394
0.0019	0.0174	0.0098	0.0399	0.0870	0.1295	0.1660	0.1966	0.2450	0.2635	0.3240	0.3339	0.4030	0.4078	0.4820	0.4842	0.5610	0.5620	0.6400	0.6404
0.0020	0.0179	0.0099	0.0401	0.0880	0.1304	0.1670	0.1975	0.2460	0.2644	0.3250	0.3348	0.4040	0.4088	0.4830	0.4852	0.5620	0.5630	0.6410	0.6414
0.0021	0.0183	0.0100	0.0403	0.0890	0.1313	0.1680	0.1983	0.2470	0.2653	0.3260	0.3357	0.4050	0.4097	0.4840	0.4862	0.5630	0.5639	0.6420	0.6424
0.0022	0.0188	0.0110	0.0423	0.0900	0.1322	0.1690	0.1991	0.2480	0.2661	0.3270	0.3367	0.4060	0.4107	0.4850	0.4871	0.5640	0.5649	0.6430	0.6434
0.0023	0.0192	0.0120	0.0443	0.0910	0.1331	0.1700	0.2000	0.2490	0.2670	0.3280	0.3376	0.4070	0.4116	0.4860	0.4881	0.5650	0.5659	0.6440	0.6444
0.0024	0.0196	0.0130	0.0461	0.0920	0.1340	0.1710	0.2008	0.2500	0.2679	0.3290	0.3385	0.4080	0.4126	0.4870	0.4891	0.5660	0.5669	0.6450	0.6454
0.0025	0.0200	0.0140	0.0479	0.0930	0.1349	0.1720	0.2016	0.2510	0.2687	0.3300	0.3394	0.4090	0.4136	0.4880	0.4901	0.5670	0.5679	0.6460	0.6464
0.0026	0.0204	0.0150	0.0496	0.0940	0.1357	0.1730	0.2025	0.2520	0.2696	0.3310	0.3403	0.4100	0.4145	0.4890	0.4910	0.5680	0.5689	0.6470	0.6474
0.0027	0.0208	0.0160	0.0513	0.0950	0.1366	0.1740	0.2033	0.2530	0.2705	0.3320	0.3412	0.4110	0.4155	0.4900	0.4920	0.5690	0.5699	0.6480	0.6484
0.0028	0.0212	0.0170	0.0530	0.0960	0.1375	0.1750	0.2042	0.2540	0.2714	0.3330	0.3422	0.4120	0.4164	0.4910	0.4930	0.5700	0.5709	0.6490	0.6494
0.0029	0.0215	0.0180	0.0546	0.0970	0.1384	0.1760	0.2050	0.2550	0.2722	0.3340	0.3431	0.4130	0.4174	0.4920	0.4940	0.5710	0.5719	0.6500	0.6504
0.0030	0.0219	0.0190	0.0561	0.0980	0.1392	0.1770	0.2058	0.2560	0.2731	0.3350	0.3440	0.4140	0.4183	0.4930	0.4950	0.5720	0.5729	0.6510	0.6514
0.0031	0.0223	0.0200	0.0576	0.0990	0.1401	0.1780	0.2067	0.2570	0.2740	0.3360	0.3449	0.4150	0.4193	0.4940	0.4959	0.5730	0.5738	0.6520	0.6524
0.0032	0.0226	0.0210	0.0591	0.1000	0.1410	0.1790	0.2075	0.2580	0.2749	0.3370	0.3458	0.4160	0.4203	0.4950	0.4969	0.5740	0.5748	0.6530	0.6534
0.0033	0.0230	0.0220	0.0606	0.1010	0.1418	0.1800	0.2083	0.2590	0.2757	0.3380	0.3468	0.4170	0.4212	0.4960	0.4979	0.5750	0.5758	0.6540	0.6544
0.0034	0.0233	0.0230	0.0620	0.1020	0.1427	0.1810	0.2092	0.2600	0.2766	0.3390	0.3477	0.4180	0.4222	0.4970	0.4989	0.5760	0.5768	0.6550	0.6553
0.0035	0.0237	0.0240	0.0634	0.1030	0.1436	0.1820	0.2100	0.2610	0.2775	0.3400	0.3486	0.4190	0.4231	0.4980	0.4999	0.5770	0.5778	0.6560	0.6563
0.0036	0.0240	0.0250	0.0648	0.1040	0.1444	0.1830	0.2108	0.2620	0.2784	0.3410	0.3495	0.4200	0.4241	0.4990	0.5008	0.5780	0.5788	0.6570	0.6573
0.0037	0.0244	0.0260	0.0661	0.1050	0.1453	0.1840	0.2117	0.2630	0.2792	0.3420	0.3505	0.4210	0.4251	0.5000	0.5018	0.5790	0.5798	0.6580	0.6583
0.0038	0.0247	0.0270	0.0675	0.1060	0.1462	0.1850	0.2125	0.2640	0.2801	0.3430	0.3514	0.4220	0.4260	0.5010	0.5028	0.5800	0.5808	0.6590	0.6593
0.0039	0.0250	0.0280	0.0688	0.1070	0.1470	0.1860	0.2134	0.2650	0.2810	0.3440	0.3523	0.4230	0.4270	0.5020	0.5038	0.5810	0.5818	0.6600	0.6603
0.0040	0.0253	0.0290	0.0701	0.1080	0.1479	0.1870	0.2142	0.2660	0.2819	0.3450	0.3532	0.4240	0.4279	0.5030	0.5048	0.5820	0.5828	0.6700	0.6703
0.0041	0.0257	0.0300	0.0713	0.1090	0.1488	0.1880	0.2150	0.2670	0.2827	0.3460	0.3542	0.4250	0.4289	0.5040	0.5058	0.5830	0.5838	0.6800	0.6803
0.0042	0.0260	0.0310	0.0726	0.1100	0.1496	0.1890	0.2159	0.2680	0.2836	0.3470	0.3551	0.4260	0.4299	0.5050	0.5067	0.5840	0.5848	0.6900	0.6902
0.0043	0.0263	0.0320	0.0738	0.1110	0.1505	0.1900	0.2167	0.2690	0.2845	0.3480	0.3560	0.4270	0.4308	0.5060	0.5077	0.5850	0.5857	0.7000	0.7002
0.0044	0.0266	0.0330	0.0751	0.1120	0.1513	0.1910	0.2175	0.2700	0.2854	0.3490	0.3570	0.4280	0.4318	0.5070	0.5087	0.5860	0.5867	0.7100	0.7102
0.0045	0.0269	0.0340	0.0763	0.1130	0.1522	0.1920	0.2184	0.2710	0.2863	0.3500	0.3579	0.4290	0.4327	0.5080	0.5097	0.5870	0.5877	0.7200	0.7202
0.0046	0.0272	0.0350	0.0775	0.1140	0.1530	0.1930	0.2192	0.2720	0.2871	0.3510	0.3588	0.4300	0.4337	0.5090	0.5107	0.5880	0.5887	0.7300	0.7302
0.0047	0.0275	0.0360	0.0787	0.1150	0.1539	0.1940	0.2201	0.2730	0.2880	0.3520	0.3597	0.4310	0.4347	0.5100	0.5118	0.5890	0.5897	0.7400	0.7401
0.0048	0.0278	0.0370	0.0798	0.1160	0.1547	0.1950	0.2209	0.2740	0.2889	0.3530	0.3607	0.4320	0.4356	0.5110	0.5126	0.5900	0.5907	0.7500	0.7501
0.0049	0.0281	0.0380	0.0810	0.1170	0.1556	0.1960	0.2217	0.2750	0.2898	0.3540	0.3616	0.4330	0.4366	0.5120	0.5136	0.5910	0.5917	0.7600	0.7601
0.0050	0.0284	0.0390	0.0822	0.1180	0.1564	0.1970	0.2226	0.2760	0.2907	0.3550	0.3625	0.4340	0.4376	0.5130	0.5146	0.5920	0.5927	0.7700	0.7701
0.0051	0.0286	0.0400	0.0833	0.1190	0.1573	0.1980	0.2234	0.2770	0.2916	0.3560	0.3635	0.4350	0.4385	0.5140	0.5156	0.5930	0.5937	0.7800	0.7801
0.0052	0.0289	0.0410	0.0844	0.1200	0.1581	0.1990	0.2243	0.2780	0.2925	0.3570	0.3644	0.4360	0.4395	0.5150	0.5166	0.5940	0.5947	0.7900	0.7901
0.0053	0.0292	0.0420	0.0855	0.1210	0.1590	0.2000	0.2251	0.2790	0.2933	0.3580	0.3653	0.4370	0.4405	0.5160	0.5175	0.5950	0.5957	0.8000	0.8001
0.0054	0.0295	0.0430	0.0866	0.1220	0.1598	0.2010	0.2260	0.2800	0.2942	0.3590	0.3663	0.4380	0.4414	0.5170	0.5185	0.5960	0.5967	0.8100	0.8101
0.0055	0.0298	0.0440	0.0877	0.1230	0.1607	0.2020	0.2268	0.2810	0.2951	0.3600	0.3672	0.4390	0.4424	0.5180	0.5195	0.5970	0.5977	0.8200	0.8201
0.0056	0.0300	0.0450	0.0888	0.1240	0.1615	0.2030	0.2276	0.2820	0.2960	0.3610	0.3681	0.4400	0.4434	0.5190	0.5205	0.5980	0.5986	0.8300	0.8300
0.0057	0.0303	0.0460	0.0899	0.1250	0.1624	0.2040	0.2285	0.2830	0.2969	0.3620	0.3691	0.4410	0.4443	0.5200	0.5215	0.5990	0.5996	0.8400	0.8400
0.0058	0.0306	0.0470	0.0910	0.1260	0.1632	0.2050	0.2293	0.2840	0.2978	0.3630	0.3700	0.4420	0.4453	0.5210	0.5225	0.6000	0.6006	0.8500	0.8500
0.0059	0.0308	0.0480	0.0920	0.1270	0.1640	0.2060	0.2302	0.2850	0.2987	0.3640	0.3709	0.4430	0.4463	0.5220	0.5235	0.6010	0.6016	0.8600	0.8600
0.0060	0.0311	0.0490	0.0931	0.1280	0.1649	0.2070	0.2310	0.2860	0.2996	0.3650	0.3719	0.4440	0.4472	0.5230	0.5244	0.6020	0.6026	0.8700	0.8700
0.0061	0.0314	0.0500	0.0942	0.1290	0.1657	0.2080	0.2319	0.2870	0.3005	0.3660	0.3728	0.4450	0.4482	0.5240	0.5254	0.6030	0.6036	0.8800	0.8800
0.0062	0.0316	0.0510	0.0952	0.1300	0.1666	0.2090	0.2327	0.2880	0.3014	0.3670	0.3738	0.4460	0.4492	0.5250	0.5264	0.6040	0.6046	0.8900	0.8900
0.0063	0.0319	0.0520	0.0962	0.1310	0.1674	0.2100	0.2336	0.2890	0.3023	0.3680	0.3747	0.4470	0.4501	0.5260	0.5274	0.6050	0.6056	0.9000	0.9000
0.0064	0.0321	0.0530	0.0973	0.1320	0.1682	0.2110	0.2344	0.2900	0.3031	0.3690	0.3756	0.4480	0.4511	0.5270	0.5284	0.6060	0.6066	0.9100	0.9100
0.0065	0.0324	0.0540	0.0983	0.1330	0.1691	0.2120	0.2353	0.2910	0.3040	0.3700	0.3766	0.4490	0.4521	0.5280	0.5294	0.6070	0.6076	0.9200	0.9200
0.0066	0.0326	0.0550	0.0993	0.1340	0.1699	0.2130	0.2361	0.2920	0.3049	0.3710	0.3775	0.4500	0.4530	0.5290	0.5304	0.6080	0.6086	0.9300	0.9300
0.0067	0.0329	0.0560	0.1003	0.1350	0.1708	0.2140	0.2370	0.2930	0.3058	0.3720	0.3785	0.4510	0.4540	0.5300	0.5313	0.6090	0.6096	0.9400	0.9400
0.0068	0.0331	0.0570	0.1013	0.1360	0.1716	0.2150	0.2378	0.2940	0.3067	0.3730	0.3794	0.4520	0.4550	0.5310	0.5323	0.6100	0.6106	0.9500	0.9500
0.0069	0.0334	0.0580	0.1023	0.1370	0.1724	0.2160	0.2387	0.2950	0.3076	0.3740	0.3803	0.4530	0.4560	0.5320	0.5333	0.6110	0.6116	0.9600	0.9600
0.0070	0.0336	0.0590	0.1033	0.1380	0.1733	0.2170	0.2395	0.2960	0.3085	0.3750	0.3813	0.4540	0.4569	0.5330	0.5343	0.6120	0.6126	0.9700	0.9700
0.0071	0.0339	0.0600	0.1043	0.1390	0.1741	0.2180	0.2404	0.2970	0.3094	0.3760	0.3822	0.4550	0.4579	0.5340	0.5353	0.6130	0.6135	0.9800	0.9800
0.0072	0.0341	0.0610	0.1053	0.1400	0.1750	0.2190	0.2412	0.2980	0.3103	0.3770	0.3832	0.4560	0.4589	0.5350	0.5363	0.6140	0.6145	0.9900	0.9900
0.0073	0.0343	0.0620	0.1063	0.1410	0.1758	0.2200	0.2421	0.2990	0.3112	0.3780	0.3841	0.4570	0.4598	0.5360	0.5373	0.6150	0.6155	1.0000	1.0000
0.0074	0.0346	0.0630	0.1072	0.1420	0.1766	0.2210	0.2429	0.3000	0.3121	0.3790	0.3850	0.4580	0.4608	0.5370	0.5382	0.6160	0.6165		
0.0075	0.0348	0.0640	0.1082	0.1430	0.1775	0.2220	0.2438	0.3010	0.3130	0.3800	0.3860	0.4590	0.4618	0.5380	0.5392	0.6170	0.6175		
0.0076	0.0351	0.0650	0.1092	0.1440	0.1783	0.2230	0.2446	0.3020	0.3139	0.3810	0.3869	0.4600	0.4628	0.5390	0.5402	0.6180	0.6185		
0.0077	0.0353	0.0660	0.1101	0.1450	0.1791	0.2240	0.2455	0.3030	0.3148	0.3820	0.3879	0.4610	0.4637	0.5400	0.5412	0.6190	0.6195		
0.0078	0.0355	0.0670	0.1111	0.1460	0.1800	0.2250	0.2463	0.3040	0.3157	0.3830	0.3888	0.4620	0.4647	0.5410	0.5422	0.6200	0.6205		

Table 10.2-3. Important Wave Parameters – Linear Wave Theory (Ref. 10-5).

Wave Parameter	Shallow Water $d < 1/20L$	Transitional Water $1/20L < d < 1/2L$	Deep Water $d > 1/2L$
Wave Length (L)	$L = T\sqrt{gD}$	$L = \dfrac{gT^2}{2\pi}\tanh\left(\dfrac{2\pi d}{L}\right)$	$L = \dfrac{gT^2}{2\pi}$
Wave Particle Horizontal Orbital Velocity (V_{orb})	$V_{orb} = \dfrac{H}{2}\sqrt{\dfrac{g}{d}}\cos\theta_w$	$V_{orb} = \dfrac{H}{2}\dfrac{gT}{L}\dfrac{\cosh\left[2\pi\dfrac{(z+d)}{L}\right]}{\cosh\left(2\pi\dfrac{d}{L}\right)}\cos\theta_w$	$V_{orb} = \dfrac{\pi H}{T}e^{\left(\frac{2\pi z}{L}\right)}\cos\theta_w$
Wave Particle Vertical Orbital Velocity $(V_{orb,v})$	$V_{orb,v} = \dfrac{H\pi}{T}\left(1+\dfrac{z}{d}\right)\sin\theta_w$	$V_{orb,v} = \dfrac{H}{2}\dfrac{gT}{L}\dfrac{\sinh\left[2\pi\dfrac{(z+d)}{L}\right]}{\cosh\left(2\pi\dfrac{d}{L}\right)}\sin\theta_w$	$V_{orb,v} = \dfrac{\pi H}{T}e^{\left(\frac{2\pi z}{L}\right)}\sin\theta_w$
Semi-Orbital Length (a)	$a = \dfrac{H}{2}\dfrac{L}{2\pi d}$	$a = \dfrac{H}{2}\dfrac{gT}{L}\dfrac{\cosh\left[2\pi\dfrac{(z+d)}{L}\right]}{\sinh\left(2\pi\dfrac{d}{L}\right)}$	$a = \dfrac{H}{2}e^{\left(\frac{2\pi z}{L}\right)}$
Semi-Orbital Width (B)	$B = \dfrac{H}{2}\left(1+\dfrac{z}{d}\right)$	$B = \dfrac{H}{2}\dfrac{gT}{L}\dfrac{\sinh\left[2\pi\dfrac{(z+d)}{L}\right]}{\sinh\left(2\pi\dfrac{d}{L}\right)}$	$B = \dfrac{H}{2}e^{\left(\frac{2\pi z}{L}\right)}$

10.2.2.3 Ship Propellers

Propeller-induced scour may occur and must be considered in the design of marine structures. The propeller of a ship or vessel produces a turbulent jet that erodes the soil if the velocities exceed the critical velocity of the soil. Propellers can produce speeds of up to 13-26 ft/s (Ref. 10-8). By comparison, tidal currents are about 3-7 ft/s. The propeller draws fluid in during motion with the flow accelerating into the jet up to a distance of approximately twice the propeller diameter downstream (Ref. 10-9). The velocity field also spreads out from the propeller in a cone shape, with the velocity decreasing with increasing distance from the propeller. The maximum velocity near the bed $(V_{b,max})$ due to the propeller can be estimated from Equation 10-4 (Ref. 10-10) using the units indicated.

$$V_{b,max}\left(\frac{ft}{s}\right) = C_1 V_0\left(\frac{ft}{s}\right)\frac{D_p(ft)}{H_p(ft)} \tag{10-4}$$

where:

V_0 = initial centerline propeller velocity defined by Equation 10-5 (Ref. 10-10) (ft/s)

C_1 = 0.22 for a non-ducted propeller and 0.30 for a ducted propeller

D_p = propeller diameter (ft)

H_p = vertical distance from the center of the propeller shaft to the channel bottom (ft)

Note that this Equation 10-4 is valid only when the diameter of the propeller is less than or equal to 1.2 times the vertical distance from the center of the propeller shaft to the channel bottom ($D_p/H_p \leq 1.2$). Also note that in Equation 10-5, the units indicated must be used for each variable.

$$V_o(ft/s) = F\left(\frac{P_d(Hp)}{D_p^2(ft)}\right)^{\frac{1}{3}}$$ (10-5)

where:

F = 9.72 for a free or non-ducted propeller and 7.68 for a propeller in a nozzle or a ducted propeller

P_d = engine power (Hp)

The location of the maximum velocity (X) is found at a horizontal distance behind the propeller equal to 4 to 10 times the height of the propeller (H_p) (Figure 10.2-13). The scour hole tends to develop symmetrically about the centerline of the propeller wash. Erosion initially takes place close to the propeller and then deposition of some of that sediment takes place at the end of the scour hole (Figure 10.2-14a).

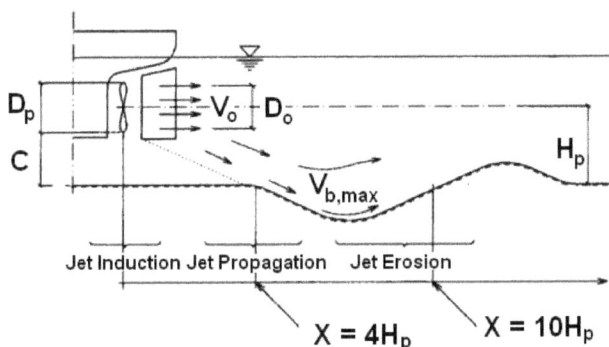

Figure 10.2-13. Schematic of unconfined propeller jet (Ref. 10-8).

10-13

When there is a quay wall (Figure 10.2-15) or other obstruction providing confinement to the propeller wash, the scour profile will change (Figure 10.2-14b). The scour depth will increase at the base of the wall and the scour hole will widen along the centerline of the wash (Ref. 10-11).

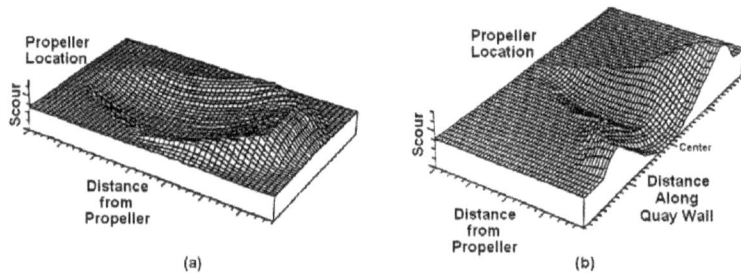

Figure 10.2-14. Scour hole produced by propeller wash (a) unconfined (b) confined (Ref. 10-11).

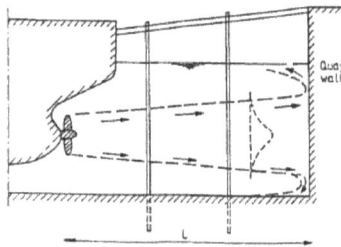

Figure 10.2-15. Confined propeller jet (Ref. 10-11).

10.2.3 The Obstacle

An obstacle located in the water will cause a disturbance in the normal flow pattern. When the water reaches the obstacle, it must flow around it. To maintain the same flow rate, the water has to accelerate around the obstacle. This acceleration can result in a local velocity which can be 1.5 times higher than the approach velocity. If the velocity exceeds the critical velocity (V_c), scour will occur around the obstacle.

If the approach velocity is lower than the critical velocity, but the local velocity at the obstacle is higher than the critical velocity, clear-water scour will occur. Clear-water scour refers to scour created by water which either does not carry any soil particles or carries a small number of soil particles which remain in suspension. If the approach velocity and the local velocity are both higher than the critical velocity, live-bed scour will occur. Live-bed scour refers to scour created by water which is carrying a significant amount of soil particles, some of which fall back on the soil surface. Typically, live-bed scour produces a scour depth smaller than clear water scour (Figure 10.2-16). This is because some of the particles in suspension during live bed scour will fall down thereby reducing the size of scour.

10-14

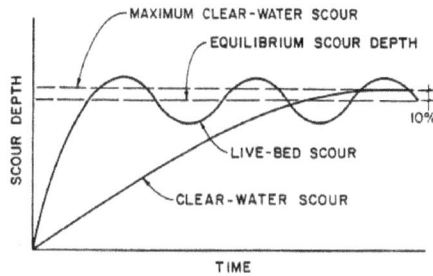

Figure 10.2-16. Clear water vs. live bed scour (Ref. 10-12).

10.2.3.1 Piers

Piers are solid objects piercing the water surface and founded in the bed soil. The reference pier in most pier scour calculations is a circular cylinder in deep water. The method to calculate pier scour in cohesionless coarse-grained soils, such as sands and gravels, is described in the Federal Highway Administration (FHWA) Hydraulic Engineering Circular No. 18 (Ref. 10-12). This method will be referred to as the "HEC-18 Sand" method throughout the chapter.

The method to calculate pier scour in cohesive fine-grained soils, such as clays and silts, was developed by the Transportation Research Board (TRB) under their National Cooperative Highway Research Program (NCHRP) in 2004. It is called the Scour Rate in Cohesive Soils–Erosion Function Apparatus Method (SRICOS-EFA) and is described in NCHRP Report 516 (Ref. 10-13). It is also part of the FHWA-HEC 18. For this chapter, the SRICOS-EFA method will be referred to as the "HEC-18 Clay" method.

For each method, correction factors must be applied to the calculated scour depth to account for differences between the pier considered and the reference pier. This can include modifications in pier shape, angle of attack, bed grain size, water depth, and pier spacing. These methods are described below.

10.2.3.1.1 Piers in Coarse-Grained Soils: Correction Factors

Before calculating the pier scour depth by the HEC-18 Sand method, described in Section 10.4.1.1, several factors that impact pier scour must be investigated. The shape and size of a pier greatly influences the flow of water around it. The most common pier shapes are circular and rectangular. Circular piers tend to have less scour than rectangular piers, with square piers experiencing the greatest scour depth (Figure 10.2-17). For a rectangular pier, L_p is the length and B_p is the width ($L_p/B_p = 1$ for a square pier).

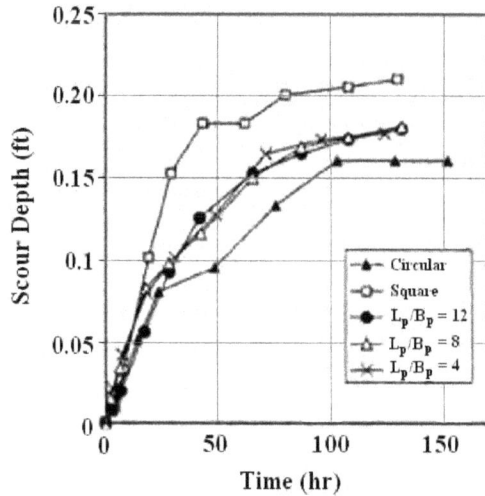

Figure 10.2-17. Scour depth versus time curves from flume tests (Ref. 10-13).

Scour calculations for pier scour are based on a circular pier. To account for the rectangular shape, a shape correction factor (K_{sh}) is applied to the calculated maximum pier scour depth ($y_{s,pier}$). Typically, rectangular piers (L_p/B_p >1) are installed with their length parallel to the major flow direction. In this case, the length of the pier (L_p) does not have a great influence (Figure 10.2-17). Therefore, for L_p/B_p >1, a good approximation for the shape factor is 1.1.

The shape of the nose for the pier also impacts scour. The nose shape is typically the first interaction between the water and the obstacle. The reference pier is circular so there is no correction factor needed for this standard case. For different pier nose shapes (Figure 10.2-18), recommended shape correction factors are given in Table 10.2-4.

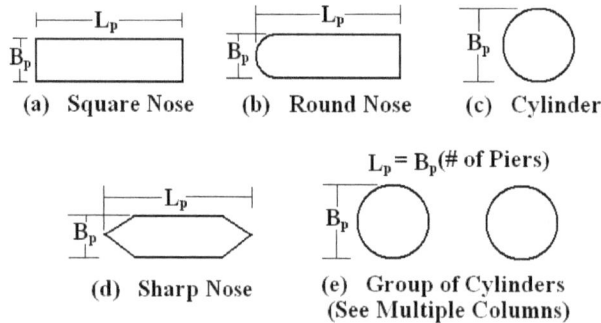

Figure 10.2-18. Common pier shapes (Ref. 10-12).

Table 10.2-4. Correction factor for pier nose shape (Ref. 10-12).

Shape of Pier Nose	Correction Factor (K_{sh})
Square	1.1
Round	1.0
Circular Cylinder	1.0
Sharp	0.9
Group of Cylinders	1.0

Piers are not always placed normal to the flow of water; sometimes, they are skewed. The attack angle (Δ) is the angle between the length direction of the rectangular pier and the flow direction (Figure 10.2-19). The correction factor for the attack angle (K_a) can be calculated according to Equation 10-6.

$$K_a = \left(\frac{L_p}{B_p} \sin \Delta + \cos \Delta \right)^{0.65}$$ (10-6)

where:

L_p = length of the pier [L]

B_p = width if the pier [L]

Δ = attack angle [deg]

The limiting L_p/B_p ratio is 16. If L_p/B_p is greater than 16, the ratio should be reduced to 16 for use in Equation 10-6.

The attack angle correction factor increases with increasing attack angle (Figure 10.2-20). The scour depth therefore increases as the pier is more skewed. For attack angles greater than 5°, K_a will dominate and K_{sh} should be taken as 1.0. Otherwise ($\Delta < 5°$), apply both factors as needed.

The condition of the seafloor or channel bed must also be taken into account (Figure 10.2-21). A correction factor for bed conditions is typically based on dune height. Dunes are repeating hills formed during sediment transport, or the movement of soil across the bed. If present, their effect must be accounted for by using a bed condition correction factor (K_b).Table 10.2-5 provides some recommended values for K_b.

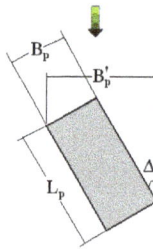

Figure 10.2-19. Attack angle (Δ) definition (Ref. 10-14).

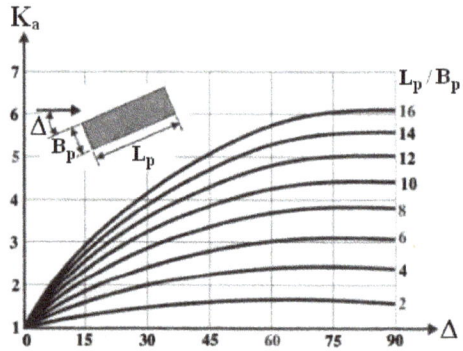

Figure 10.2-20. Correction factor (K_a) for attack angle (Ref. 10-15).

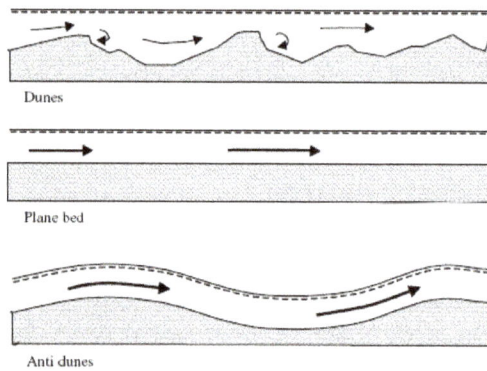

Figure 10.2-21. Bed forms (Ref. 10-16).

Table 10.2-5. Correction factor for bed conditions (Ref. 10-12).

Bed Conditions	Dune Height (ft)	K_b
Clear-Water Scour	N/A	1.1
Plane Bed and Anti-Dune Flow	N/A	1.1
Small Dunes	2-10	1.1
Medium Dunes	10-30	1.1-1.2
Large Dunes	≥ 30	1.3

A correction factor for armoring by bed material size (K_{ar}) is another factor that must be taken into account when considering pier scour in coarse-grained soils. This factor is beneficial as it reduces the scour depth. It accounts for the fact that as finer particles erode, the coarser particles form an armor on the bed and decrease scour. If the median diameter of the bed material (D_{50}) is less than 0.0066 feet or the diameter of the bed particles for which 95% are smaller (D_{95}) is less than 0.067 feet, then K_{ar} is equal to unity. If these conditions are not met, then K_{ar} is found according to Equation 10-7. The minimum value for K_{ar} is 0.4.

$$K_{ar} = 0.4 \left[\frac{V_1 - 0.645 \left(\dfrac{D_{50}}{B_p} \right)^{0.053} (11.17) y_1^{1/6} D_{50}^{1/3}}{(11.17) y_1^{1/6} D_{50}^{1/3} - 0.645 \left(\dfrac{D_{95}}{B_p} \right)^{0.053} (11.17) y_1^{1/6} D_{95}^{1/3}} \right]^{0.15} \geq 0.4 \qquad (10\text{-}7)$$

where:

V_1 = average velocity upstream of the pier [L/T]

B_p = width if the pier [L]

y_1 = average water depth upstream of the pier [L]

The scour depth also depends on the width of the pier in relation to the water depth. If the following conditions are met, a wide pier in shallow water correction factor (K_{ws}) should be applied (Equation 10-8): the water depth (y) is less than $0.8B_p$ ($y/B_p < 0.8$), the width of the pier is greater than $50D_{50}$ ($B_p/D_{50} > 50$), and the Froude number (Fr) is less than 1.

$$K_{ws} = \begin{cases} 2.58 \left(\dfrac{y}{B_p} \right)^{0.34} Fr_1^{0.65} & for \ \dfrac{V}{V_c} < 1 \\[3mm] 1.0 \left(\dfrac{y}{B_p} \right)^{0.13} Fr_1^{0.25} & for \ \dfrac{V}{V_c} \geq 1 \end{cases}$$

(10-8)

where:

y = water depth [L]

B_p = width if the pier [L]

F_1 = upstream Froude number (Equation 10-9)

V = average velocity [L/T]

V_c = critical velocity [L/T]

The upstream Froude number is given by Equation 10-9:

$$Fr = \frac{V}{\sqrt{gy}}$$

(10-9)

where:

V = average velocity [L/T]

g = acceleration due to gravity [L/T^2]

y = water depth [L]

The subscript for the Froude number in Equation 10-8 (Fr_1) simply means that in the Froude number calculation (Equation 10-9), y_1, the average water depth upstream and V_1, the original approach velocity are used.

The Froude number is a dimensionless value that describes the ratio of inertial forces to gravitational forces. If Fr is less than 1, the flow is considered subcritical; if Fr is equal to 1, the flow is critical; if Fr is greater than 1, the flow is supercritical. Subcritical flow refers to a calm flow in which the velocity is small and the water depth is high. Conversely, supercritical flow refers to a rapid flow in which the velocity is high and the water depth is low. Critical flow occurs when both the velocity and water depth are critical.

10.2.3.1.2 Piers in Fine-Grained Soils: Correction Factors

For fine-grained soils, corrections for water depth, pier spacing, and pier shape must be taken into account. The pier nose shape correction factor (K_{sh}) for the HEC-18 Clay method is the same as in HEC-18 Sand (Table 10.2-4). With respect to water depth, when the ratio of the water depth (y) to the pier width (B_p), is less than 1.62, the water is considered shallow (Ref. 10-13). A shallow water correction factor (K_w) must be used in this case (Equation 10-10). Note that scour for piers located in deep water (y/B_p >1.62) is independent of the water depth (K_w = 1).

$$K_w = 0.85 \left(\frac{y}{B_p} \right)^{0.34} \quad \text{for} \quad \frac{y}{B_p} < 1.62 \qquad (10\text{-}10)$$

where:

y = water depth [L]

B_p = width of the pier [L]

Another correction exists for the pier spacing effect (K_{sp}). Scour depth increases as piers get closer together. This causes a contraction in the normal flow as the water has to move between piers. The spacing correction factor is given by Equation 10-11.

$$K_{sp} = \frac{W_p}{W_p - n_p B_p} \qquad (10\text{-}11)$$

where:

W_p = width of the channel without the piers [L]

n_p = number of piers

B_p = width of the pier [L]

10.2.3.2 Contractions

A contraction causes the area of flow to reduce (Figure 10.2-22). To maintain the same flow rate, the water must accelerate through the contraction. If the increased velocity exceeds the critical velocity (V_c), contraction scour will occur.

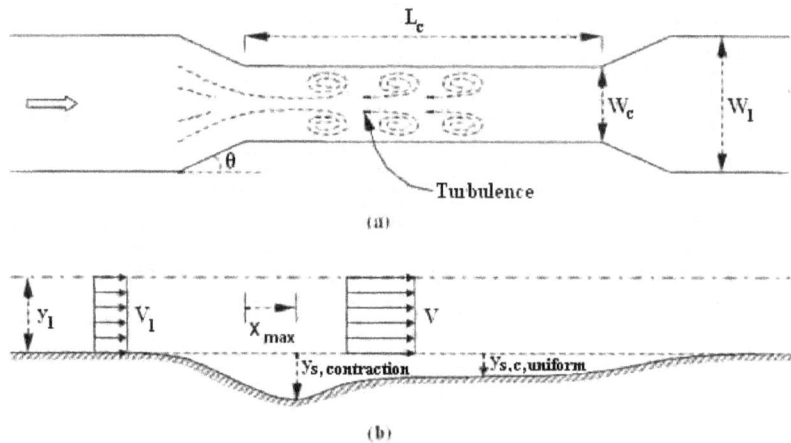

Figure 10.2-22. Schematic of contraction scour (Ref. 10-13).

The critical velocity can be estimated by Equation 10-12 for coarse grained soils (using the units indicated) or by Figure 10.2-4 for all soils. Note that the maximum contraction scour depth $(y_{s,contraction})$ will occur at a location X_{max} from the beginning of the contracted section (Figure 10.2-22).

$$V_c \left(\frac{ft}{s} \right) = 11.17 \left(y_1(ft) \right)^{\frac{1}{6}} \left(D_{50}(ft) \right)^{\frac{1}{3}}$$

(10-12)

where:

y_1 = average water depth upstream of the pier (ft)

D_{50} = median grain size of the soil (ft)

For coarse-grained soils, HEC-18 Sand does not suggest any correction factors to their proposed scour equations. The HEC-18 Clay method, however, does recommend correction factors for the transition angle (K_θ) and the contraction length (K_L). Tests to investigate the influence of the transition angle on scour have shown that the scour depth is independent of the transition angle $(K_\theta = 1)$ (Ref. 10-13).

The transition angle (θ) does have an impact on the location of maximum scour (X_{max}) though. The location of maximum scour for a transition angle $(X_{max}(\theta))$ can be expressed as the location of maximum scour for no contraction $(X_{max}(90°))$ multiplied by a transition angle correction factor $(K_{\theta/Xmax})$. The transition angle correction factor for location of maximum scour $(K_{\theta/Xmax})$ is given by Equation 10-13.

$$K_{\theta/X_{\max}} = \frac{X_{\max}(\theta)}{X_{\max}(90^o)} = 1 + \frac{1}{2\tan\theta} \qquad\qquad (10\text{-}13)$$

Generally, the length of the contraction (L_c) has no impact on the scour depth or its location ($K_L = 1$). When the length is smaller than ¼ of the contraction width (W_c), however, increasing the predicted $y_{s,contraction}$ and X_{max} values is necessary. A correction factor for this case has not been introduced. Further research is needed to obtain the contraction length factor (K_L).

10.2.3.3 Abutments and Walls

Abutments can be considered as acting like half of a wide pier, which also creates significant contraction for bridges. Therefore, the size and shape of the abutment, along with its contraction ratio, are important factors in abutment scour development. There are three common types of abutments: vertical wall abutments, wing-wall abutments, and spill through abutments (Figure 10.2-23). Typically, the blunter the face of the abutment, the deeper the abutment scour hole will be (Ref. 10-15). Therefore, spill-through abutments tend to produce less scour than vertical and wing-wall abutments.

To account for the shape of the abutment, a shape factor (K_s) or correction must be made in the HEC-18 Sand method (Table 10.2-6). Reference 10-17 also recommends values for various abutment shapes (Figure 10.2-23).

Abutment model	Abutment shape	Shape factor, K_s
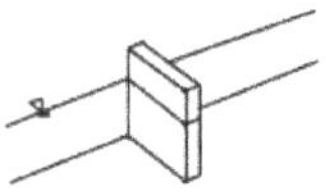	Vertical-wall	1·00
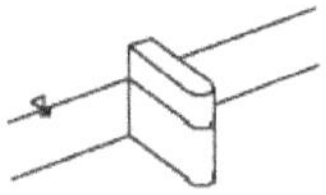	Semicircular ended	0·75
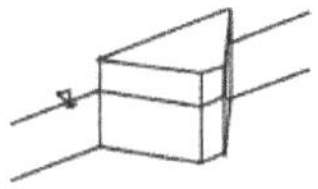	45° wing-wall	0·75
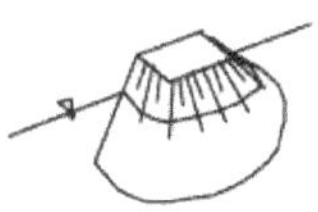	Spill-through with slope horizontal : vertical 0·5 : 1 1 : 1 1·5 : 1	0·60 0·50 0·45

Figure 10.2-23. Abutment types (Ref. 10-17).

Table 10.2-6. Correction factor for abutment shape (Ref. 10-12).

Abutment Shape	Shape Factor, K_s
Vertical Wall	1.00
Vertical Wall with Wing Walls	0.82
Spill-Through	0.55

As with pier scour, the alignment of the abutment (K_ψ) must also be considered (Equation 10-14). In Equation 10-14, ψ is the angle of the abutment [deg].

$$K_\psi = \left(\frac{\psi}{90^o} \right)^{0.13}$$

(10-14)

An abutment that is perpendicular to the flow has an angle (ψ) equal to 90°. The angle of the abutment (ψ) is less than 90° if the abutment points downstream and greater than 90° if the abutment points upstream. The HEC-18 Clay method is currently limited to pier and contraction scour so there is no verified method to calculate abutment scour in cohesive soils. The HEC-18 Sand method does have equations for abutment scour in cohesionless soils.

Walls typically serve as protection structures. For stable bank protection, vertical walls or bulkheads are common. For flood protection, seawalls are usually chosen. When a vertical wall is parallel to the flow, local velocities will often increase at the wall face producing an increase in boundary shear stress (Ref. 10-12). The increase in velocity at the wall boundary is due to a decrease in roughness as the wall is typically smoother than the natural soil. For example, a natural sand channel may have a Manning's roughness coefficient (n) of approximately 0.025 whereas a vertical concrete wall may have a roughness of about 0.015 (Ref. 10-18). The resulting increase in shear stress at the wall boundary will cause scour along the wall. Scour will continue until the increase in flow area causes the velocity to equal the average channel velocity. Conversely, if the wall is rougher than the soil bottom, then deposition can occur near the wall.

In coastal environments, a seawall is often used to protect the shoreline. Scour around the toe of seawalls can cause a dislocation of the foundation material, geotechnical instability, and modification of wave patterns in front of the wall. If the scour becomes large enough, a major failure can occur. Typically, seawalls are vertically faced, but they can also be inclined. Theoretically, reducing the slope of the seawall will produce less scour at the toe of the seawall (Ref. 10-19). Scour calculated for a vertical wall will therefore be a conservative estimate for a sloping seawall under the same conditions.

10.2.3.4 Pipelines and Cables

Pipelines and cables are typically laid along the bed of a body of water, but they can also be buried or trenched. Pipelines often transport oil and gas from offshore platforms, while cables typically anchor platforms to the seafloor. For pipelines and cables resting on the surface of the seafloor, scour due to both waves and currents can undermine the underlying soil, causing the pipeline to be suspended in free span (Figure 10.2-24). As indicated in the figure, scour will not occur across an entire pipeline. Some spans of pipeline will experience scour, while other spans may experience burial from the soil removed during scour.

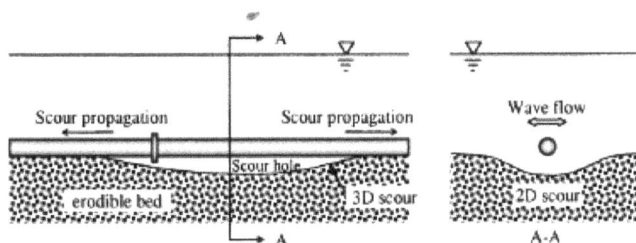

Figure 10.2-24. Pipeline scour (Ref. 10-20).

The conditions of the water flow can impact the resulting pipeline scour. Under a current, the velocity of interest is the velocity flowing across the top of the pipeline (V_{pl}). Under both current and waves, the velocity of interest is the maximum orbital velocity at the seafloor ($V_{orb,bm}$).

Water depth is an important parameter in pipeline scour. More scour is expected in shallow waters as compared to deep waters. The embedment depth (e) of the pipeline, measured from the surface of the seafloor to the bottom of the pipeline (see Figure 10.2-25), is also a factor. If the depth of embedment is greater than or equal to half of the diameter of the pipeline ($e \geq D_{pl}/2$), scour will not develop (Ref. 10-21). It is often costly to install the pipeline to this embedment depth, so a cost-benefit analysis is often needed. If the pipeline is less than half buried, scour must be evaluated.

There are three stages of pipeline scour: the onset of scour, tunnel erosion, and lee-wake erosion. The onset of scour begins as a pressure difference forms between the upstream and downstream sides of the pipeline. This pressure difference is the result of the current flowing across the pipeline.

Seepage is induced in the soil underneath the pipeline because of this pressure difference. The surface of the soil immediately downstream of the pipe will begin to rise. As this continues, a point will be reached where the risen sand on the downstream side, mixed with water, breaks off and boils up; this is called piping. The result is a small gap that forms below the pipeline and the seafloor.

Reference 10-6 describes the criterion for the onset of scour due to a current (Equation 10-15, Figure 10.2-25). Note that Equation 10-15 uses English units.

$$V_{pl,c}^2\left(\frac{ft^2}{s^2}\right) = 0.0023\exp\left[9\left(\frac{e(ft)}{D_{pl}(ft)}\right)^{0.5}\right]g\left(\frac{ft^2}{s^2}\right)D_{pl}(ft)(1-n_s)\left[\left(\frac{\gamma_s\left(\frac{lb}{ft^3}\right)}{\gamma_w\left(\frac{lb}{ft^3}\right)}\right)-1\right] \quad (10\text{-}15)$$

where:

$V_{pl,c}$ = critical velocity of the coarse-grained soil for pipeline scour to be initiated (ft/s)

e = embedment depth of the pipeline (ft)

D_{pl} = pipeline diameter (ft)

g = acceleration due to gravity (ft/s^2)

n_s = soil porosity

γ_s = specific weight of the soil grains (lb/ft^3)

γ_w = specific weight of water (lb/ft^3)

If the velocity flowing across the top of the pipeline (V_{pl}) exceeds the critical velocity ($V_{pl,c}$) found by Equation 10-15, then scour will occur. The criterion for the onset of scour due to a current can also be expressed graphically (Figure 10.2-25).

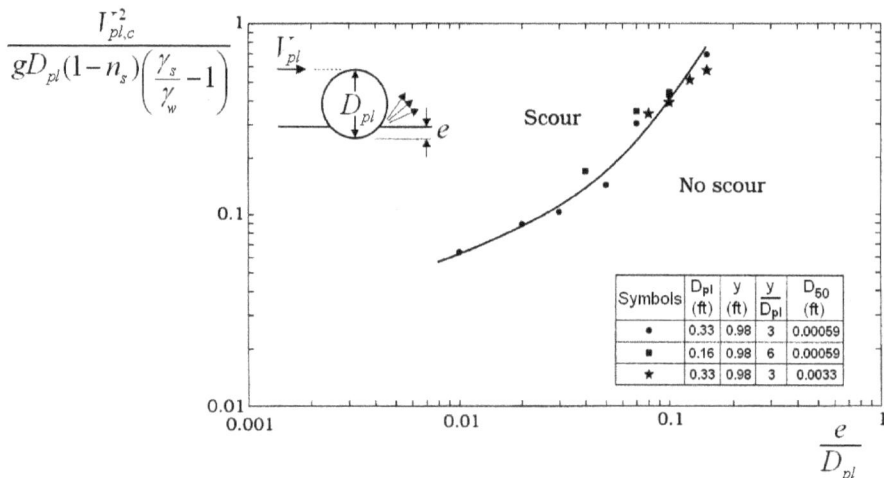

Figure 10.2-25. Onset of scour due to currents (Ref. 10-22).

In this figure (Figure 10.2-25), $V_{pl,c}$ is the critical velocity of the coarse-grained soil for pipeline scour to be initiated, g is the acceleration due to gravity, D_{pl} is the pipeline diameter, D_{50} is the mean grain size, n_s is the soil porosity, γ_s is the specific weight of the soil grains, γ_w is the specific weight of water, y is the water depth, and e is the embedment depth. If the velocity flowing across the top of the pipeline (V_{pl}) exceeds the critical velocity ($V_{pl,c}$) found either by Figure 10.2-25 or Equation 10-15, then scour will occur.

Note that the above figures and equations for pipeline scour are only valid for cohesionless soils (i.e., coarse-grained soils). The permeability and soil properties of cohesive soils such as clays will influence the scour process. For example, an impermeable soil would not experience the onset of scour because seepage would not occur.

If the pipeline is exposed to waves, the onset of scour will be determined differently. First, the velocity that should be considered is the maximum orbital velocity at the bed ($V_{orb,bm}$) rather than the undisturbed flow velocity at the top of the pipeline (V_{pl}) as used in the current only formulation. The maximum orbital velocity at the bed can be calculated using Table 10.2-2.

Knowing the embedment depth ratio (e/D_{pl}) and the Keulegan-Carpenter (KC) number (Equation 10-3), the critical velocity for pipelines under wave forces ($V_{orb,bm,c}$) can be found by using Figure 10.2-26. If the velocity across the pipeline ($V_{orb,bm}$) exceeds the critical velocity ($V_{orb,bm,c}$) found in Figure 10.2-26, scour will occur.

$$\frac{V_{m,c}^2}{gD_{pl}(1-n_s)\left(\dfrac{\gamma_s}{\gamma_w}-1\right)}$$

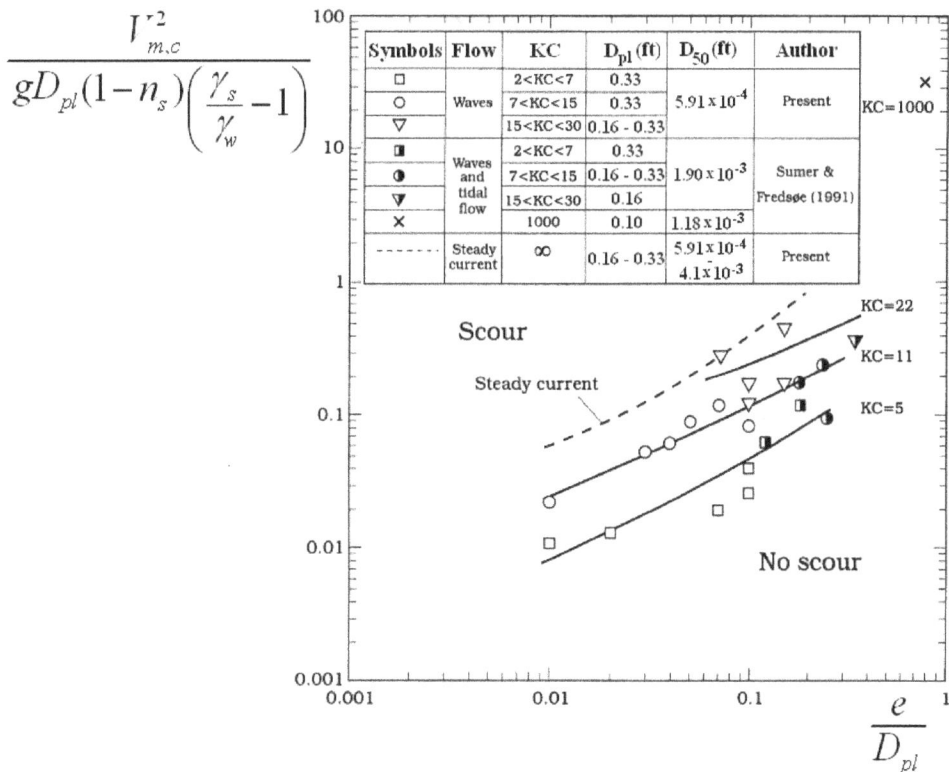

Figure 10.2-26. Onset of scour due to waves (Ref. 10-22).

10-27

The onset of scour is typically followed by tunnel erosion. Tunnel erosion occurs as a result of high shear stresses on the bed that develop when water flows through the small gap that formed between the pipeline and the soil during the onset of scour. This increase in shear stress at the bed leads to further erosion below the pipeline (Ref. 10-23). Scour depth can substantially increase during this stage. If the flow depth is greater than 3.5 times the pipe diameter ($y > 3.5D_{pl}$), however, there is no tunnel erosion (Ref. 10-24).

The final stage of scour under pipelines is termed lee-wake erosion. This stage is governed by vortex shedding (Ref. 10-25). The vortex shedding begins when the gap between the pipeline and the bed reaches a certain depth (Ref. 10-6). Under a current, the sediment transport will greatly increase on the lee side of the pipeline, resulting in lee-wake erosion. This will continue until equilibrium occurs. Under waves, erosion will take place on both sides of the pipeline due to the oscillatory motion of the water.

10.2.3.5 *Footings*

Footings are shallow foundations that do not pierce the water surface. They can either rest on the seafloor or penetrate the seafloor. Scour for footings must be corrected for factors such as shape, size, angle of attack, and bed conditions in the same manner as for piers. Like pipelines, footings are also susceptible to burial.

Observations on scour at small footings and objects on the seafloor (Ref. 10-26) show that objects placed in shallow ocean water depths of less than 30 feet are buried very quickly. At depths of around 60 feet, no burial was observed, but scour around footings caused settlement and tilting. The critical velocity for scour around footings can be approximated by using Figure 10.2-4 for all soils. Higher velocity currents will bury the footings more rapidly. Lower velocity currents are not expected to result in significant scour under the footing.

10.3 ESTIMATING GENERAL SCOUR

General scour refers to the change in seafloor level that would occur even if the structure was not present. It can also be referred to as bed degradation or bed aggradation. General scour can be broken down into short-term and long-term scour. Short-term general scour occurs as a result of a single event, such as a hurricane or flood. Long-term general scour refers to the progressive change of the bed due to natural hydrological or geomorphological changes over a long period of time.

The best method for estimating long term scour at a location is from historical data for the site. Often, however, records are not kept concerning the seafloor elevation at a particular site. In the absence of this data, general scour can be roughly estimated according to empirical correlations, discussed below.

There are several methods available to estimate the magnitude or depth of general scour. The best approach is to use a range of these methods. Field observations and engineering judgment will dictate the initial quantitative estimate of general scour depth.

The maximum general scour (y_{gs}) can be roughly estimated (±20%) by Equation 10-16 (Ref. 10-26).

$$y_{gs} = 1.15H_e - 4.1 \qquad (10\text{-}16)$$

where H_e is the annual extreme wave height (Equation 10-17, Ref. 10-27), which is the significant wave height exceeded 12hr/yr.

$$H_e = \overline{H_s} + 5.6\sigma_d \approx 4.5\overline{H_s} \qquad (10\text{-}17)$$

where:

$\overline{H_s}$ = average monthly significant wave height [L]

σ_d = standard deviation of the monthly average significant wave heights [L]

H_s = significant wave height = average of the highest one-third of the waves [L]

As scour occurs, the soil is moved and accumulated in another location. The maximum bar height is the highest elevation of this marine sediment deposition. If the maximum bar height in the area of interest is known, the magnitude of maximum scour can be estimated as twice the maximum bar height, but may be as much as three times the maximum bar height in some locations (Ref. 10-28).

General scour can also be determined from the erosion function found during EFA testing (Figure 10.2-7). The shear stress on the bed, under current only, can be estimated using Equation 10-18 (Ref. 10-29).

$$\tau_{bc} = \rho C_D V_b^2 \qquad (10\text{-}18)$$

where:

ρ = mass density of water [FT^2/L^4]

C_D = drag coefficient computed using Equation 10-19 (Ref. 10-29)

V_b = mean depth water velocity of the current [L/T]

$$C_D = \left[\frac{0.4}{1 + \ln\left(\dfrac{z_o}{y}\right)} \right]^2$$

(10-19)

where:

z_0 = hydraulic roughness length [L]

y = water depth [L]

The hydraulic roughness length (z_0) is the distance, or elevation, from the bed corresponding to an extrapolated zero velocity (Figure 10.3-1). In smooth flow, the velocity is zero at the bed (z_0 = 0); for rough flow, however, the roughness length will increase depending on the velocity distribution. For coarse-grained soils, the hydraulic roughness length is estimated according to Equation 10-20 (Ref. 10-29). In Equation 10-20, k_s represents the Nikuradse equivalent bed roughness, which is given by Equation 10-21.

$$z_0 = \frac{k_s}{30}$$

(10-20)

$$k_s = \alpha_s D_{xs}$$

(10-21)

where:

α_s = a proportionality constant (Table 10.3-1)

D_{xs} = diameter of the bed material corresponding to a percent finer equal to $xs\%$ [L]

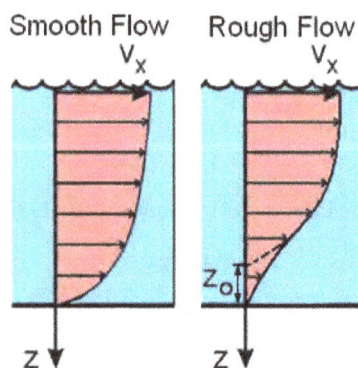

Figure 10.3-1. Hydraulic roughness length definition.

Table 10.3-1. Nikuradse equivalent bed roughness (After Ref. 10-30).

Investigator	D_{xs}	a_s
Ackers & White (1973)	D_{35}	1.23
Hammond et al. (1984)	D_{50}	6.6
Engelund & Hansen (1967)	D_{65}	2.0
Lane & Carlson (1953)	D_{75}	3.2
Gladki (1979)	D_{80}	2.5
Richardson & Davis (2001) - Sand	D_{84}	1.0
Richardson & Davis (2001) - Gravel	D_{84}	3.5
Simons & Richardson (1966)	D_{85}	1
Hoffmans & Verheij (1997)	D_{90}	3.0

The bed shear stress under waves only (τ_{bw}) can be estimated using Equation 10-22 (Ref. 10-29).

$$\tau_{bw} = \frac{1}{2}\rho f_w V_{orb,bm}^2 \tag{10-22}$$

where:

ρ = mass density of water [FT^2/L^4]

f_w = wave friction factor given by Equation 10-23 (Ref. 10-31)

$V_{orb,bm}$ = maximum orbital velocity at the bed (Table 10.2-3) [L/T]

and:

$$f_w = 0.04\left(\frac{a}{k_s}\right)^{-\frac{1}{4}} \ for \ \frac{a}{k_s} > 50 \tag{10-23}$$

where:

a = semi-orbital length of the wave (Table 10.2-3) [L]

k_s = Nikuradse equivalent bed roughness (Equation 10-21)

There are a number of other equations available to estimate the wave friction factor (f_w) based on different parameters such as the wave period, dynamic viscosity, near-bed orbital diameter of wave motion, etc. (Ref. 10-32).

According to Reference 10-29, the mean bed shear stress due to both currents and waves ($\tau_{bcw,m}$) is found by Equation 10-24. Equation 10-25 presents the maximum bed shear stress due to both currents and waves ($\tau_{bcw,max}$).

$$\tau_{bcw,m} = \tau_{bc}\left[1+1.2\left(\frac{\tau_{bw}}{\tau_{bc}+\tau_{bw}}\right)^{3\,2}\right] \tag{10-24}$$

$$\tau_{bcw,max} = \left[\left(\tau_{bcw,m}+\tau_{bw}\cos\phi\right)^2+\left(\tau_{bw}\sin\phi\right)^2\right]^{\frac{1}{2}} \tag{10-25}$$

where:

τ_{bc} = shear stress due to currents only (Equation 10-18) [F/L^2]

τ_{bw} = shear stress due to waves only (Equation 10-22) [F/L^2]

ϕ = angle between the direction of wave propagation and the direction of the current

Once the shear stress is found, the corresponding erosion rate ($\dot{Z}$) can be determined from the erosion function (Figure 10.2-7). If the shear stress is less than the critical shear stress (τ_c), there will be no scour ($\dot{Z} = 0$). The depth of general scour is then the product of the erosion rate and the time over which the shearing occurs ($y_{gs} = \dot{Z}\Delta t$).

10.4 ESTIMATING LOCAL SCOUR

Local scour refers to scour due to a structure founded on the seafloor or channel bed. The typical obstacles that are encountered in marine environments have been discussed in Section 10.2.3. This section will focus on how to calculate the magnitude of scour at each type of obstacle. Two methods will be discussed: HEC-18 Sand and HEC-18 Clay (also known as SRICOS-EFA). HEC-18 Sand is for coarse-grained soils whereas HEC-18 Clay is for fine-grained soils.

10.4.1 Pier Scour

Piers are considered to be either simple or complex. A simple pier is a single solid pier which is made of a continuous structure from the bottom to the top, such as a cylinder. A complex pier is made up of several components from the bottom to the top. These components can be a pile group with a pile cap on top of it and a large column on top of the pile cap.

The maximum depth of scour is defined as the depth of scour reached when the water velocity flows over the soil for an infinitely long period of time (z_{max}). The final depth of scour is the scour depth reached at the end of the storm event being considered (z_{final}). In coarse-grained soils, one storm (say 24 hours) is usually enough to create the maximum depth of scour. Therefore, in coarse-grained soils there is no need to distinguish between z_{max} and z_{final}. In fine-grained soils, however, the same storm may only generate a fraction of that maximum depth scour, and z_{final} may be much less than z_{max}. One can be very conservative and use the value of z_{max} calculated even for fine-grained soils. However, this may be costly and a method exists to predict z_{final} for a given velocity and given storm duration for pier scour predictions (HEC-18 Clay).

10.4.1.1 Simple Pier Scour in Coarse-Grained Soils (HEC-18 Sand)

For coarse-grained soils, HEC-18 Sand estimates the maximum depth of scour ($y_{s,pier}$) at piers by Equation 10-26.

$$y_{s,pier} = 2B_p K_{sh} K_a K_b K_{ar} K_{ws} \left(\frac{y_1}{B}\right)^{0.35} Fr_1^{0.43} \tag{10-26}$$

where:

B_p = pier diameter [L]

K_{sh} = shape correction factor (Table 10.2-4)

K_a = angle of attack correction factor (Equation 10-6, Figure 10.2-20)

K_b = bed condition correction factor (Table 10.2-5)

K_{ar} = armoring correction factor (Equation 10-7)

K_{ws} = wide pier in shallow water correction factor (Equation 10-18)

y_1 = the average water depth in the upstream main channel [L]

Fr_1 = Froude number directly upstream of the pier (Equation 10-9)

10.4.1.2 Simple Pier Scour in Fine-Grained Soils (HEC-18 Clay)

For fine-grained soils, the HEC-18 Clay method defines the maximum depth of scour for a circular pier (in feet) according to Equation 10-27.

$$y_{s,pier}(ft) = \left(5.91 \times 10^{-4}\right) K_w K_{sp} K_{sh} (\text{Re})^{0.635} \tag{10-27}$$

where:

Re = Reynolds number = $V_1B'_p/v$

V_1 = depth average approach velocity [L/T]

B'_p = pier projection width (Figure 10.2-19) [L]

v = kinematic viscosity of the water [L^2/T]

K_w = shallow water correction factor (Equation 10-10)

K_{sp} = pier spacing correction factor (Equation 10-11)

K_{sh} = shape correction factor (Table 10.2-4)

This equation is equally valid for coarse-grained soils as well (Ref. 10-13). The main difference between the HEC-18 Sand method and the HEC-18 Clay method is that the HEC-18 Clay method utilizes a time rate of scour effect. This is a significant advantage of the HEC-18 Clay method. The rate of scour has an important impact on scour prediction. The depth of scour found using Equation 10-27 is the maximum depth of scour (z_{max}) that can occur. The final scour depth after one storm (z_{final}), however, may not reach the maximum depth (Figure 10.4-1). It is therefore advantageous to determine the time rate of scour for cases where the soil erodes slowly, specifically for fine-grained soils. The following sections are concerned with the maximum depth of scour. The method to calculate the depth of scour after a storm event is given in Section 10.4.8.

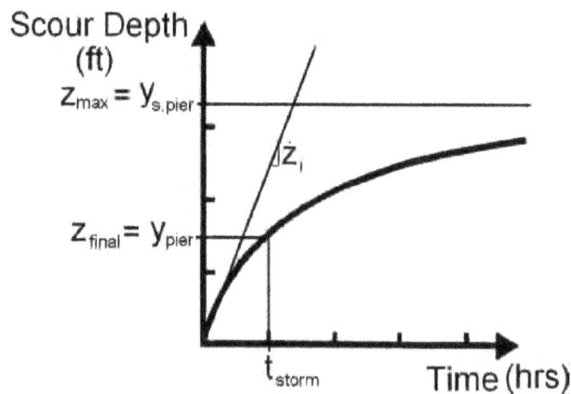

Figure 10.4-1. Time rate of scour.

10-34

10.4.1.3 Complex Pier Scour in Coarse-Grained Soils (HEC-18 Sand)

For complex pier foundations (Figure 10.4-2), the maximum total scour is a superposition of the scour created by the pier ($y_{s,pier}$), the pile cap ($y_{s,pilecap}$), and the pile group ($y_{s,pilegroup}$). The same equation for simple pier scour (Equation 10-26) is used to find the pier scour component (Equation 10-28), except a correction factor ($K_{h,pier}$) is applied to account for both the height of the pier stem above the bed (h_l) and the shielding effect from the pile cap overhang (f).

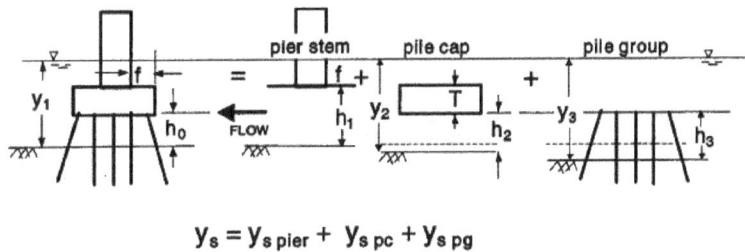

$$y_s = y_{s\,pier} + y_{s\,pc} + y_{s\,pg}$$

Figure 10.4-2. Complex pier scour (Ref. 10-12).

$$y_{s,pier} = K_{h,pier}\left[2BK_{sh}K_aK_bK_{ar}K_{ws}\left(\frac{y_1}{B_p}\right)^{0.35}Fr_1^{0.43}\right]$$

(10-28)

where:

$K_{h,pier}$	=	complex pier correction factor (Equation 10-29)
B_p	=	pier diameter [L]
K_{sh}	=	shape correction factor (Table 10.2-4)
K_a	=	angle of attack correction factor (Equation 10-6, Figure 10.2-20)
K_b	=	bed condition correction factor (Table 10.2-5)
K_{ar}	=	armoring correction factor (Equation 10-7)
K_{ws}	=	wide pier in shallow water correction factor (Equation 10-18)
y_1	=	the average water depth in the upstream main channel [L]
Fr_1	=	Froude number directly upstream of the pier (Equation 10-9)

$$K_{h,pier} = \left(.4075 - .0669\frac{f}{B_p}\right) - \frac{h_1}{B_p}\left(.4271 - .0778\frac{f}{B_p}\right)$$

$$+ \left(\frac{h_1}{B_p}\right)^2 \left(.01615 - .0455\frac{f}{B_p}\right) - \left(\frac{h_1}{B_p}\right)^3 \left(.0269 - .012\frac{f}{B_p}\right)$$

(10-29)

where:

f = pilecap overhang [L]

B_p = pier diameter [L]

h_1 = the height of the pier stem above the bed [L]

The scour due to the pile cap is found in the same way as for a pier. The difference lies in the inputs for the variables. If the pile cap is above the bed, its width (B_{pc}) needs to be reduced to an equivalent full depth solid pier width (B^*_{pc}) (Equation 10-30).

$$B^*_{pc} = B_{pc}\,\exp\left[-2.705 + 0.51Ln\left(\frac{t_{pc}}{y_2}\right) - 2.783\left(\frac{h_2}{y_2}\right)^3 + \frac{1.751}{\exp\left(\frac{h_2}{y_2}\right)}\right]$$

(10-30)

where:

B_{pc} = width of the pile cap [L]

t_{pc} = thickness of the pile cap exposed to the flow [L]

y_2 = adjusted flow depth = $y_1 + 0.5(y_{s,pier})$ [L]

y_1 = upstream water depth [L]

$y_{s,pier}$ = maximum pier scour depth [L]

h_2 = adjusted pile cap height = $h_{o,pc} + 0.5(y_{s,pier})$ [L]

$h_{o,pc}$ = actual pile cap height [L]

The pile cap scour, $y_{s,pilecap}$, can then be found by using Equation 10-31. Note that the quantity V_2 in Fr_2 is the adjusted flow velocity for the pile cap ($V_2 = V_1(y_1/y_2)$), and that V_1 is the approach velocity.

$$\frac{y_{s,pilecap}}{y_2} = 2K_{sh}K_aK_bK_{ar}K_{ws}\left(\frac{B^*_{pc}}{y_2}\right)^{0.65} Fr_2^{0.43}$$ (10-31)

where:

K_{sh} = shape correction factor (Table 10.2-4)

K_a = angle of attack correction factor (Equation 10-6, Figure 10.2-20)

K_b = bed condition correction factor (Table 10.2-5)

K_{ar} = armoring correction factor (Equation 10-7)

K_{ws} = wide pier in shallow water correction factor (Equation 10-18)

B^*_{pc} = equivalent full depth solid pier width (Equation 10-30)[L]

y_2 = adjusted flow depth = $y_1 + 0.5(y_{s,pier})$ [L]

y_1 = upstream water depth [L]

$y_{s,pier}$ = maximum pier scour depth [L]

Fr_2 = adjusted Froude number for the pile cap (Equation 10-9)

If the pile cap is on or below the bed, it can be considered as a footing. The scour due to the pile cap in this case will need to be redefined (Equation 10-32). Note that the quantity V_f in Fr_f is the average velocity in the flow zone below the top of the pile cap or footing (Equation 10-33).

$$\frac{y_{s,pilecap}}{y_f} = 2K_{sh}K_aK_bK_{ar}K_{ws}\left(\frac{B_{pc}}{y_f}\right)^{0.65} Fr_f^{0.43}$$ (10-32)

where:

K_{sh} = shape correction factor (Table 10.2-4)

K_a = angle of attack correction factor (Equation 10-6, Figure 10.2-20)

K_b = bed condition correction factor (Table 10.2-5)

K_{ar} = armoring correction factor (Equation 10-7)

K_{ws} = wide pier in shallow water correction factor (Equation 10-18)

B_{pc} = width of the pilecap [L]

y_f = distance from the bed to the top of the footing [L]

Fr_f = Froude number for the footing (Equation 10-9)

$$V_f = V_2 \left[\dfrac{Ln\left(10.93\dfrac{y_f}{k_s}+1\right)}{Ln\left(10.93\dfrac{y_2}{k_s}+1\right)} \right]$$

(10-33)

where:

V_2 = adjusted flow velocity for the pile cap = $V_1(y_1/y_2)$ [L/T]

V_1 = approach velocity [L/T]

y_1 = upstream water depth [L]

y_2 = adjusted flow depth = $y_1 + 0.5(y_{s,pier})$ [L]

$y_{s,pier}$ = maximum pier scour depth [L]

y_f = distance from the bed to the top of the footing [L]

k_s = grain roughness of the bed (Equation 10-21)

K_{ws} = wide pier in shallow water correction factor (Equation 10-18)

B_{pc} = width of the pilecap [L]

y_f = distance from the bed to the top of the footing [L]

For sand size and gravel size particles, k_s is equal to D_{84} and $3.5D_{84}$, respectively, where D_{84} is the diameter of the bed material corresponding to a percent finer equal to 84% (Ref. 10-12). The total scour depth for this type of pier configuration (pier with a footing) will include only the pier scour and the pile cap scour components.

If several piles are present under the pile cap, the scour relating to the pile group will need to be found ($y_{s,pilegroup}$). Determining scour for the pile group requires reducing the pile group to an equivalent pier by finding an equivalent width (B^*_{pg}) as in Equation 10-34.

$$B^*_{pg} = B_{proj} K_{sp,pg} K_m$$

(10-34)

where:

B_{proj} = projected width of the pile group [L]

$K_{sp,pg}$ = coefficient for pile spacing (Equation 10-35)

K_m = coefficient for number of aligned rows (Equation 10-36); Note: K_m = 1, if the pile group is skewed or staggered

$$K_{sp,pg} = 1 - \frac{4}{3} \left[1 - \frac{1}{\left(\dfrac{B_{proj}}{B_p} \right)} \right] \left[1 - \left(\frac{S_p}{B_p} \right)^{-0.6} \right]$$

(10-35)

$$K_m = 0.9 + 0.1 m_p - 0.0714(m_p - 1) \left[2.4 - 1.1 \left(\frac{S_p}{B_p} \right) + 0.1 \left(\frac{S_p}{B_p} \right)^2 \right]$$

(10-36)

where:

S_p = pile spacing [L]

m_p = number of rows in the pile group (Figure 10.4-3)

B_p = pile diameter [L]

B_{proj} = projected width of the pile group [L]

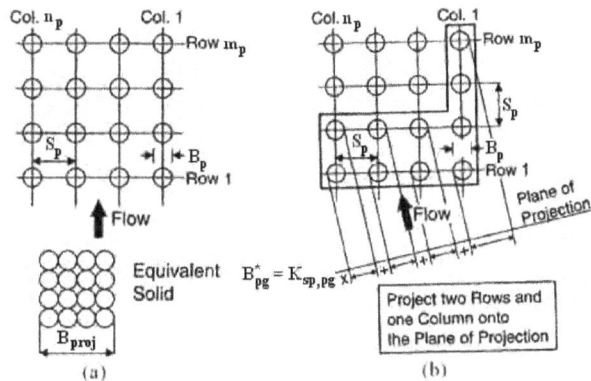

Figure 10.4-3. Projected width of piles in pile group (a) aligned with flow, and (b) not aligned with flow (Ref. 10-12).

The depth of scour due to the pile group ($y_{s,pilegroup}$) can then be found according to Equation 10-37. Note that the quantity V_3 in Fr_3 is the adjusted flow velocity ($V_3 = V_1(y_1/y_3)$), and that V_1 is the approach velocity.

$$\frac{\left(y_s\right)_{pilegroup}}{y_3} = K_{h,pg}\left[2K_{sh}K_b K_{ar}K_{ws}\left(\frac{B^*_{pg}}{y_3}\right)^{0.65} Fr_3^{0.43}\right]$$

(10-37)

where:

y_3 =   adjusted flow depth = $y_1 + 0.5(y_{s,pier}) + 0.5(y_{s,pilecap})$ [L]

y_1 =   upstream water depth [L]

$y_{s,pier}$ =   maximum pier scour depth [L]

$y_{s,pilecap}$ =   maximum pile cap scour depth [L]

$K_{h,pg}$ =   pile group height factor (Equation 10-38)

K_{sh} =   shape correction factor (Table 10.2-4)

K_b =   bed condition correction factor (Table 10.2-5)

K_{ar} =   armoring correction factor (Equation 10-7)

K_{ws} =   wide pier in shallow water correction factor (Equation 10-18)

B^*_{pg} =   equivalent width of the pile group (Equation 10-34) [L]

Fr_3 =   adjusted pile group Froude number (Equation 10-9)

$$K_{h,pg} = \left[3.08\left(\frac{h_3}{y_3}\right) - 5.23\left(\frac{h_3}{y_3}\right)^2 + 5.25\left(\frac{h_3}{y_3}\right)^3 - 2.10\left(\frac{h_3}{y_3}\right)^4\right]^{\frac{1}{0.65}}$$

(10-38)

where:

h_3 =   height of the pile group considering pier and pile cap scour = $h_{o,pg} + y_3 - y_1$ [L]

$h_{o,pg}$ =   actual pile group height [L]

y_3 =   adjusted flow depth = $y_1 + 0.5(y_{s,pier}) + 0.5(y_{s,pilecap})$ [L]

y_1 =   upstream water depth [L]

$y_{s,pier}$ =   maximum pier scour depth [L]

$y_{s,pilecap}$ =   maximum pile cap scour depth [L]

10.4.1.4 Wave-Induced Pier Scour

The above equations were for piers experiencing a current only. Waves can also have an impact on scour development around piers. Scour under waves only is considerably smaller than scour under current only (Ref. 10-33). Wave-induced scour involves erosion due to vortex shedding. For a/B_p less than 0.2, the effects of waves are minimal. Scour for this case is similar to the case with a current velocity. As the wave length increases or the size of the structure decreases, vortices begin to form behind the structure and the scour depth increases.

There are two important parameters necessary in describing wave-induced pier scour: first, the a/B_p parameter where B_p is the pier diameter and a is the wave semi-orbital length (Table 10.2-3) and second, the Keulegan-Carpenter (KC) number (Equation 10-7). For KC less than 6, there is no vortex shedding behind the pier. Wave-induced scour for KC greater than or equal to 6 is given by Equation 10-39 (Ref. 10-2). Note that in Equation 10-39, $y_{s,pier}$ is the maximum pier scour.

$$y_{s,pier,wave} = y_{s,pier}\left[1 - \exp\left[-0.03(KC - 6)\right]\right]$$

(10-39)

10.4.2 Contraction Scour

The maximum depth of scour is defined as the depth of scour reached when the water velocity flows over the soil for an infinitely long period of time (z_{max}). The final depth of scour is the scour depth reached at the end of the storm event being considered (z_{final}). In coarse-grained soils, one storm (say 24 hours) is usually enough to create the maximum depth of scour. Therefore in coarse-grained soils there is no need to distinguish between z_{max} and z_{final}. In fine-grained soils, however, the same storm may only generate a fraction of that maximum depth scour and z_{final} may be much less than z_{max} (Figure 10.4-1). One can be very conservative and use the value of z_{max} calculated even for fine-grained soils. However, this may be costly and a method exists to predict z_{final} for a given velocity and given storm duration for contraction scour predictions. The following sections are concerned with the maximum depth of scour. The method to calculate the depth of scour after a storm event is given in Section 10.4.8.

10.4.2.1 Contraction Scour in Coarse-Grained Soils (HEC-18 Sand)

Before calculating contraction scour, the velocity upstream of the contraction and in the contracted zone should be obtained. This can be done by using a 1-dimensional flow simulation program such as HEC-RAS (Ref. 10-34). The upstream velocity is compared to the critical velocity to determine if there will be clear-water scour or live-bed scour. This is important because contraction scour will be estimated differently depending on which type of scour is present. Clear-water contraction scour ($y_{s,contraction,cw}$) can be found using Equation 10-40.

$$y_{s,contraction,cw} = \left(\frac{0.0077Q^2}{D_m^{2/3}W_c^2} \right)^{\frac{3}{7}} - y_o \qquad (10\text{-}40)$$

where:

Q = discharge through the contraction [L³/T]

D_m = diameter of the smallest non-transportable particle in the bed martial
= $1.25D_{50}$ [L]

D_{50} = median diameter of the bed material [L]

W_c = bottom width of the contraction section [L]

y_o = initial water depth for the existing bed depth [L]

If piers are present within the contraction, the bottom width of the contraction section (W_c) will be the total width less any pier widths. Live bed contraction scour ($y_{s,contraction,lb}$) is estimated differently (Equation 10-41).

$$y_{s,contraction,lb} = y_1 \left[\left(\frac{Q_2}{Q_1} \right)^{\frac{6}{7}} \left(\frac{W_1}{W_c} \right)^{k_1} \right] - y_o \qquad (10\text{-}41)$$

where:

y_1 = average water depth upstream [L]

Q_2 = flow in the contracted channel [L³/T]

Q_1 = flow in the upstream channel [L³/T]

W_1 = bottom width of the upstream channel [L]

W_c = bottom width of the main channel in the contracted section [L]

k_1 = an exponent (Table 10.4-1)

y_o = initial water depth for the existing bed depth [L]

Table 10.4-1. Exponent for live-bed contraction scour (Ref. 10-12).

V^*/ω	k_1	Mode of Bed Material Transport
< 0.50	0.59	Mostly contact bed material discharge
0.50 – 2.0	0.64	Some suspended bed material discharge
> 2.0	0.69	Mostly suspended bed material discharge

In Table 10.4-1, V^* is the shear velocity in the upstream section (Equation 10-42), and ω is the fall velocity of the bed material (Figure 10.4-4), defined as the velocity at which a sediment particle falls through a column of still water (Ref. 10-12).

$$V^* = \left(\frac{\tau_b}{\rho} \right)^{\frac{1}{2}} = \left(g y_1 S_1 \right)^{\frac{1}{2}}$$

(10-42)

where:

τ_b = shear stress on the bed [F/L^2]

ρ = mass density of the water [FT^2/L^4]

g = acceleration due to gravity [L/T^2]

y_1 = upstream water depth [L]

S_1 = slope of the energy grade line of the main channel

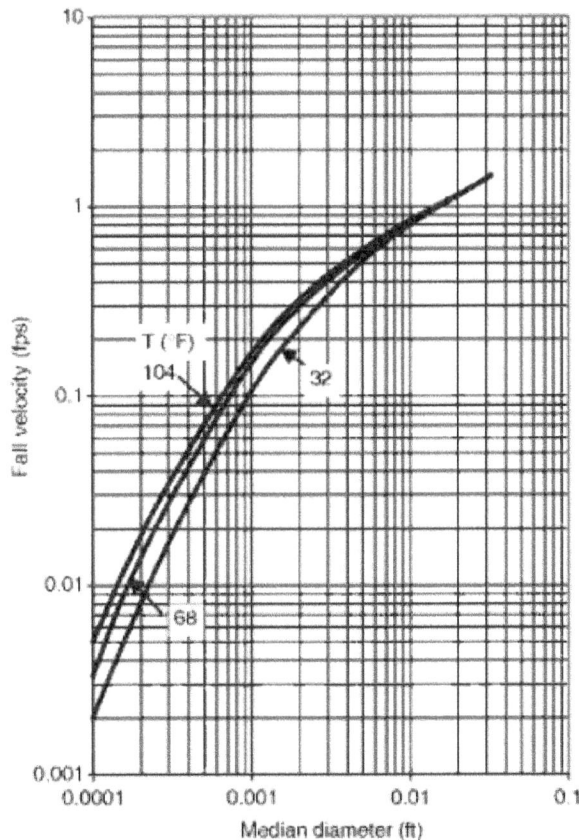

Figure 10.4-4. Particle fall velocity (Ref. 10-35).

The slope of the energy grade line (S_f) can be found according to Manning's equation (Equation 10-43) using the units noted.

$$V\left(\frac{ft}{s}\right) = \frac{1.49\left(R_h(ft)\right)^{\frac{2}{3}} S^{\frac{1}{2}}}{n\left(\dfrac{s}{ft^{\frac{1}{3}}}\right)}$$

(10-43)

where:

V	=	the average velocity (ft/s)
R_h	=	hydraulic radius = A/P (ft)
A	=	cross-sectional area of flow (ft^2)
P	=	wetted perimeter (ft)
S	=	slope of the energy grade line
n	=	Manning's roughness coefficient (s/ft$^{1/3}$)

Average values for Manning's roughness coefficient for different soil types are given in Table 10.4-2. The units of Manning's coefficient in this chapter are in s/ft$^{1/3}$, but in the literature, n is typically given in units of s/m$^{1/3}$. To convert the given n values in units of s/m$^{1/3}$ to units of s/ft$^{1/3}$, divide by 1.49.

Table 10.4-2. Average values for Manning's coefficient (Ref. 10-36).

Description	n (s/m$^{1/3}$)	n (s/ft$^{1/3}$)
Clay	0.023	0.015
Sand	0.020	0.013
Gravel	0.030	0.020
Rock	0.040	0.027

10.4.2.2 Contraction Scour in Fine-Grained Soils (HEC-18 Clay)

The HEC-18 Clay method (Ref. 10-13) predicts maximum contraction scour according to Equation 10-44. Equation 10-44 indicates that if the critical shear stress in cohesive soils is large enough, no scour will take place.

$$y_{s,contraction} = K_\theta K_L 1.90 y_1 \left[\frac{1.38 \left(V_1 \frac{W_1}{W_c} \right)}{\sqrt{gy_1}} \left(\frac{\tau_c}{\rho} \right)^{\frac{1}{2}} - \frac{1}{gny_1^{\frac{1}{3}}} \right] \qquad \text{(10-44)}$$

where:

K_θ = correction factor for the influence of the transition angle = 1 (Ref. 10-13)

K_L = correction factor for the contraction length = 1 (Ref. 10-13)

y_1 = upstream water depth [L]

V_1 = average upstream velocity [L/T]

W_1 = bottom width of the upstream channel [L]

W_c = bottom width of the main channel in the contracted section [L]

τ_c = critical stress [F/L^2]

ρ = mass density of the water [FT^2/L^4]

g = acceleration due to gravity [L/T^2]

n = Manning's roughness coefficient from Table 10.4-2 (T/L$^{1/3}$)

The location of maximum contraction scour (X_{max}) is given by Equation 10-45 (Figure 10.2-22). In Equation 10-45, W_c is the bottom width of the main channel in the contracted section [L], and W_1 is the bottom width of the upstream channel [L].

$$X_{max} = W_c \left(2.25 \frac{W_c}{W_1} + 0.15 \right) \qquad \text{(10-45)}$$

10.4.3 Propeller Induced Scour

Scour due to propellers ($y_{s,propeller}$) in feet is found using Equation 10-46, developed in Reference 10-11. In Equation 10-46, t is time, and Γ and Ω are given by Equations 10-47 and 10-48, respectively. Note: The variables in Equations 10-47 and 10-48 must be in consistent units.

$$y_{s,propeller}(ft) = 0.128\Omega \left\{ \ln \left[t(\text{sec}) \right] \right\}^{\Gamma} \qquad \text{(10-46)}$$

$$\Gamma = Fr_0^{-0.53} \left(\frac{D_p}{D_{50}} \right)^{-0.48} \left(\frac{C}{D_{50}} \right)^{0.94} \qquad (10\text{-}47)$$

$$\Omega = \Gamma^{-6.38} \qquad (10\text{-}48)$$

where:

Fr_0 = densimetric Froude number (Equation 10-49)

D_p = propeller diameter [L]

D_{50} = median grain size [L]

C = distance between the propeller tip and the seabed (Figure 10.2-13) [L]

$$Fr_0 = \frac{V_o}{\sqrt{gD_{50}\left(\frac{\Delta\rho}{\rho}\right)}} \qquad (10\text{-}49)$$

where:

V_o = initial centerline propeller velocity (Equation 10-5) [L/T]

g = acceleration due to gravity [L/T²]

D_{50} = median grain size [L]

ρ = mass density of the water [FT²/L⁴]

$\Delta\rho$ = difference between the soil density (ρ_s) and the water density = $\rho_s - \rho$ [FT²/L⁴]

Equation 10-46 is valid only for those seabed depths below the propeller (C) in the range of 0.5 to 2.5 times the propeller diameter (D_p). The equation provides scour as a function of time. The maximum propeller scour depth will occur as the time (t) goes to infinity. The location of maximum scour from the propeller (X_m) is estimated according to Equation 10-50 (Ref. 10-11), where C is the clearance distance between the propeller tip [L] and the seabed and Fr_0 is the densimetric Froude number (Equation 10-49).

$$X_m = CFr_0^{0.94} \qquad (10\text{-}50)$$

The presence of a rudder, used for steering, will impact scour. The rudder splits the jet formed by the propeller into two streams (Ref. 10-6). Based on a study presented in Reference 10-37, scour will change from the calculated propeller scour (Equation 10-46) depending on the rudder angle (ζ, Figure 10.4-5).

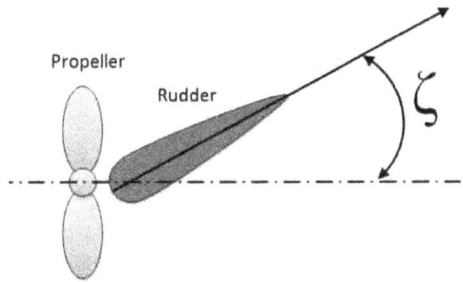

Figure 10.4-5. Definition of rudder angle.

The equation used to calculate the scour due to a propeller with a rudder is given by Equation 10-51. For a rudder angle of 0°, the maximum scour depth increases by about 25% compared to the scour depth due to a propeller with no rudder (Ref. 10-37).

$$y_{s,rudder} = y_{s,propeller} \left[0.75 - 0.07 Fr_o + 0.02 \left(\frac{D_p}{D_{50}} \right) - 0.15 \left(1 + \zeta \right) \right]$$

(10-51)

where:

$y_{s,propeller}$ = the scour due to an unconfined propeller only (Equation 10-46) [L]

Fr_o = densimetric Froude number (Equation 10-49)

D_p = propeller diameter [L]

D_{50} = median grain size [L]

ζ = rudder angle [radians]

10.4.4 Abutment and Wall Scour

10.4.4.1 Abutment Scour in Coarse-Grained Soils (HEC-18 Sand)

There are two methods available to calculate the depth of abutment scour. If the ratio of projected abutment length (L') to the upstream flow depth (y_1) is greater than 25 ($L'/y_1 > 25$), then the HIRE abutment scour equation (Equation 10-52) should be used (Ref. 10-12). Equation 10-52 is a result of field observations of scour in the Mississippi River.

$$\frac{y_{s,abutment}}{y_1} = 4 Fr_1^{0.33} \frac{K_s}{0.55} K_\psi$$

(10-52)

where:

y_1 = average upstream water depth [L]

10-47

Fr_1 = Froude number upstream of the abutment (Equation 10-9)

K_s = abutment shape factor (Table 10.2-6)

K_ψ = correction factor for angle of attack (Equation 10-14)

If the ratio of projected abutment length (L') to the upstream flow depth (y_1) is less than 25 ($L'/y_1 < 25$), then the Froehlich abutment scour equation (Equation 10-53) should be used (Ref. 10-12). Note that in both Equations 10-52 and 10-53, there is no difference in the calculation between live-bed scour and clear-water scour.

$$\frac{y_{s,abutment}}{y_a} = 2.27 K_s K_\psi \left(\frac{L'}{y_a} \right) \left(\frac{V_e}{(gy_a)^{\frac{1}{2}}} \right)^{0.61} + 1 \tag{10-53}$$

where:

y_a = average depth of flow on the floodplain [L]

K_s = abutment shape factor (Table 10.2-6)

K_ψ = correction factor for angle of attack (Equation 10-14)

L' = length of the abutment projected normal to flow [L]

V_e = velocity of the flow in the obstructed section [L/T]

g = acceleration due to gravity [L/T^2]

10.4.4.2 Seawall Scour

A basic rule of thumb for scour at vertical seawalls, based on laboratory testing and field observation, is that the maximum scour depth is less than or equal to the height of the unbroken deepwater wave height (H_o) (Ref. 10-38). Reference 10-19 recommends using an equation (Equation 10-54) based on incident deepwater significant wave height (H_s), deepwater wave length (L_d), and water depth at the wall (v).

$$y_{s,seawall} = H_s \sqrt{22.72 \left(\frac{y}{L_d} \right) + 0.25} \tag{10-54}$$

Equation 10-54 is valid only where $-0.011 \leq y/L_d \leq 0.045$ and $0.015 \leq H_s/L_d \leq 0.04$. For cases outside of these bounds, Reference 10-19 suggests estimating maximum scour using the rule of thumb ($y_{s,seawall} \leq H_o$).

10.4.4.3 Propeller Induced Scour at Quay Walls

The jet produced by propellers causes scour whether the wash is confined or unconfined (Figure 10.2-14). Quay walls serve to confine and interfere with the propeller wash causing an increased depth of scour (Ref. 10-6). The influence of the quay wall depends on the distance between the wall and the face of the propeller (X_w). As the distance X_w increases, the scour at the quay wall will decrease. The maximum propeller induced scour at a quay wall can be found by using Equation 10-55.

$$y_{s,quaywall} = S_a \left[1.18 \left(\frac{X_w}{X_m} \right)^{-0.2} - 1 \right] + y_{s,propeller} \qquad (10\text{-}55)$$

where:

S_a	=	maximum equilibrium scour depth in the unconfined scour case measured from the centerline of the propeller shaft [L] $= y_{s,propeller} + H_p$
$y_{s,propeller}$	=	maximum scour for the unconfined case measured from the initial seafloor elevation (Equation 10-46) [L]
H_p	=	distance from the propeller shaft to the channel bottom [L]
X_w	=	distance between the wall and the face of the propeller [L]
X_m	=	distance from maximum unconfined propeller scour to the propeller (Equation 10-50) [L]

In the case where a rudder is present, propeller induced scour at quay walls is calculated according to Equation 10-56. <u>Note</u>: The variables in Equation 10-56 must be in consistent units.

$$y_{s,quaywall,rudder} = D_p \left[2.3 \times 10^{-4} \left(\frac{C}{D_{50}} \right)^{0.581} \left(\frac{D_p}{D_{50}} \right)^{0.427} \left(\frac{X_w}{D_p} \right)^{-0.052} (1+\zeta)^{-0.772} Fr_o^{4.403} \right] \qquad (10\text{-}56)$$

where:

D_p	=	propeller diameter [L]
C	=	clearance distance between the propeller tip and the seabed [L]
D_{50}	=	median grain size [L]
X_w	=	distance between the wall and the face of the propeller [L]
ζ	=	rudder angle [radians]
Fr_o	=	densimetric Froude number (Equation 10-49)

10.4.5 Pipeline and Cable Scour

For clear-water scour under pipelines and cables, under current only, Reference 10-21 suggests using Equation 10-57, as proposed by Reference 10-24.

$$y_{s,pipeline,cw} = \frac{D_{pl}}{2}\left[\left(\frac{k_s}{12D_{pl}}\right)Ln\left(\frac{6D_{pl}}{k_s}\right)\right]\left(\frac{V}{V_c}\right)$$

(10-57)

where:

D_{pl}	=	pipe diameter [L]
V	=	depth averaged velocity [L/T]
V_c	=	critical velocity [L/T]
k_s	=	effective bed roughness (k_s = 3D$_{90}$ for clays, from Equation 10-21 and Table 10.3-1) [L]
D_{90}	=	diameter of the bed material of which 90% are smaller [L]

In the case of live-bed scour, a simple equation is proposed by Reference 10-39 that relates scour under current only to the pipe diameter (Equation 10-58). In Equation 10-58, D_{pl} is the pipeline diameter.

$$y_{s,pipeline,lb}(ft) = 0.6D_{pl}(ft) \pm 0.1D_{pl}(ft)$$

(10-58)

In the presence of wave action, scour under pipelines is a function of the pipe diameter (D_{pl}), the Keulegan-Carpenter (KC) number (Equation 10-3), and the embedment depth (e) (Equation 10-59, Ref. 10-40). The embedment depth accounts for partial embedment of the pipeline due to burial. If the embedment ratio (e/D_{pl}) is greater than 0.5, no scour occurs, and Equation 10-59 is invalid.

$$y_{s,pipeline,wave} = 0.1D_{pl}\sqrt{KC}\left(1-1.4\frac{e}{D_{pl}}\right)+\frac{e}{D_{pl}}$$

(10-59)

10.4.6 Scour for Structures Piercing Water Surface

Piers, offshore platforms, and offshore wind turbines are types of marine structures that pierce the water surface. The method to calculate scour for structures piercing the water surface is the same as the methods used to calculate pier scour in Section 10.4.1 (Equations 10-26 and 10-27). Correction factors must be applied to account for structures that are not piers.

10.4.7 Scour for Structures Resting on the Seafloor

Sometimes, structures placed on the seafloor do not pierce the water surface. Scour will still occur in these cases because there is still an obstacle to the flow at the soil-water interface. For footings or other structures resting on the seafloor (not including pipelines and cables), calculations for scour can make use of the complex pier scour equation, which includes contributions from the pier or column, the pile cap, and the pile group (Section 10.4.1.3). For structures resting on the seafloor with a shallow foundation, the column and pile group contributions will be omitted from the complex pier scour equation. Pipelines and cables are other structures resting on the seafloor. Scour for these structures is found differently as in Section 10.4.5.

10.4.8 Time for Scour Development

The time rate of scour is an important part of scour predictions. The time rate method describes the relationship between the scour depth and the time during which the water has been flowing at a given velocity over the soil. If a storm occurs, scour will develop around the foundation of the structure. If the storm is short, the full amount of scour may not be realized, and the scour depth is limited to z_{final} (Section 10.4.1.2). If the storm is long, the full amount of scour may be realized, and the maximum depth of scour z_{max} is reached (Section 10.4.1.2). It is therefore advantageous to determine the time rate of scour for cases where the soil erodes slowly.

Clear-water scour will reach equilibrium slower than live-bed scour (Figure 10.2-16). While clear-water scour will approach a maximum scour value, live-bed scour oscillates as removal and deposition of the bed material continually occurs. The time rate of scour will also differ for the type of structure involved. The scour depth is related to time according to Equation 10-60 (Ref. 10-41).

$$z_{final}(t) = \frac{t}{\dfrac{1}{\dot{z}_i} + \dfrac{t}{z_{max}}}$$ (10-60)

where:

z_{final} = the final scour depth after time t [L]

$\dot{z}_i$ = initial rate of scour (Figure 10.2-7, Figure 10.4-1) [L/T]

z_{max} = maximum magnitude of scour [L]

The initial scour rate can be found using the initial shear stress at the soil-water interface together with the erosion function. This erosion function can be obtained from a laboratory test such as the EFA test (Ref. 10-42). Once the shear stress at the bed is calculated, the initial scour rate can be found on the erosion function (Figure 10.2-7).

For piers, the initial shear stress used in the HEC-18 Clay method is the maximum hydraulic shear stress exerted by the water on the riverbed ($\tau_{max,pier}$) around the pier. This can be calculated according to Equation 10-61 (Ref. 10-13).

$$\tau_{max,pier} = k_w k_{sp} k_{sh} k_a \times 0.094 \rho V_1^2 \left[\frac{1}{\log \text{Re}} - \frac{1}{10} \right] \tag{10-61}$$

where:

k_w = correction factor for the effect of water depth (Equation 10-62)

k_{sp} = correction factor for the effect of pier spacing (Equation 10-63)

k_{sh} = correction factor for the effect of pier shape (Equation 10-64)

k_a = correction factor for the effect of attack angle (Equation 10-65)

ρ = mass density of water [FT^2/L^4]

V_1 = upstream velocity [L/T]

Re = Reynolds number = $V_1 B_p / v$

B_p = pier width [L]

v = kinematic viscosity of the water [L^2/T]

The correction factors used in Equation 10-61 are given by Equations 10-62 through 10-65, as follows:

$$k_w = 1 + 16 e^{-4\frac{y}{B_p}} \tag{10-62}$$

$$k_{sp} = 1 + 5 e^{-1.1\frac{S_p}{B_p}} \tag{10-63}$$

$$k_{sh} = 1.15 + 7 e^{-4\frac{L_p}{B_p}} \tag{10-64}$$

$$k_a = 1 + 1.5 \left(\frac{\Delta}{90} \right)^{0.57} \tag{10-65}$$

where:

y =   water depth [L]

B_p =   pier width [L]

S_p =   pier spacing [L]

L_p =   pier length [L]

Δ =   attack angle of the pier [deg]

For contractions, the initial shear stress used in the HEC-18 Clay method is the maximum hydraulic shear stress exerted by the water on the riverbed ($\tau_{max,contraction}$) in the contraction. This can be calculated according to Equation 10-66 (Ref. 10-13).

$$\tau_{max,contraction} = k_{c-R}k_{c-\theta}k_{c-y}k_{c-Lc}\gamma_w\, n^2 V^2 R_h^{-\frac{1}{3}}$$

(10-66)

where:

k_{c-R} =   correction factor for the contraction ratio (Equation 10-67)

$k_{c-\theta}$ =   correction factor for the contraction transition angle (Equation 10-68)

k_{c-y} =   correction factor for the contraction water depth ($k_{c-y} \approx 1$, Ref. 10-13)

k_{c-Lc} =   correction factor for the contraction length (Equation 10-69)

γ_w =   unit weight of water [F/L^3]

n =   Manning's roughness coefficient from Table 10.4-2 (T/L$^{1/3}$)

V_1 =   upstream velocity [L/T]

R_h =   hydraulic radius = A/P

A =   cross-sectional area of flow [L^2]

P =   wetted perimeter [L]

The correction factors used in Equation 10-66 are given by Equations 10-67 through 10-69, as follows:

$$k_{c-R} = 0.62 + 0.38\left(\frac{W_1}{W_c}\right)^{1.75}$$

(10-67)

$$k_{c-\theta} = 1 + 0.9\left(\frac{\theta}{90}\right)^{1.5}$$

(10-68)

$$k_{c-Lc} = \begin{cases} 1 & for \quad k_{c-Lc} \geq 0.35 \\ 0.77 + 1.36\left(\dfrac{L_c}{W_1 - W_c}\right) - 1.98\left(\dfrac{L_c}{W_1 - W_c}\right)^2 & for \quad k_{c-L} < 0.35 \end{cases}$$

(10-69)

where:

W_1 = width of the upstream channel [L]

W_c = width of the contraction [L]

θ = transition angle [deg]

L_c = contraction length [L]

The initial scour rate depends on the same factors that affect scour magnitude including pier shape, angle of attack, bed grain size, water depth, and pier spacing. The scour rate will decrease with an increase in pier width and flow depth; it will increase with an increase in contraction and the sharpness of pier corners (Ref. 10-13). The HEC-18 Clay method (SRICOS-EFA Method) incorporates time rate of scour development in addition to determining scour depth. The available program is simple, free, and will calculate scour depth and rate for given input geometries and flow conditions.

10.5 SCOUR COUNTERMEASURES

Countermeasures can be used to reduce or minimize the effects of scour at a structure. The following sections discuss the various countermeasure approaches, as well scour monitoring instrumentation.

10.5.1 Countermeasure Design Concepts and Approach

Scour countermeasures can be divided into three main categories: armor, hydraulic control, and grade control. Armoring works to protect the bed or seafloor soil from erosion. The flow is not significantly changed by this type of countermeasure. In narrow channels, however, armor can serve to contract the channel even further leading to increased contraction scour. Hydraulic control involves altering the flow. This can be accomplished by reducing the flow velocity or changing the flow path. Grade control is another type of scour countermeasure. It involves realigning existing beds to reduce scour at the structure. The most commonly used scour countermeasures are presented in this section.

10.5.1.1 Riprap

Riprap is the most common type of armoring protection used for scour. It consists of placing a layer of rock around the structure at the soil-water interface to reduce erosion. The riprap must be big enough to withstand the flow and any wave forces, but small enough to avoid loss of the underlying material through void spaces. This loss of material can be circumvented by placing a sand filter or geosynthetic filter between the bed material and the riprap or by grouting the riprap. The disadvantage of grouting is that flexibility is sacrificed. Partial grouting may be a better solution.

It is advantageous to use a well graded riprap because the different sizes will interlock together producing a more stable solution. This will only work if, during placement of the riprap, the graded rocks are uniformly distributed. After any increase in flow, such as during a storm, an inspection should be conducted to ensure the stability of the riprap.

The size of the riprap depends on the structure being protected. For example, riprap used for pier scour will be different than riprap used for abutment scour. There are several methods available to size riprap for different types of structures. The U.S. Army Corps of Engineers published a chapter on riprap protection (Ref. 10-43). NCHRP Report No. 568 also provides riprap design criteria specifically for bridges (Ref. 10-44). For example, the median diameter of riprap ($D_{50,riprap}$) for piers in fresh water can be evaluated according to Equation 10-70 using the units indicated.

$$D_{50,riprap}(ft) = \frac{0.692 \left(KV(ft) \right)^2}{2g \left(\dfrac{ft}{s^2} \right)(S_s - 1)}$$

(10-70)

where:

K =   coefficient for pier shape (1.5 for round-nose pier, 1.7 for rectangular pier)

V =   average approach velocity in line with the pier (ft)

g =   acceleration due to gravity (ft/s^2)

S_s =   specific gravity of the riprap (≈ 2.65)

10.5.1.2 Filters

Filters are made of either granular or geosynthetic materials. They are extremely important for adequate performance of both armoring and hydraulic control countermeasures. They prevent erosion of soil through voids in the armoring and relieve hydrostatic pressure within the soil.

Granular filters are created by placing layers of soil on top of the natural soil before placing the riprap. Each layer is made of soil coarser than the previous layer with the finest layer being at the bottom, closest to the natural soil. The largest layer could be as big as riprap material. Grading criteria for this type of material is available through the National Engineering Handbook (Ref. 10-45).

The size of the openings in the geosynthetic filter, also known as filter fabric, is an important design consideration. If the openings are too large, piping can occur and soil can migrate. If the openings are too small, hydrostatic pressures can increase leading to instability. Further guidelines for designing for erosion protection are available through Reference 10-46, an Army and Air Force Technical Manual on the engineering use of geotextiles.

10.5.1.3 Artificial Seaweed

Artificial seaweed acts as a hydraulic control scour countermeasure. Artificial seaweed comes as a mat consisting of many buoyant polypropylene fronds. The seaweed fronds work to lower the current velocity around their location. They also tend to encourage deposition of sediment rather than erosion. Artificial seaweed is not suitable in areas with intense and frequent wave action. Otherwise, it is useful for reducing local scour around any offshore structure, including pipelines.

10.5.1.4 Peripheral Skirt

Peripheral skirts do not prevent scouring; rather, they prevent scour from undermining the foundation. Skirts are placed vertically around the footing and penetrate the seafloor. As scour occurs, they prevent the structure from rocking and instability. It is therefore important that the skirt penetrate below the estimated maximum depth of scour.

10.5.1.5 Articulated Mats

Articulated mats are an armoring countermeasure for permanent scour protection (Ref. 10-47). This type of scour protection is the easiest and most positive method for offshore structures (Ref. 10-48). Articulated mats are typically made of concrete and come in separate blocks that are bound together. The blocks can be any size and shape. Some types of blocks are shaped so that they interlock with other blocks. Other types do not interlock, but are connected using cables. Commonly, a filter is placed under articulated mats as a backing. Design methods for articulated mats are available. FHWA's HEC-23 provides design guidelines for an articulated concrete block system (Ref. 10-49).

10.5.2 Scour Monitoring and Instrumentation

In some cases, the scour countermeasure is simply monitoring the site for potential scour problems. Visual inspection is not reliable, especially during storms. Instrumentation installed around the structure and the bed more accurately depicts the scour conditions during any flow condition. There are two categories of instrumentation available for scour monitoring: portable instruments and fixed instruments.

10.5.2.1 Portable Instrumentation

Portable instruments are not fixed to the structure; rather, they are brought to a site each time a measurement is necessary, such as after a storm event. This type of instrumentation can be used at different locations at a structure or it can also be moved to a different site completely.

There are three classifications of portable instruments: physical probing, sonar, and geophysical (Ref. 10-50). Physical probes, such as sounding poles and sounding weights, extend the reach of inspectors, allowing them to measure scour depths. The length of the probes is the limiting factor in their selection. Sonar instruments quantify the scour at a site by sending an acoustic pulse from a generating transducer to the bed and measuring the elapsed time for the pulse to travel to the bed and back to the transducer. Geophysical instruments such as fathometers and ground-penetrating radar use wave propagation and reflection measurements. Similar to sonar, geophysical instruments transmit a signal into the water which is reflected back when it reaches a material with different physical properties.

10.5.2.2 Fixed Instrumentation

When frequent measurements are necessary, fixed instruments are employed. They are either attached to the structure or installed around the structure in the channel bed as permanent fixtures for scour monitoring. Measurements can be taken daily, weekly, or continuously for a record of scour over time.

There are four classifications of recommended fixed instruments: sonars, magnetic sliding collars, float-out devices, and sounding rods (Ref. 10-51). Fixed sonar instruments are the most commonly used and work in the same manner as their portable instrument counterpart except that they are mounted directly on the structure. Magnetic sliding collars are devices attached to the structure that pierce the bed. As scour occurs, the magnetic sliding collar will slide down the device as much as the soil has been eroded. Float-out devices are initially buried near the structure in the bed to a predetermined depth. If scour becomes too great, the float-out device will be exposed and will float to the top of the water. Float-outs are typically used to monitor when a scour critical depth has been reached, not to obtain a scour profile with time. Sounding rods are rod shaped physical probes with a foot at the end that rests on the bed. When the soil is eroded, the rod will drop to the new bed elevation. It is important that the sounding rods do not penetrate the bed under its own weight as this will skew the measurements.

Other fixed instruments that are currently in use are tilt and vibration sensors and time domain reflectometers (TDR). Tilt and vibration sensors are installed on the structure and measure both movement and rotation. While they do not directly measure scour, settlement or movement of the structure can indicate erosion of the supporting soil. Time domain reflectometers are embedded in the soil. They measure the time it takes for an electromagnetic pulse to reach the soil-water interface and reflect back to the surface. The part of the electromagnetic pulse energy that is not initially reflected back will continue to propagate to either the boundary between soil layers or to the end of the reflectometer. The difference in time between the two pulses can then be equated to distance. If the time between pulses changes, the soil elevation has changed.

10.5.2.3 Selecting Instrumentation

The plan of action should incorporate which type of instrument should be used, the number of instruments that should be placed, the frequency of data collection, and the accuracy needed. Determining which instrumentation is appropriate for a given site depends on the soil, the water, and the obstacle. There are also limitations within each type of instrument that must be evaluated before selection. Ultimately, engineering judgment will play a role in the instrumentation selection.

There are advantages and disadvantages for each instrument. The first decision is whether the instruments will be portable or fixed (Table 10.5-1). Sometimes, having both types of instruments for a site is most advantageous. If portable instrumentation is chosen, then the type of portable instrument should be selected (Table 10.5-2). Similarly, if fixed instrumentation is chosen, then the type of fixed instrument should be selected (Table 10.5-3). There are a number of instruments available within each type of portable or fixed instruments. Careful review of the advantages and disadvantages of each instrument should be made for each site.

Table 10.5-1. Portable versus fixed instrumentation (Ref. 10-50).

Instrument	Advantages	Disadvantages
Portable	Point measurement or complete mapping, use at many bridges	Labor intensive, special platforms often required
Fixed	Continuous monitoring, low operational cost, easy to use	Maximum scour not at instrument location, maintenance/loss of equipment

Table 10.5-2. Portable instrumentation summary (Ref. 10-50).

Instrument	Best Application	Advantages	Limitations
Physical Probes	Small bridges and channels (up to reach of probe)	Simple technology	Accuracy, high flow application
Sonar	Larger bridges and channels (beyond reach of physical probes)	Point data or complete mapping, accurate	High flow application
Geophysical	Fresh water and soils other than dense, moist clays	Forensic evaluation	Specialized training required, labor intensive, costly

Table 10.5-3. Fixed instrumentation summary (Refs. 10-51 and 10-50).

Instrument	Best Application	Advantages	Limitations
Sonar	Coastal regions	Can track both scour and deposition, continuous measurement	Debris, high sediment or air entrapment
Magnetic Sliding Collar	Fine-bed channels	Simple, mechanical device	Can only monitor maximum scour depth – No deposition recorded, debris
Float-out Devices	Dry river beds, new construction, before placing riprap	Lower cost, easy to install	Battery life – No continuous measurements
Sounding Rods	Coarse-bed channels	Simple, mechanical device	Must ensure no penetration under its own weight or from vibrations due to flowing water
Tilt and Vibration Sensors	Any structure	Measure movement and rotation of structure	False scour readings due to structural movement not related to soil
Time Domain Reflectometry (TDR)	Any condition	Allows processes affecting sediment to be correlated to scour	Debris, organic material

10.6 EXAMPLE PROBLEMS

10.6.1 Problem 1 – General Scour

10.6.1.1 Problem Statement

The amount of general scour at a site is important in understanding the seasonal fluctuations of the seafloor. At a particular site, the average monthly significant wave height ($\overline{H_s}$) is estimated at 3 feet. Find the amount of general scour.

10.6.1.2 Problem Solution

Problem 10.6-1

ANALYTICAL PROCEDURES	COMPUTATIONS
37. From Equation 10-17, the annual extreme wave height (H_e) is approximated as: $H_e = \overline{H_s} + 5.6\sigma_d \approx 4.5\overline{H_s}$	$H_e \approx (4.5)(3\,ft) = 13.5\,ft$
38. The maximum general scour at this site can now be estimated according to Equation 10-16 as: $y_{gs} = 1.15H_e - 4.1$	$y_{gs} = (1.15)(13.5\,ft) - 4.1 = 11.425\,ft$
SUMMARY	
The amount of general scour at this site is 11.43 ft.	

10.6.2 Problem 2 – Pier Scour Example

10.6.2.1 Problem Statement

A square pier, with sides 10 feet in length, is located in 100 ft of deep water with an approaching velocity of 10 ft/s (angle of attack = 0°), as shown in Figure 10.6-1. The upstream channel width is 150 feet, and the bed condition is plane. The water temperature is 20°C.

The pier is located in water with waves only. The waves have a wave length of 500 feet, a wave period of 15 seconds, and a wave height of 3 feet. Also, EFA testing was conducted on a clay sample from the site. The results of this testing include the erosion function shown in Figure 10.6-2.

Based on the site data given above, find:

   (a) The simple pier scour in coarse-grained soils.

   (b) The simple pier scour in fine-grained soils.

   (c) The depth of scour for simple pier scour in fine-grained soils after the first 24 hour storm.

   (d) The wave-induced pier scour.

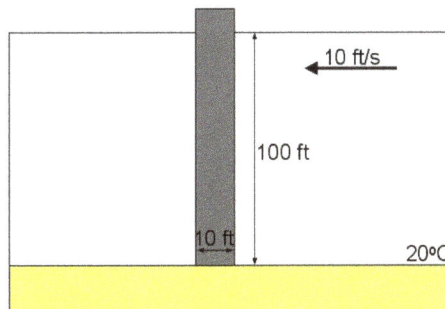

Figure 10.6-1. Pier scour example.

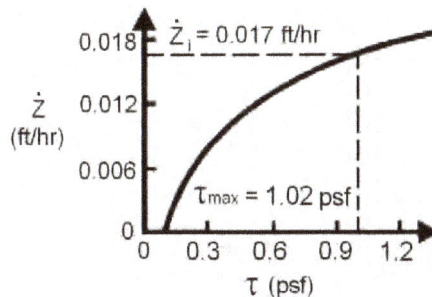

Figure 10.6-2. Time rate of pier scour example.

10.6.2.2 Problem Solution

The analytical and computational procedures for the problem's solution are shown below. They follow the methods presented in this chapter for computation of pier scour.

Problem 10.6-2

ANALYTICAL PROCEDURES	COMPUTATIONS
Part (a): Computation of simple pier scour in coarse-grained soils	
1. Assume that the bed soil is a sand with a D_{50} of 0.02 ft and a D_{95} of 0.071 ft. Since the soil is coarse-grained, the HEC-18 Sand method should be used.	$D_{50} = 0.02\,ft$ $D_{95} = 0.071\,ft$
2. Before calculating the local pier scour using Equation 10-26, the correction factors K_{sh}, K_a, K_b, K_{ar}, and K_{ws} must be determined. Because D_{50} is greater than 0.0066 ft and D_{95} is greater than 0.067 ft, K_{ar} must be calculated according to Equation 10-7, shown below: $$K_{ar} = 0\,4\left[\frac{V_1 - 0\,645\left(\dfrac{D_{50}}{B_p}\right)^{0.053}(11\;17)y_1^{1/6}D_{50}^{1/3}}{(11\;17)y_1^{1/6}D_{50}^{1/3} - 0\,645\left(\dfrac{D_{95}}{B_p}\right)^{0.053}(11\;17)y_1^{1/6}D_{50}^{1/3}}\right]^{0.15} \geq 0\,4$$	$K_{sh} = 1.1$ (from Table 10.2-4) $K_a = 1.0$ (from Equation 10-6 for $\Delta = 0°$) $K_b = 1.1$ (from Table 10.2-5) $K_{ar} = 0.5$ (from Equation 10-7)
3. K_{ws} is the wide pier in shallow water correction factor; it is only applied when certain conditions are met, including if $y/B_p < 0.8$. Since $y/B_p = 10$, K_{ws} is not applied (K_{ws} equal to unity). The Froude number (Fr_l) for local pier scour is found by using Equation 10-9. $$Fr = \frac{V}{\sqrt{gy}}$$	$K_{ws} = 1.0$ $$Fr_1 = \frac{V_1}{\sqrt{gy_1}} = \frac{10}{\sqrt{(32.2)(100)}} = 0.18$$

Problem 10.6-2

ANALYTICAL PROCEDURES	COMPUTATIONS
4. The local pier scour using HEC-18 Sand can now be found by using Equation 10-26. $$y_{s,pier} = 2B_p K_{sh} K_a K_b K_{ar} K_{ws} \left(\frac{y_1}{B_p}\right)^{0.35} Fr_1^{0.43}$$	$$y_{s,pier} = 2(10)(1.1)(1.0)(1.1)(0.5)(1.0)\left(\frac{100}{10}\right)^{0.35}(0.18)^{0.43}$$ $$y_{s,pier} = 12.96\,ft$$
Part (b): Computation of simple pier scour in fine-grained soils	
5. Now assume that the bed soil is clay. Since the soil is fine-grained, the HEC-18 Clay method should be used to estimate pier scour (Equation 10-27). Before calculating the local pier scour using Equation 10-27, the correction factors K_w, K_{sp}, and K_{sh} must be determined.	According to Section 10.2.3.1.2, scour for piers located in deep water ($y/B_p >1.62$) is independent of the water depth, therefore, $K_w = 1$ Also, in this example, there is only a single pier. Pier spacing is therefore not a concern. $K_{sp} = 1$ $K_{sh} = 1.1$ (from Table 10.2-4)
6. The Reynolds number Re' must then be calculated. Then the maximum local depth of pier scour using HEC-18 Clay is found using Equation 10-27. $$Re' = \frac{VB}{v}$$ $$y_{s,pier} = (5.91 \times 10^{-4}) K_w K_{sp} K_{sh} (Re')^{0.635}$$	At 20°C (68°F), $v \approx 1.08 \times 10^{-5}$ ft²/s $$Re' = \frac{(10)(10)}{1.08 \times 10^{-5}} = 9259259$$ $$y_{s,pier} = (5.91 \times 10^{-4})(1.0)(1.0)(1.1)(9259259)^{0.635}$$ $$y_{s,pier} = 17.25\,ft$$
Part (c): Computation of the time rate of scour for simple pier scour in fine-grained soils	
7. The pier scour value found in part (b) is the max scour depth for fine-grained soil. However, in fine-grained soils, the final scour that occurs after a single storm can be much less if the soil erodes very slowly. Part (c) will explore how to find this final scour value after a 24 hour storm.	

Problem 10.6-2

ANALYTICAL PROCEDURES	COMPUTATIONS
8. Before calculating the amount of scour resulting from the 24 hour storm (Equation 10-60), first compute the correction factors k_w (Equation 10-62) and k_{sh} (Equation 10-64). No correction for pier spacing (k_{sp}) and angle of attack (k_a) are needed in this problem as there is only one pier and the angle of attack is 0°. $$k_w = 1 + 16e^{-4\frac{y}{B_p}}$$ $$k_{sh} = 1.15 + 7e^{-4\frac{L_p}{B_p}}$$	$$k_w = 1 + 16e^{-4\frac{100}{10}} = 1.0$$ $$k_{sh} = 1.15 + 7e^{-4\frac{10}{10}} = 1.28$$ In this example, there is only a single pier. Pier spacing is, therefore, not a concern, and: $$k_{sp} = 1$$ Likewise, because the angle $K_a = 1.0$ (from Equation 10-6 for $\Delta = 0°$)
9. Knowing these correction factors, the maximum shear stress around the pier ($\tau_{max,pier}$) is calculated as shown below. Note that the mass density of the water and the viscosity of the water will change depending on temperature and salt concentration in the water. $$\tau_{max,pier} = k_w k_{sp} k_{sh} k_a \times 0.094 \rho V_1^2 \left[\frac{1}{\log \text{Re}} - \frac{1}{10} \right]$$	$$\tau_{max,pier} = (1.0)(1.0)(1.28)(1.0) \times$$ $$0.094(1.94)(10)^2 \left[\frac{1}{\log(9259259)} - \frac{1}{10} \right]$$ $$\tau_{max,pier} = 1.02 \, psf$$
10. Since the maximum hydraulic shear stress around the pier has been calculated (1.02 psf), the initial scour rate ($\dot{z}_i$) can be found using the erosion function (Figure 10.6-2) and is equal to 0.017 ft/hr.	$\dot{z}_i = .017$ ft/hr

Problem 10.6-2

ANALYTICAL PROCEDURES	COMPUTATIONS
11. The maximum scour at the site was calculated as 17.25 ft in the previous example (Sect. 10.6.2.2) using HEC-18 Clay. The first storm event for the pier lasted 24 hours. The amount of scour resulting from this storm can be found using Equation 10-60, shown below, where $z(t)$ is the scour depth as a function of time, z_{max} is the maximum scour depth ($y_{s,pier}$), and $\dot{z}_i$ is the initial scour rate (found from Figure 10.6-2). $$z(t) = \dfrac{t}{\dfrac{1}{\dot{z}_i} + \dfrac{t}{z_{max}}}$$	$$z(t) = \dfrac{24}{\dfrac{1}{0.017} + \dfrac{24}{17.25}}$$ $z(24hrs) = 0.4\,ft$

<table>
<tr><td colspan="2" align="center">Part (d): Computation of the wave-induced pier scour</td></tr>
<tr>
<td>12. First, it must be determined whether the pier is located in shallow water, transitional water, or deepwater. The pier is located in a water depth (d) of 100 feet. This is less than $L/2$, which is equal to 250 feet. It is larger than $L/20$, however, so the pier is located in transitional water. The maximum orbital velocity at the bed ($V_{orb,bm}$) can then found using Table 10.2-3.

$$V_{orb,bm} = \frac{H}{2}\frac{gT}{L}\frac{\cosh\left[2\pi\dfrac{(z+d)}{L}\right]}{\cosh\left(2\pi\dfrac{d}{L}\right)}\cos\theta_w$$</td>
<td>$$V_{orb,bm} = \frac{3}{2}\frac{(32.2)(15)}{500}\frac{\cosh\left[2\pi\dfrac{(-100+100)}{500}\right]}{\cosh\left(2\pi\dfrac{100}{500}\right)}\cos(0)$$

$V_{orb,bm} = 0.76\dfrac{ft}{s}$</td>
</tr>
</table>

Problem 10.6-2

ANALYTICAL PROCEDURES	COMPUTATIONS
13. The Keulegan-Carpenter number (KC) must now be found (Equation 10-3). $$KC = \frac{V_{orb,bm}T}{B_p}$$	$$KC = \frac{(0.76)(15)}{10} = 1.14$$ Since KC is less than 6, waves will not impact the local pier scour (Ref. 10-33). It can be considered negligible ($y_{s,pier,wave} = 0$).

SUMMARY

a. The simple pier scour for the case of coarse-grained soils is 12.96 ft.

b. The simple pier scour for the case of fine-grained soils is 17.25 ft.

c. The scour depth at the pier is 0.4 ft after the first 24 hour storm. This is a significant difference from the maximum scour possible at the site. Subsequent storms and other events will further the scour at the pier. In fine-grained soil, the final scour depth after a series of storms or a velocity hydrograph can be found using the SRICOS-EFA Method.

d. The wave-induced pier scour is 0 ft (no scour).

10.6.3 Problem 3 – Contraction Scour Example

10.6.3.1 Problem Statement

A contracted channel has an upstream channel width (W_1) of 100 feet and a contracted channel width (W_c) of 75 feet as shown in Figure 10.6-3. The length of contraction (L_c) is 20 feet. The transition angle for the contraction is 90° and the approach velocity is 10 ft/s. The upstream water depth (y_1) and the water depth in the contracted section (y_o) are both 25 feet, prior to any scour. The discharge (Q) is 25,000 ft³/s through the channel. The water temperature is 20°C.

Also EFA testing was conducted on a soil sample at the site. The results of this test include the erosion function shown in Figure 10.6-4.

Based on the site data given above, find:

   (a) The contraction scour in coarse-grained soils.

   (b) The contraction scour in fine-grained soils.

   (c) The depth of contraction scour in fine-grained soils after the first 24 hour storm.

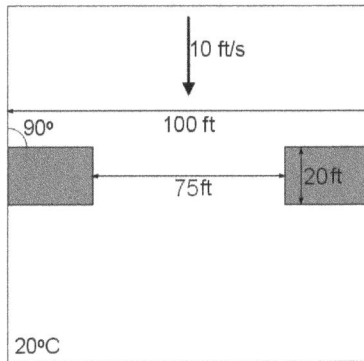

Figure 10.6-3. Contraction scour example.

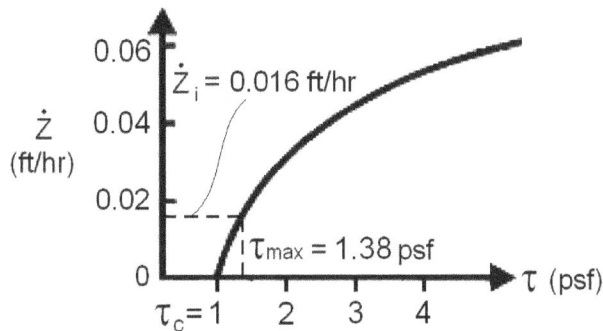

Figure 10.6-4. Time rate of contraction scour example.

10.6.3.2 Problem Solution

The analytical and computational procedures for the problem's solution are shown below. They follow the methods presented in this chapter for computation of pier scour.

Problem 10.6-3

ANALYTICAL PROCEDURES	COMPUTATIONS
Part (a): Computation of contraction scour in coarse-grained soils	
1. Assume that the soil is coarse grained with a D_{50} of 0.03 feet. To use the HEC-18 Sand method in this case, a distinction must be made as to whether clear-water scour or live-bed scour will develop.	$D_{50} = 0.02\,ft$
2. The critical velocity (V_c) is determined by Equation 10-12, shown below. $$V_c = 11.17 y_1^{1/6} D_{50}^{1/3}$$	$V_c = 11.17(25)^{1/6}(0.03)^{1/3}$ $V_c = 5.94\,ft/s$ The approach velocity (10 ft/s) is greater than the critical velocity (5.94 ft/s), therefore, live-bed scour will result.
3. Find the fall velocity of the bed material (ω) according to Figure 10.4-4 using the known value of D_{50}, and the shear velocity in the upstream section (V^*) from Equation 10-42. $$V^* = \left(g y_1 S_1\right)^{\frac{1}{2}}$$ The slope of the energy grade line (S_1) is found using Manning's equation (Equation 10-43), which is shown below. $$V_1 = \frac{1.49 R_{h,1}^{\frac{2}{3}} S_1^{\frac{1}{2}}}{n}$$ $$\Rightarrow S_1 = \frac{0.45 V_1^2 n^2}{R_{h,1}^{\frac{4}{3}}} = \frac{0.45 V_1^2 n^2}{\left(\frac{A_1}{P_1}\right)^{\frac{4}{3}}} = \frac{0.45 V_1^2 n^2}{\left(\frac{y_1 W_1}{2 y_1 + W_1}\right)^{\frac{4}{3}}}$$	$\omega = 1.4$ ft/s. From Table 10.4-2, $n = 0.02$ s/m$^{1/3}$ for sand <u>Note:</u> Since Manning's equation (Equation 10-43) already incorporates the constant 1.49, use the n values given in Table 10.4-2 with the units of s/m$^{1/3}$. The 1.49 will automatically convert the units of n to s/ft$^{1/3}$. $S_1 = \dfrac{0.45(10)^2 (0.02)^2}{\left[\dfrac{(25)(100)}{2(25)+(100)}\right]^{\frac{4}{3}}} = 0.000423$

Problem 10.6-3

ANALYTICAL PROCEDURES	COMPUTATIONS
4. Now that S_1 has been found, the shear velocity can be calculated (Equation 10-42). $V^* = \left(gy_1S_1\right)^{\frac{1}{2}}$ Find V^*/ω	$V^* = \left[(32.2)(25)(0.000423)\right]^{\frac{1}{2}} = 0.58\dfrac{ft}{s}$ $V^*/\omega = .58/1.4 = .41$
5. Find the value of k_1 from Table 10.4-1, using the computed value of V^*/ω.	For $V^*/\omega = 0.41$, $k_1 = .59$
6. Using Equation 10-41, the local live-bed contraction scour can be calculated. $y_{s,contraction,lb} = y_1\left[\left(\dfrac{Q_2}{Q_1}\right)^{\frac{6}{7}}\left(\dfrac{W_1}{W_c}\right)^{k_1}\right] - y_o$	$y_{s,contraction,lb} = (25)\left[(1)^{\frac{6}{7}}\left(\dfrac{100}{75}\right)^{0.59}\right] - 25$ $y_{s,contraction,lb} = 4.62\,ft$
Part (b): Computation of contraction scour in fine-grained soils	
7. Now assume the soil is a fine-grained clay. Find the critical shear stress (τ_c) using Figure 10.6-4. Also find the correction factors for transition angle (K_θ) and contraction length (K_L) and Manning's coefficient (n) based on the guidance for HEC-18 Clay method (Section 10.4.2.2). From Table 10.4-2, Manning's roughness (n) is equal to 0.023 for clay. Remember that in English units, n must be divided by a factor of 1.49 ($n/1.49$) to keep units consistent. In SI units, n is divided by a factor of 1 ($n/1 = n$).	From Figure 10.6-4, $\tau_c = 1.0\ psf$ Based on the guidance in Section 10.4.2.2, $K_\theta = 1$ $K_L = 1$ From Table 10.4-2 for clay, $n = 0.023\ s/m^{1/3}$ Note: Remember that in English units, n must be divided by a factor of 1.49 ($n/1.49$) to keep units consistent. In SI units, n is divided by a factor of 1 ($n/1 = n$).

Problem 10.6-3

ANALYTICAL PROCEDURES	COMPUTATIONS
8. The local contraction scour can now be calculated according to Equation 10-44, as shown below. $$y_{s,contraction} = K_\theta K_L 1.90 y_1 \left[\frac{1.38 \left(V_1 \dfrac{W_1}{W_c} \right)}{\sqrt{gy_1}} - \frac{\left(\dfrac{\tau_c}{\rho} \right)^{\frac{1}{2}}}{gny_1^{\frac{1}{3}}} \right]$$	$$y_{s,contraction} = (1)(1)(1.9)(25)\left\{ \frac{1.38\left[(10)\left(\dfrac{100}{75}\right)\right]}{\sqrt{(32.2)(25)}} - \frac{\left[\dfrac{1.0}{1.94}\right]^{\frac{1}{2}}}{(32.2)\left(\dfrac{0.023}{1.49}\right)(25)^{\frac{1}{3}}} \right\}$$ $y_{s,contraction} = 7.34\,ft$
Part (c): Computation of the depth of contraction scour in fine-grained soils after the first 24 hour storm	
9. The maximum hydraulic shear stress in the contraction zone ($\tau_{max,contraction}$) must first be calculated in order to find the initial scour rate (Equation 10-66). To do so, find the values of the correction factors for the contraction ratio (k_{c-R}), transition angle ($k_{c-\theta}$), and contraction length (k_{c-Lc}) using Equations 10-67, 10-68, and 10-69, respectively. The depth of the water has a negligible influence, so the correction for water depth, k_{c-y} = 1. $$k_{c-R} = 0.62 + 0.38\left(\frac{W_1}{W_c}\right)^{1.75}$$ $$k_{c-\theta} = 1 + 0.9\left(\frac{\theta}{90}\right)^{1.5}$$ $$k_{c-Lc} = \begin{cases} 1 & for \quad k_{c-Lc} \geq 0.35 \\ 0.77 + 1.36\left(\dfrac{L_c}{W_1 - W_c}\right) - 1.98\left(\dfrac{L_c}{W_1 - W_c}\right)^2 & for \quad k_{c-L} < 0.35 \end{cases}$$ $$\tau_{max,contraction} = k_{c-R}k_{c-\theta}k_{c-y}k_{c-Lc}\gamma n^2 V^2 R_h^{-\frac{1}{3}}$$ $$= k_{c-R}k_{c-\theta}k_{c-y}k_{c-Lc}\gamma n^2 V^2 \left(\frac{y_1 W_1}{2y_1 + W_1}\right)^{-\frac{1}{3}}$$	$$k_{c-R} = 0.62 + 0.38\left(\frac{100}{75}\right)^{1.75} = 1.25$$ $$k_{c-\theta} = 1 + 0.9\left(\frac{90}{90}\right)^{1.5} = 1.9$$ $$k_{c-Lc} = 0.77 + 1.36\left(\frac{20}{100-75}\right) - 1.98\left(\frac{20}{100-75}\right)^2 = 0.59 \geq 0.35$$ $$k_{c-Lc} = 1$$ $$\tau_{max,contraction} = (1.25)(1.9)(1)(1)(62.4)\left(\frac{0.023}{1.49}\right)^2 (10)^2 \left[\frac{(25)(100)}{2(25)+(100)}\right]^{-\frac{1}{3}}$$ $$\tau_{max,contraction} = 1.38\,psf$$

Problem 10.6-3

ANALYTICAL PROCEDURES	COMPUTATIONS
10. Since the maximum hydraulic shear stress in the contraction has been calculated (1.38 psf), the initial scour rate ($\dot{z}_i$) can be found using the erosion function (Figure 10.6-4). The maximum scour at the site was calculated as 7.34 feet in the previous example (Sect. 10.6.3.2) using HEC-18 Clay. The storm event lasted 24 hours. The amount of scour resulting from this storm can be found using Equation 10-60, shown below, where $z(t)$ is the scour depth as a function of time, z_{max} is the maximum scour depth ($y_{s,contraction}$), and $\dot{z}_i$ is the initial scour rate (found from Figure 10.6-4). $$z(t) = \dfrac{t}{\dfrac{1}{\dot{z}_i} + \dfrac{t}{z_{max}}}$$	From Figure 10.6-4, for $\tau_{max,contraction} = 1.38\ psf$, $\dot{z}_i = 0.016$ ft/hr $$z(t) = \dfrac{24}{\dfrac{1}{0.016} + \dfrac{24}{7.34}}$$ $z(24hrs) = 0.36\ ft$

SUMMARY

a. The contraction scour in coarse-grained soil is 4.62 ft.

b. The contraction scour in fine-grained soil is 7.34 ft.

c. The scour depth in the contraction zone is 0.36 ft after the first 24 hour storm. This is a significant difference from the maximum scour possible at the site. Subsequent storms and other events will further the scour at the pier. In fine-grained soil, the final scour depth after a series of storms or a velocity hydrograph can be found using the SRICOS-EFA Method.

10.6.4 Problem 4 – Abutment Scour in Coarse-Grained Soils Example

10.6.4.1 Problem Statement

A concrete wing wall abutment has a projected abutment length of 5 feet for both the right abutment (labeled R on Figure 10.6-5) and the left abutment (labeled L on Figure 10.6-5). The average flow depth and upstream water depth are 15 ft. The left abutment is aligned at a 60° angle pointing downstream. The right abutment is aligned at a 120° angle pointing upstream. The velocity of the flow in the obstructed section is 10 ft/s. Calculate the scour depth at both the right and left abutments.

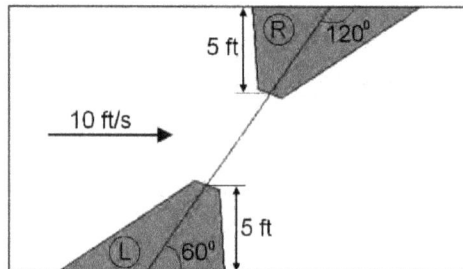

Figure 10.6-5. Abutment scour example.

10.6.4.2 Problem Solution

The analytical and computational procedures for the problem's solution are shown below. They follow the methods presented in this chapter for computation of abutment scour.

Problem 10.6-4

ANALYTICAL PROCEDURES	COMPUTATIONS
1. According to Section 10.4.4.1, there are two methods available to calculate the depth of abutment scour. If the ratio of projected abutment length (L') to the upstream flow depth (y_1) is greater than 25 ($L'/y_1 > 25$), then the HIRE abutment scour equation is used. If $L'/y_1 < 25$, the Froehlich abutment scour equation is used. Compute L'/y_1.	For both abutments, $L'/y_1 = 5/15 = 1/3 < 25$ The Froehlich abutment scour equation will be applied (Equation 10-53).

Problem 10.6-4

ANALYTICAL PROCEDURES	COMPUTATIONS
2. Since the abutment is a wing wall abutment, a shape factor (K_s) must be applied. The angle of attack also impacts local scour. To account for the alignment, a correction factor (K_ψ) must also be applied (Equation 10-14). For the left abutment, the angle of abutment alignment (ψ) is 60°. For the right abutment, the angle of abutment alignment is 120°.	From Table 10.2-6, $K_s = 0.82$ From Equation 10-14, $$K_{\psi,L} = \left(\frac{60^o}{90^o}\right)^{0.13} = 0.95$$ $$K_{\psi,R} = \left(\frac{120^o}{90^o}\right)^{0.13} = 1.04$$
3. Now, the local abutment scour can be calculated according to the Froehlich equation (Equation 10-53). For the left abutment: $$\frac{y_{s,abutment,L}}{y_a} = 2.27 K_s K_\psi \left(\frac{L'}{y_a}\right)\left(\frac{V_e}{(gy_a)^{\frac{1}{2}}}\right)^{0.61} + 1$$	$$\frac{y_{s,abutment,L}}{y_a} = 2.27(0.75)(0.95)\left(\frac{5}{15}\right)\left[\frac{10}{(32.2*15)^{\frac{1}{2}}}\right]^{0.61} + 1$$ $y_{s,abutment,L} = 1.33(15)$ $y_{s,abutment,L} = 19.95\,ft$
4. For the right abutment: $$\frac{y_{s,abutment,R}}{y_a} = 2.27 K_s K_\psi \left(\frac{L'}{y_a}\right)\left(\frac{V_e}{(gy_a)^{\frac{1}{2}}}\right)^{0.61} + 1$$	$$\frac{y_{s,abutment,R}}{y_a} = 2.27(0.82)(1.04)\left(\frac{5}{15}\right)\left[\frac{10}{(32.2*15)^{\frac{1}{2}}}\right]^{0.61} + 1$$ $y_{s,abutment,R} = 1.40 y_a = 1.40(15)$ $y_{s,abutment,R} = 21\,ft$

<u>SUMMARY</u>

The right abutment experiences more scour than the left abutment in this equation. This is because the right abutment is angled upstream. The left abutment has a slight "shielding" effect since it is angled downstream.

1. $y_{s,abutment,L}$ = 19.95 ft

2. $y_{s,abutment,R}$ = 21 ft

10.6.5 Problem 5 – Seawall Scour in Coarse-Grained Soils Example

10.6.5.1 Problem Statement

A vertical seawall is placed along the shoreline, as shown in Figure 10.6-6. The water level is 5 feet at the location of the wall. The deepwater significant wave height (H_s), unbroken wave height (H_o), and wave length (L_d) are 4 feet, 2 feet, and 250 feet, respectively. Find the maximum seawall scour.

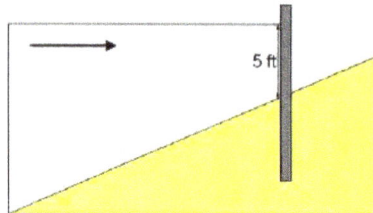

Figure 10.6-6. Seawall scour example.

10.6.5.2 Problem Solution

The analytical and computational procedures for the problem's solution are shown below. They follow the methods presented in this chapter for computation of seawall scour.

Problem 10.6-5

ANALYTICAL PROCEDURES	COMPUTATIONS
1. As a rule of thumb, the maximum seawall scour is less than or equal to the unbroken deepwater wave height, H_o. Equation 10-54 can be used to more accurately estimate the maximum scour as long as $-0.011 \leq y/L_d \leq 0.045$ and $0.015 \leq H_s/L_d \leq 0.04$. $$y_{s,seawall} = H_s \sqrt{22.72\left(\frac{y}{L_d}\right) + 0.25}$$	From the problem statement, $H_s = 4\ ft$, $H_o = 2\ ft$, and $L_d = 250\ ft$ $y/L_d = 5/250 = 0.02$ $H_s/L_d = 4/250 = .016$ Equation 10-54 is valid since y/L_d and H_s/L_d fall within the constraints. $$y_{s,seawall} = 4\sqrt{22.72\left(\frac{5}{250}\right) + 0.25}$$ $y_{s,seawall} = 3.36\ ft$
SUMMARY	
The maximum seawall scour is 3.36 ft.	

10.6.6 Problem 6 – Pipeline Scour Example

10.6.6.1 Problem Statement

A 2.5 foot diameter pipeline is laid along the seafloor which is at a depth (d) of 100 feet, as shown in Figure 10.6-7. Under its own weight, it embeds 0.5 feet. The pipeline experiences wave action with the waves having a wave length (L) of 500 feet, a wave period (T) of 10 seconds, and a wave height (H) of 5 feet. Find the maximum pipeline scour.

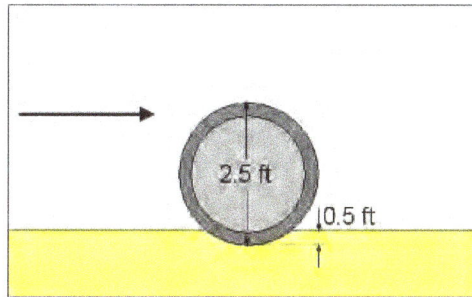

Figure 10.6-7. Pipeline scour example.

10.6.6.2 Problem Solution

The analytical and computational procedures for the problem's solution are shown below. They follow the methods presented in this chapter for computation of pipeline scour.

Problem 10.6-6

ANALYTICAL PROCEDURES	COMPUTATIONS
1. The pipeline is located in a water depth (d) of 100 feet. This is less than $L/2$, which is equal to 250 feet. It is larger than $L/20$, however, so the pipe is therefore located in transitional water. The maximum orbital velocity at the bed ($V_{orb,bm}$) can be found using Table 10.2-3. $$V_{orb,bm} = \frac{H}{2}\frac{gT}{L}\frac{\cosh\left[2\pi\frac{(z+d)}{L}\right]}{\cosh\left(2\pi\frac{d}{L}\right)}\cos\theta_w$$	$$V_{orb,bm} = \frac{5}{2}\frac{(32.2)(10)}{500}\frac{\cosh\left[2\pi\frac{(-100+100)}{500}\right]}{\cosh\left(2\pi\frac{100}{500}\right)}\cos(0)$$ $$V_{orb,bm} = 0.85\frac{ft}{s}$$

Problem 10.6-6

ANALYTICAL PROCEDURES	COMPUTATIONS
2. The embedment ratio (e/D_{pl}) is less than 0.5, so scour will occur. Since the pipeline is in the presence of wave action, the maximum depth of scour is calculated according to Equation 10-59. $$y_{s,pipeline,wave} = 0.1 D_{pl}\sqrt{KC}\left(1-1.4\frac{e}{D_{pl}}\right)+\frac{e}{D_{pl}}$$ $$= 0.1 D_{pl}\sqrt{\frac{V_{orb,m}T}{D_{pl}}}\left(1-1.4\frac{e}{D_{pl}}\right)+\frac{e}{D_{pl}}$$	$$y_{s,pipeline,wave} = 0.1(2.5)\sqrt{\frac{(0.85)(10)}{2.5}}\left[1-1.4\left(\frac{0.5}{2.5}\right)\right]+\left(\frac{0.5}{2.5}\right)$$ $$y_{s,pipeline,wave} = 0.53\,ft$$

SUMMARY

The maximum depth of pipeline scour is 0.53 ft.

10.6.7 Problem 7 – Propeller Induced Scour Example

10.6.7.1 Problem Statement

A vessel's propeller has a diameter (D_p) of 8 feet and an initial centerline velocity (V_o) of 15 ft/s, as shown in Figure 10.6-8. The propeller has a clearance distance to the seabed (C) of 15 feet. The seabed soil $(D_{50} = 0.005$ ft) and the seawater have mass densities of 3.6 slugs/ft^3 and 2 slugs/ft^3, respectively. The ship's propellers remain in motion at this one site for 1 hour. Find the scour due to the propeller.

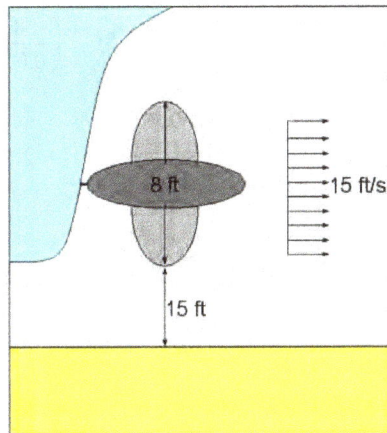

Figure 10.6-8. Propeller scour example.

10.6.7.2 Problem Solution

The analytical and computational procedures for the problem's solution are shown below. They follow the methods presented in this chapter for computation of propeller induced scour.

Problem 10.6-7

ANALYTICAL PROCEDURES	COMPUTATIONS
1. Equation 10-46 can be used to compute the propeller induced scour, as long as the seabed depth below the propeller (C) is in the range of 0.5 to 2.5 times the propeller diameter (D_p). Is $0.5D_p < C < 2.5D_p$?	$0.5(8\ ft) = 4\ ft$ $2.5\ (8\ ft) = 20\ ft$ YES, C is between 4 and 20 feet, so Equation 10-46 can be applied.

Problem 10.6-7

ANALYTICAL PROCEDURES	COMPUTATIONS
2. Before calculating the propeller induced scour, the densimetric Froude number (Fr_o) must first be found (Equation 10-49). Then, both Γ and Ω need to be estimated (Equations 10-47 and 10-48). $$Fr_o = \dfrac{V_o}{\sqrt{gD_{50}\left(\dfrac{\Delta\rho}{\rho}\right)}}$$ $$\Gamma = Fr_o^{-0.53}\left(\dfrac{D_p}{D_{50}}\right)^{-0.48}\left(\dfrac{C}{D_{50}}\right)^{0.94}$$ $$\Omega = \Gamma^{-6.38}$$	$$Fr_o = \dfrac{15}{\sqrt{(32.2)(0.005)\left(\dfrac{3.6-2}{2}\right)}} = 41.8$$ $$\Gamma = (41.8)^{-0.53}\left(\dfrac{8}{0.005}\right)^{-0.48}\left(\dfrac{15}{0.005}\right)^{0.94} = 7.4$$ $$\Omega = (7.4)^{-6.38} = 2.8\times10^{-6}$$
3. The propeller induced scour is then found according to Equation 10-46. Since the amount of time the propeller is moving is known ($t = 3,600s$), scour can be directly calculated. If the time was unknown and a maximum asymptotic value is wanted, try different time intervals until the calculated scour for one time step is within a chosen accuracy value from the previous time step. $$y_{s,propeller}(ft) = 0.128\Omega\left\{\ln\left[t(\sec)\right]\right\}^{\Gamma}$$	$$y_{s,propeller} = 0.128(2.8\times10^{-6})\left[\ln(3600)\right]^{7.4}$$ $$y_{s,propeller} = 2.05\,ft$$

SUMMARY
The maximum depth of propeller induced scour is 2.05 ft.

10.6.8 Problem 8 – Footing Scour with Skirt Countermeasure Example

10.6.8.1 Problem Statement

A 50 foot square footing is going to be placed on the seafloor normal to the direction of flow. The footing extends 5 feet above the plane bed seafloor (D_{50} = 0.006 ft, D_{84} = 0.05 ft, and D_{95} = 0.065 ft) with no embedment. The water depth at this location is 100 ft and the current is moving at 5 ft/s. There is no wave action. Find the depth to which a skirt must be placed around the perimeter of the footing to avoid the footing being undermined by scour (Figure 10.6-9).

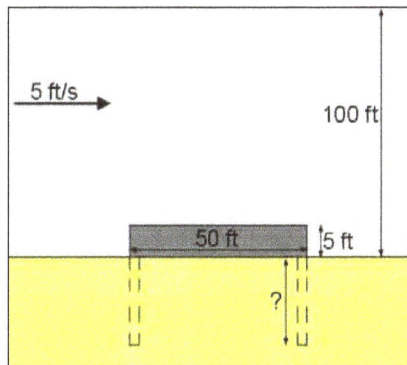

Figure 10.6-9. Footing scour example.

10.6.8.2 Problem Solution

Peripheral skirts must be placed below the estimated scour depth to ensure foundation stability. The method used to calculate footing scour is the same as that for pilecap/footing scour for a complex pier (Section 10.4.1.3), except there is no pier scour and no pile group scour.

Problem 10.6-8

ANALYTICAL PROCEDURES	COMPUTATIONS
1. The footing scour can be estimated using Equation 10-32. But first, any necessary correction factors must be found. The correction factors for pier nose shape (K_{sh}), angle of attack (K_a), bed condition (K_b), armoring (K_{ar}), and wide pier in shallow water (K_{ws}) are found according to the guidelines in Section 10.2.3.1.1.	For square footing, Table 10.2-4 gives: K_{sh} = 1.1 For flow normal to the footing: K_a = 1.0 For a plane bed, Table 10.2-5 gives: K_b = 1.1 Since D_{50} < 0.0066 ft there is no correction for armoring: K_{ar} = 1.0 The water depth > 0.8 (footing width), so the footing is not considered wide in shallow water: K_{ws} = 1.0

Problem 10.6-8

ANALYTICAL PROCEDURES	COMPUTATIONS
2. Second, the Froude number for this case (Fr_f, Equation 10-9) must be calculated using the average velocity of flow (V_f) at the exposed footing (Equation 10-33). $$Fr_f = \frac{V_f}{\sqrt{gy_f}}$$ $$V_f = V_2 \frac{\left[Ln\left(10.93\frac{y_f}{k_s}+1\right)\right]}{\left[Ln\left(10.93\frac{y_2}{k_s}+1\right)\right]}$$	$$V_f = 5\left\{\frac{\left[Ln\left(10.93\left(\frac{5}{0.05}\right)+1\right)\right]}{\left[Ln\left(10.93\left(\frac{100}{0.05}\right)+1\right)\right]}\right\} = 3.50\frac{ft}{s}$$ $$Fr_f = \frac{3.50}{\sqrt{(32.2)(5)}} = 0.28$$
3. Finally, the footing scour can be estimated using Equation 10-32. $$y_{s,footing} = 2y_f K_{sh}K_a K_b K_{ar}K_{ws}\left(\frac{B_{pc}}{y_f}\right)^{0.65} Fr_f^{0.43}$$	$$y_{s,footing} = 2(5)(1.1)(1.0)(1.1)(1.0)(1.0)\left(\frac{50}{5}\right)^{0.65}(0.28)^{0.43}$$ $$y_{s,footing} = 31.27 ft$$

<u>SUMMARY</u>

The skirt must be placed to a depth of at least 31.27 ft around the footing. If riprap or other scour countermeasures are also used at this site, the scour depth, and thus the skirt depth, will be decreased.

10.7 REFERENCES

10-1.   Karrock.   "Hjulström's Diagram."   Wikipedia: The Free Encyclopedia Website. http://en.wikipedia.org/wiki/File:Hjulströms_diagram_en.PNG (October 25, 2009).

10-2.   Briaud, J.-L., 2008, "Case Histories in Soil and Rock Erosion: Woodrow Wilson Bridge, Brazos River Meander, Normandy Cliffs, and New Orleans Levees," 9th Ralph B. Peck Lecture, *Journal of Geotechnical and Geoenvironmental Engineering*, ASCE.

10-3.   Cao, Y., Wang, J., Briaud, J.-L., Chen, H.C., Li, Y., Nurtjahyo, P., 2002, "Erosion Function Apparatus Overview and Discussion of Influence Factors on Cohesive Soil Erosion Rate in EFA Test," *Proceedings of the First International Conference on Scour of Foundations*, Texas A&M University, Dpt. Of Civil Engineering, College Station, TX, 2002

10-4.   Martin, J.L., McCutcheon, S.C., 1999, Hydrodynamics and Transport for Water Quality Modeling, CRC Press.

10-5.   Demerbilek, Z., Vincent, L., 2006, Water Wave Mechanics, Coastal Engineering Manual, Part 2, Chapter 1, Engineer Manual 1110-2-1100, U.S. Army Corps of Engineers, Washington, DC.

10-6.   Sumer, B.M., Fredsøe, J., 2002, The Mechanics of Scour in the Marine Environment, World Scientific, Advanced Series on Ocean Engineering, Vol. 17, River Edge, NJ.

10-7.   Dey, S., Sumer, B.M., Fredsøe, J., 2006, "Control of Scour at Vertical Circular Piles under Waves and Current," *Journal of Hydraulic Engineering*, ASCE, March 2006.

10-8.   Thoresen, C.A., 2003, Port Designer's Handbook: Recommendations and Guidelines, Thomas Telford, Ltd., London, UK.

10-9.   Tsinker, G.P., 1995, Marine Structures Engineering: Specialized Applications, Chapman & Hall, New York, NY.

10-10.   Blaaw, H.G., van de Kaa, E.J., 1978, "Erosion of Bottom and Banks Caused by the Screw Race of Maneuvering Ships," Publication No. 202, Delft Hydraulics Laboratory, Delft, The Netherlands, presented at the 7th International Harbor Congress, Antwerp, May 22-26.

10-11.   Hamill, G.A., Johnston, H.T., Stewart, D.P., 1999, "Propeller Wash Scour near Quay Walls," *Journal of Waterway, Port, Coastal, and Ocean Engineering*, Vol. 125, Issue 4, pp. 170-175, July/August 1999.

10-12.   Richardson, E.V., Davis, S.R., 2001, Evaluating Scour at Bridges, Fourth Edition, Hydraulic Engineering Circular No. 18, Report No. FHWA NHI 01-001 HEC-18, May 2001.

10-13.   Briaud, J.-L., Chen, H.C., Li, Y., Nurtjahyo, P., Wang, J., 2004, Pier and Contraction Scour in Cohesive Soils, NCHRP Report 516, Washington, DC.

10-14. Wang, J., 2004, The SRICOS-EFA Method for Complex Pier and Contraction Scour, Ph.D. Dissertation, Texas A&M University, College Station, TX, May 2004.

10-15. Melville, B.W., Coleman, S.E., 2000, Bridge Scour, Water Resources Publication, Highlands Ranch, CO.

10-16. DHI, 2007, "MIKE 21C River Morphology: A Short Description," DHI Water & Environment.

10-17. Barbhuiya, A.K., Dey, S., 2004, "Local Scour at Abutments: A Review," Sadhana, Vol. 29, Part 5, Springer India, October 2004.

10-18. Arcement, JR., G.J., Schneider, V.R., 1984, "Guide for Selecting Manning's Roughness Coefficients for Natural Channels and Flood Plains," United States Geological Survey Water-supply Paper 2339, Metric Version, USGS.

10-19. Fowler, J.E., 1992, Scour Problems and Methods for Prediction of Maximum Scour at Vertical Seawalls, Coastal Engineering Research Center, Vicksburg, MS.

10-20. Liang, D., Cheng, L., 2005, "Numerical Model for Wave-Induced Scour Below a Submarine Pipeline," Journal of Waterway, Port, Coastal, and Ocean Engineering. Vol. 131, Issue 5, pp. 193-202, September/October 2005.

10-21. Burcharth, H.F., Hughes, S.A., 2006, Fundamentals of Design – Part 3, Coastal Engineering Manual, Part 6, Chapter 5, Engineer Manual 1110-2-1100, U.S. Army Corps of Engineers, Washington, DC.

10-22. Sumer, B.M., Truelsen, C., Sichmann, T., Fresøe, J., 2001, "Onset of Scour Below Pipelines and Self-burial," Coastal Engineering, Vol. 42, issue 4, pp. 313-335, April 2001.

10-23. Liang, D., Cheng, L., Li, F., 2005, "Numerical Modeling of Flow and Scour Below a Pipeline in Currents: Part II. Scour simulation," Coastal Engineering, Elsevier, Vol. 52, pp. 43-62.

10-24. Hoffmans, G.J.C.M., Verheij, H.J., 1997, Scour Manual, A.A. Balkema Publishers, Netherlands.

10-25. Hosseini, D., Hakimzadeh, H., Ghiassi, R., 2004, "Scour Below Submarine Pipeline due to Currents," Proceedings of 2nd International Conference on Scour and Erosion, Meritus Mandarin, Singapore, November 14-17, 2004.

10-26. Nordin, C.F., Richardson, E.V., 1967, "The Use of Stochastic Models in Studies of Alluvial Channel Processes," Proceedings of the 12th IAHR Conference, Vol. 2, pp. 96-102, Ft. Collins, CO.

10-27. Shore Protection Manual, 1977, US Army Engineer Waterways Experiment Station, Coastal Engineering Research Centre, US Government Printing Office, Washington, DC.

10-28. Dewall, A.E., Christenson, J.A., 1979, Guidelines for Predicting Maximum Nearshore Sand Level Changes on Unobstructed Beaches, Army Coastal Engineering Research Center, Unpublished Report, Fort Belvoir, VA.

10-29. Reeve, D., Chadwick, A., 2004, Coastal Engineering: Processes, Theory and Design, Taylor & Francis.

10-30. Yen, B.C.,ed., 1992, Channel Flow Resistance: Centennial of Manning's Formula, Water Resources Publications, LLC, Highlands Range, CO.

10-31. Fredsøe, J., Deigaard, R., 1992, *Mechanics of Coastal Sediment Transport*, Singapore: World Scientific.

10-32. Le Roux, J.P., 2003, "Wave Friction Factor as Related to the Shields Parameter for Steady Currents," Sedimentary Geology, Vol. 155, Issues 1-2, pp. 37-43, January 2003.

10-33. McConnell, K., Allsop, W., Cruickshank, I., 2004, Piers, Jetties and Related Structures Exposed to Waves: Guidelines for Hydraulic Loadings, Thomas Telford.

10-34. Brunner, G.W., 2002, "HEC-RAS River Analysis System Hydraulic Reference Manual," Version 3.1, Report No. CPD-69, U.S. Army Corps of Engineers, Institute for Water Resources, Hydrologic Engineering Research Center, Davis, CA 95616.

10-35. Akan, A.O., 2006, Open Channel Hydraulics, Butterworth-Heinemann/Elsevier.

10-36. Caltrans (California Department of Transportation), 2006, Highway Design Manual, Chapter 860.

10-37. Hamill, G.A., McGarvey, J.A., 1997, "The Influence of a Ship's Rudder on the Scouring Action of a Propeller Wash," Proceedings of the 7th International Offshore and Polar Engineering Conference (ISOPE), Vol. IV, pp. 754-757, May 25-30, 1997, Honolulu, HI.

10-38. Shore Protection Manual, 1984, 4th ed., 2 Vols, US Army Engineer Waterways Experiment Station, Coastal Engineering Research Centre, US Government Printing Office, Washington, DC.

10-39. Sumer, B.M., Fredsøe, J., 1992, "A Review of Wave/Current-Induced Scour Around Pipelines," Proceedings of the 23rd International Coastal Engineering Conference, Vol. 3, ASCE, pp. 2839-2852.

10-40. Klomb, W.H.G., Tonda, P.L., 1995, "Pipeline Cover Stability," *Proceedings of the 5th International Conference on Offshore Mechanics and Arctic Engineering*, American Society of Mechanical Engineering, Vol. 2, pp. 15-22.

10-41. Briaud J.-L., Ting F. C. K., Chen H. C., Gudavalli R., Perugu S., Wei G., 1999, "SRICOS: Prediction of Scour Rate in Cohesive Soils at Bridge Piers," Journal of Geotechnical and Geoenvironmental Engineering, Vol. 125, No.4, pp. 237-246, ASCE, Reston, Virginia.

10-42. Briaud J.-L., Ting F., Chen H.C., Cao Y., Han S.-W., Kwak K., 2001, "Erosion Function Apparatus for Scour Rate Predictions", Journal of Geotechnical and Geoenvironmental Engineering, Vol. 127, No.2, pp. 105-113, Feb. 2001, ASCE, Reston, Virginia.

10-43. USACE (U.S. Army Corps of Engineers), 1994, Engineering and Design – Hydraulic Design of Flood Control Channels, EM 1110-2-1601.

10-44. Lagasse, P.F., Clopper, P.E., Zevenbergen, L.W., Ruff, J.F., 2006, Riprap Design Criteria, Recommended Specifications, and Quality Control, NCHRP Report 568, Washington, DC.

10-45. USDA (United States Department of Agriculture), 1994, Gradation Design of Sand and Gravel Filters, National Engineering Handbook, Ch. 26, Part 633.

10-46. Army and Air Force Technical Manual, 1995, Engineering Use of Geotextiles, Joint Departments of the Army and Air Force, Army TM 5-818-8, Air Force AFJMAN 32-1030.

10-47. Anderson, D.W., 2003, "Use of Articulating Concrete Block Revetment Systems for Stream Restoration and Stabilization Projects," *Proceedings of the 2003 ASABE Annual International Meeting*, Las Vegas, NV, July 27-30, 2003.

10-48. Gerwick, Jr., B.C., 2000, Construction of Marine and Offshore Structures, Second Edition, CRC Press, Francis & Taylor Group.

10-49. Lagasse, P.F., Zevenbergen, L.W., Schall, J.D., Clopper, P.E., 2001, Bridge Scour and Stream Instability Countermeasures: Experience, Selection, and Design Guidance, Second Edition, Hydraulic Engineering Circular No. 23, Report No. FHWA NHI 01-003 HEC-23, March 2001.

10-50. Lagasse, P.F., Schall, J.D., Richardson, E.V., 2001, Stream Stability at Highway Structures, Third Edition, Hydraulic Engineering Circular No. 20, Report No. FHWA NHI 01-002 HEC-20, March 2001.

10-51. Hunt, B., (2008), Practices for Monitoring Scour Critical Bridges, NCHRP Project 20-5, Topic 36-02, Transportation Research Board, Washington, D.C.

10.8 SYMBOLS

A	Cross-sectional area of flow [L^2]
a	Wave semi-orbital length [L]
B	Wave semi-orbital width [L]
B_p	Width/Diameter of pier [L]
B'_p	Width of pier normal to the flow [L]
B_{pc}	Width of pile cap [L]
B^*_{pc}	Equivalent full depth solid pier width for pile cap [L]
B^*_{pg}	Equivalent full depth solid pier width for pile group [L]
B_{proj}	Projected width of pile group [L]
C	Clearance distance between the propeller tip and the seabed [L]
C_1	Coefficient for maximum velocity at the bed due to a propeller
C_D	Drag coefficient
$C.G.$	Center of gravity
D	Diameter of structure [L]
D_o	Diameter of initial centerline propeller velocity [L]
D_{50}	Mean grain size [L]
$D_{50,riprap}$	Median diameter of riprap [L]
D_{84}	Diameter of soil of which 84% of the particles are smaller [L]
D_{90}	Diameter of soil of which 90% of the particles are smaller [L]
D_{95}	Diameter of soil of which 95% of the particles are smaller [L]
D_m	Diameter of the smallest non-transportable particle in the bed material [L]
D_p	Propeller diameter [L]
D_{pl}	Pipeline/Cable diameter [L]
D_{xs}	Diameter of soil of which xs% of the particles are smaller [L]
d	Water depth [L]
$d\gamma$	Change in shear strain [L/L]
dt	Change in time [T]
dV_x	Change in horizontal velocity [L/T]
dx	Change in horizontal displacement [L]

dz	Change in vertical direction [L]
e	Embedment depth of pipeline/cable [L]
F	Coefficient for initial centerline propeller velocity
Fr	Froude number
Fr_o	Densimetric Froude number
Fr_1	Upstream Froude number
Fr_2	Adjusted Froude number for pile cap
Fr_3	Adjusted Froude number for pile group
Fr_f	Froude number for footing
f	Pile cap overhang [L]
f_{ci}	Forces at contacts between soil particles [F]
f_{ei}	Electrical forces between soil particles [F]
f_w	Wave friction factor
g	Acceleration due to gravity [L/T^2]
H	Wave height [L]
H_o	Unbroken deepwater wave height [L]
H_e	Annual extreme wave height [L]
H_p	Distance from the center of the propeller shaft to channel bottom [L]
$\overline{H_s}$	Average monthly significant wave height [L]
H_s	Significant wave height [L]
$h_{o,pc}$	Actual pile cap height [L]
$h_{o,pg}$	Actual pile group height [L]
h_1	Height of the pier stem above the bed [L]
h_2	Adjusted pile cap height [L]
h_3	Adjusted pile cap height [L]
K	Coefficient for pier shape for riprap
K_a	Attack angle correction factor (Figure 10.2-23)
K_{ar}	Bed armoring correction factor
K_b	Bed condition correction factor (Table 10.2-5)
$K_{h,pier}$	Complex pier scour correction factor

$K_{h,pg}$	Pile group height correction factor
K_L	Contraction length correction factor
K_m	Number of aligned rows correction factor for a pile group
K_ψ	Abutment alignment correction factor
K_s	Abutment shape correction factor (Table 10.2-6, Figure 10.2-24)
K_{sh}	Pier shape correction factor (Table 10.2-4)
K_{sp}	Pier spacing correction factor
$K_{sp,pg}$	Pile spacing correction factor for a pile group
K_θ	Transition angle correction factor
$K_{\theta/Xmax}$	Transition angle correction factor for location of maximum scour
K_w	Shallow water correction factor
K_{ws}	Wide pier in shallow water correction factor
KC	Keulegan-Carpenter number
k_1	Exponent for live-bed contraction scour (Table 10.4-1)
k_a	Shear stress correction factor for attack angle
$k_{c-\theta}$	Shear stress correction factor for contraction transition angle
k_{c-Lc}	Shear stress correction for contraction length
k_{c-R}	Shear stress correction factor for contraction ratio
k_{c-y}	Shear stress correction factor for contraction water depth
k_s	Nikuradse equivalent bed roughness [L]
k_{sh}	Shear stress correction factor for pier shape
k_{sp}	Shear stress correction factor for pier spacing
k_w	Shear stress correction factor for water depth
L	Wave length [L]
L'	Length of the abutment projected normal to flow [L]
L_c	Length of contraction [L]
L_d	Deepwater wave length [L]
L_p	Length of pier [L]
L_o	Deep water wave length [L]
m_p	Number of rows in the pile group

N	Design life [T]
n	Manning's roughness coefficient (Table 10.4-2) [$T/L^{1/3}$]
n_p	Number of piers
n_s	Soil porosity
P	Wetted perimeter [L]
P_d	Engine power [L·F/T]
PI	Plasticity index
Q	Flow rate/Discharge (ft^3/s)
Q_1	Flow rate in the upstream channel [L^3/T]
Q_2	Flow rate in the contracted channel [L^3/T]
R_h	Hydraulic radius [L]
Re	Reynolds number
S	Slope of energy grade line [L/L]
S_1	Slope of energy grade line upstream in main channel [L/L]
S_a	Maximum equilibrium scour depth for unconfined propeller scour [L]
S_p	Pile spacing [L]
S_s	Specific gravity
s_u	Undrained shear strength [F/L^2]
T	Wave period [T]
T_r	Return period [T]
t	Time [T]
t_{pc}	Thickness of the pile cap exposed to flow [L]
u_w	Water pressure around soil particle [F/L^2]
V	Average water velocity [L/T]
V^*	Shear velocity upstream [L/T]
V_o	Initial centerline propeller velocity [L/T]
V_1	Average upstream velocity/Approach velocity [L/T]
V_2	Adjusted flow velocity for pile cap [L/T]
V_3	Adjusted flow velocity for pile group [L/T]
V_b	Average current velocity near the bed [L/T]

$V_{b,max}$	Maximum velocity at the bed due to a propeller [L/T]
V_c	Critical velocity [L/T]
V_e	Velocity of the flow in the obstructed section for abutment scour [L/T]
V_f	Average velocity in the flow zone below the top of the pile cap/footing [L/T]
V_m	Maximum velocity [L/T]
V_{orb}	Horizontal orbital velocity of wave [L/T]
$V_{orb,bm}$	Maximum orbital velocity at the bed [L/T]
$V_{orb,bm,c}$	Critical maximum orbital velocity at the bed [L/T]
$V_{orb,v}$	Vertical orbital velocity of wave [L/T]
V_{pl}	Undisturbed flow velocity at the top of the pipeline (ft/s)
$V_{pl,c}$	Critical undisturbed flow velocity at the top of the pipeline (ft/s)
V_x	Horizontal water velocity [L/T]
V_y	Vertical water velocity [L/T]
W	Weight of soil particle [F]
W_1	Width of upstream channel [L]
W_c	Width of contraction [L]
W_p	Width of channel without piers [L]
w	Water content
X	Horizontal location of maximum velocity behind the propeller [L]
X_m	Distance of the maximum unconfined scour from the propeller [L]
X_{max}	Location of maximum contraction scour depth as measured from the beginning of the contracted section [L]
$X_{max}(\theta)$	Location of maximum scour for a transition angle θ [L]
$X_{max}(90°)$	Location of maximum scour for no contraction [L]
X_w	Distance of the quay wall from the face of the propeller [L]
y	Water depth [L]
y_o	Initial water depth for the existing bed depth [L]
y_1	Average upstream water depth [L]
y_2	Adjusted flow depth for pile cap [L]
y_3	Adjusted flow depth for pile group [L]

y_a	Average depth of flow on the floodplain [L]
y_f	Distance from the bed to the top of the footing [L]
y_{gs}	Maximum general scour [L]
y_{pier}	Pier scour depth [L]
$y_{s,abutment}$	Maximum abutment scour depth [L]
$y_{s,contraction}$	Maximum contraction scour depth [L]
$y_{s,contraction,cw}$	Maximum clear-water contraction scour depth [L]
$y_{s,contraction,lb}$	Maximum live-bed contraction scour depth [L]
$y_{s,c,uniform}$	Uniform contraction scour depth [L]
$y_{s,pier}$	Maximum pier scour depth [L]
$y_{s,pier,wave}$	Maximum wave-induced pier scour depth [L]
$y_{s,pilecap}$	Maximum pile cap scour depth [L]
$y_{s,pilegroup}$	Maximum pile group scour depth [L]
$y_{s,pipeline,cw}$	Maximum clear-water pipeline scour depth [L]
$y_{s,pipeline,lb}$	Maximum live-bed pipeline scour depth [L]
$y_{s,pipeline,wave}$	Maximum wave-induced pipeline scour depth [L]
$y_{s,propeller}$	Propeller induced scour depth [L]
$y_{s,quaywall}$	Maximum propeller induced scour depth at quay wall [L]
$y_{s,quaywall,rudder}$	Maximum rudder and propeller induced scour depth at quay wall [L]
$y_{s,rudder}$	Rudder and propeller induced scour depth [L]
$y_{s,seawall}$	Maximum seawall scour depth [L]
$\dot{Z}$	Erosion rate [L/T]
z	Measure of depth [L]
z_o	Hydraulic roughness length [L]
z_{final}	The final depth of scour after a storm event [L]
$\dot{z}_i$	Initial rate of scour [L/T]
z_{max}	Maximum magnitude of scour [L]
α_s	Proportionality constant for Nikuradse bed roughness (Table 10.3-1)
Γ	Propeller scour coefficient
γ	Shear strain [L/L]

γ_s	Specific weight of the soil grains [F/L³]
γ_w	Specific weight of water [F/L³]
Δ	Attack angle [rad]
$\Delta\rho$	Difference between the soil density and the water density [F·T²/L⁴]
$\Delta\sigma$	Turbulent fluctuation of the net uplift normal stress [F/L²]
$\Delta\tau$	Turbulent fluctuation of the hydraulic shear stress [F/L²]
Δt	Change in time [T]
ζ	Rudder angle [rad]
θ	Transition angle [deg]
θ_w	Angle of velocity particle around its orbit [rad]
μ	Dynamic viscosity of water [F·T/L²]
ν	Kinematic viscosity of water [L²/T]
ρ	Mass density of water [F·T²/L⁴]
ρ_s	Soil density [F·T²/L⁴]
σ_d	Standard deviation of the monthly average significant wave heights [L]
τ	Shear stress [F/L²]
τ_b	Bed shear stress [F/L²]
τ_{bc}	Bed shear stress under current only [F/L²]
$\tau_{bcw,m}$	Mean bed shear stress due to both currents and waves [F/L²]
$\tau_{bcw,max}$	Maximum bed shear stress due to both currents and waves [F/L²]
τ_{bw}	Bed shear stress under waves only [F/L²]
τ_c	Critical shear stress [F/L²]
$\tau_{max,contraction}$	Maximum hydraulic shear stress at a contraction [F/L²]
$\tau_{max,pier}$	Maximum hydraulic shear stress at a pier [F/L²]
ϕ	Angle between direction of wave propagation and direction of the current [rad]
χ	Angle of slope [rad]
ψ	Angle of abutment alignment [deg]
Ω	Propeller scour coefficient
ω	Fall velocity of bed material) [L/T]

www.ingramcontent.com/pod-product-compliance
Lightning Source LLC
Chambersburg PA
CBHW081225220326
41598CB00037B/6878

* 9 7 8 1 7 8 2 6 6 0 5 1 4 *